PYTHON FOR CHEMISTS

This accessible and self-contained guide provides a comprehensive introduction to the popular programming language Python, with a focus on applications in chemistry and chemical physics. Ideally suited to students and researchers of chemistry learning to employ Python for problem-solving in their research, this fast-paced primer first builds a solid foundation in the programming language before progressing to advanced concepts and applications in chemistry. The required syntax and data structures are established, and then applied to solve problems computationally. Popular numerical packages are described in detail, including NumPy, SciPy, Matplotlib, SymPy, and pandas. End of chapter problems are included throughout, with worked solutions available within the book. Additional resources, datasets, and Jupyter Notebooks are provided on a companion website, allowing readers to reinforce their understanding and gain confidence applying their knowledge through a hands-on approach.

CHRISTIAN HILL is a physicist and physical chemist with over 25 years' experience in scientific programming, data analysis, and database design in atomic and molecular physics. Currently the Head of the Atomic and Molecular Data Unit at the International Atomic Energy Agency, Vienna, he has previously held positions at the University of Cambridge, the University of Oxford, and University College London.

PYTHON FOR CHEMISTS

CHRISTIAN HILL
International Atomic Energy Agency

CAMBRIDGE
UNIVERSITY PRESS

Shaftesbury Road, Cambridge CB2 8EA, United Kingdom

One Liberty Plaza, 20th Floor, New York, NY 10006, USA

477 Williamstown Road, Port Melbourne, VIC 3207, Australia

314–321, 3rd Floor, Plot 3, Splendor Forum, Jasola District Centre,
New Delhi – 110025, India

103 Penang Road, #05–06/07, Visioncrest Commercial, Singapore 238467

Cambridge University Press is part of Cambridge University Press & Assessment,
a department of the University of Cambridge.

We share the University's mission to contribute to society through the pursuit of
education, learning, and research at the highest international levels of excellence.

www.cambridge.org
Information on this title: www.cambridge.org/9781009102049

DOI: 10.1017/9781009106696

© Christian Hill 2024

This publication is in copyright. Subject to statutory exception and to the provisions
of relevant collective licensing agreements, no reproduction of any part may take
place without the written permission of Cambridge University Press & Assessment.

First published 2024

A catalogue record for this publication is available from the British Library

A Cataloging-in-Publication data record for this book is available from the Library of Congress

ISBN 978-1-009-10204-9 Paperback

Cambridge University Press & Assessment has no responsibility for the persistence
or accuracy of URLs for external or third-party internet websites referred to in this
publication and does not guarantee that any content on such websites is, or will
remain, accurate or appropriate.

Contents

Acknowledgments

For Emma, Charlotte and Laurence

Many people have helped in many different ways in the preparation of this book. In addition to my family, special thanks are due to Milo Shaffer, Alison Whiteley, Chris Pickard, Helen Reynolds, Lianna Ishihara and Natalie Haynes, who are so much better about deadlines than I am.

1
Introduction

1.1 About This Book

This book aims to provide a resource for students, teachers and researchers in chemistry who want to use Python in their work. Over the last 10 years, the Python programming language has been widely adopted by scientists, who appreciate its expressive syntax, gentle learning curve and numerous packages and libraries which facilitate numerical work.

The book is composed of relatively short chapters, each with a specific job. Mostly, these jobs fall into one of two categories: to act as a tutorial on a specific part of the Python language or one of its libraries, or to demonstrate the application of Python to a particular concept in chemistry. For students and teachers, these example applications are chosen to go beyond what can be reasonably achieved with a pencil, paper and a calculator: A brief overview of the chemical concepts is usually given, but there is no in-depth tutorial on these topics. Rather, it is assumed that the reader has some familiarity with the topic and wishes to use Python to solve larger or more complex problems than those usually presented in textbooks. For example, the chapter on Hückel molecular orbital theory (Chapter 33) outlines the assumptions behind this approach to modeling the electronic structure of organic molecules and then demonstrates the use of Python in determining the π molecular orbitals of benzene, which (using an unsymmetrized basis) involves a 6×6 matrix determinant: not a problem to be solved by hand.

Researchers are also increasingly using Python in their work as a computational tool, to manage and transform data, to produce publication-quality figures and visualizations, and even (using the JupyterLab package) as a replacement for laboratory notebooks, and to disseminate data and reproducible analysis. Indeed, Jupyter Notebook was listed in an article in the journal *Nature* as one of the ten computer codes that have transformed science.[1]

---

[1] J. M. Perkel, *Nature* **589**, 344–348 (2021).

Creating high-quality reports requires some knowledge of Markdown and LaTeX to write equations, tables, and chemical reactions: These are described in Chapter 14. There are further chapters on chemical databases and formats, peak-finding, linear and nonlinear least-squares fitting and symbolic computing with SymPy.

The examples are provided on the website `https://scipython.com/chem/` in the form of downloadable Jupyter Notebooks (see Chapter 13), and are supplemented by some exercises (with solutions at the back of the book).

1.2 About Python

Python is a powerful, general-purpose programming language that is well-suited to many of the tasks of scientific computing. It is a "high-level language" in that the programmer does not have to manage the fundamental operations of data type declarations, memory management and so on. In contrast to languages such as C and Fortran, for which the user must pass their code through a "compiler" to generate executable machine code before it is executed, Python programs are compiled automatically into "bytecode" (a kind of intermediate representation between its source and the machine code executable by the processor) by the Python interpreter. This makes the process of code development much quicker: There is a single step to code execution, and any errors are reported in messages returned are generally helpful and specific.

Python and its associated libraries are free and open source, in contrast to comparable commercial languages such as Mathematica and MATLAB. It is available for every major computer operating system, including Windows, Unix, Linux and macOS. It is a highly modular language: A core functionality is provided with the Python distribution itself, but there is a large number of additional modules and packages that extend its functionality. The most notable of these, in the context of scientific computing, are as follows:[2]

- NumPy: a package implementing mathematical algorithms for fast numerical computing, including support for vectors, matrices and multi-dimensional arrays – see Chapters 9 and 18.
- SciPy: a library of scientific computing algorithms for optimization, root-finding, linear algebra, integration, interpolation, signal processing and the numerical solution of ordinary differential equations – see Chapters 21, 22 and 25.
- Matplotlib: a package for visualizing and plotting data, with the ability to generate high-resolution, publication-quality graphs and charts – see Chapter 10.

---

[2] These packages are sometimes collectively referred to as the Python scientific computing "stack."

- pandas: a library providing high-level data structures for manipulating tabular data (`DataFrames` and `Series`), popular with data scientists – see Chapter 31.
- SymPy: a library for symbolic computation, with support for arithmetic, algebra and calculus – see Chapter 35.
- Jupyter: a suite of applications comprising a platform for interactive computing allowing scientists to share code and data analysis in a way that promotes reproducibility and collaboration – see Chapter 13.

However, Python programs will generally not execute as fast as those written in compiled languages: for heavily numerical work, even Python code using the NumPy and SciPy libraries (which call pre-compiled C routines from Python) will not run as fast as code written in, for example, C, C++ or Fortran. It is also hard to obfuscate the source code of a Python program: to some extent, an open-source philosophy is built-in to the Python ecosystem.

1.3 Installing Python

The official website of Python, `www.python.org`, contains full and easy-to-follow instructions for downloading Python. However, there are several full distributions which include the NumPy, SciPy and Matplotlib libraries to save you from having to download and install these yourself:

- *Anaconda* is available for free (including for commercial use) from `www.anaconda.com/distribution` This distribution includes its own well-documented package manager that can be used to install additional packages, either using a dedicated application or the command-line `conda` command.
- *Enthought Deployment Manager (EDM)* is a similar distribution with a free version and various tiers of paid-for versions including technical support and development software. It can be downloaded from `https://assets.enthought.com/downloads/`.

In most cases, one of these distributions should be all you need. There are some platform-specific notes below.

The source code (and binaries for some platforms) for the NumPy, SciPy, Matplotlib, pandas, SymPy and Jupyter packages are available separately at:

- NumPy: `https://github.com/numpy/numpy`
- SciPy: `https://github.com/scipy/scipy`
- Matplotlib: `https://matplotlib.org/users/installing.html`
- pandas: `https://pandas.pydata.org/`
- SymPy: `www.sympy.org/`
- Jupyter Notebook and JupyterLab: `https://jupyter.org/`

1.3.1 Windows

Windows users have a couple of further options for installing Python and its libraries: *Python(x,y)* (`https://python-xy.github.io`) and *WinPython* (`https://winpy thon.github.io/`). Both are free.

1.3.2 macOS

macOS, being based on Unix, comes with Python, usually an older version of Python 3 accessible from the Terminal application as `python3`. You must not delete or modify this installation (it's needed by the operating system), but you can follow the instructions above for obtaining a distribution with a more recent version of Python 3. macOS does not have a native *package manager* (an application for managing and installing software), but the two popular third-party package managers, Homebrew (`https://brew.sh/`) and MacPorts (`www.macports.org`), can both supply the latest version of Python 3 and its packages if you prefer this option.

1.3.3 Linux

Almost all Linux distributions these days come with Python 3 but the Anaconda and Enthought distributions both have versions for Linux. Most Linux distributions come with their own software package managers (e.g., `apt` in Debian and `rpm` for RedHat). These can be used to install more recent versions of Python 3 and its libraries, though finding the necessary package repositories may take some research on the Internet. Be careful not to replace or modify your system installation as other applications may depend on it.

1.4 Code Editors

Although Python code can be successfully written in any text editor, most programmers favor one with syntax highlighting and the possibility to define macros to speed up repetitive tasks. Popular choices include:

- Visual Studio Code, a popular, free and open-source editor developed by Microsoft for Windows, Linux and macOS;
- Sublime Text, a commercial editor with per-user licensing and a free-evaluation option;
- Vim, a widely used, cross-platform keyboard-based editor with a steep learning curve but powerful features; the more basic vi editor is installed on almost all Linux and Unix operating systems;

- Emacs, a popular alternative to Vim;
- Notepad++, a free Windows-only editor;
- SciTE, a fast, lightweight source code editor;
- Atom, another free, open-source, cross-platform editor.

Beyond simple editors, there are fully featured integrated development environments (IDEs) that also provide debugging, code-execution, code-completion and access to operating-system commands and services. Here are some of the options available:

- Eclipse with the PyDev plugin, a popular free IDE (`www.eclipse.org/ide/`);
- JupyterLab, an open-source browser-based IDE for data science and other applications in Python (`https://jupyter.org/`);
- PyCharm, a cross-platform IDE with commercial and free editions (`www.jetbrains.com/pycharm/`);
- PythonAnywhere, an online Python environment with free and paid-for options (`www.pythonanywhere.com/`);
- Spyder, an open-source IDE for scientific programming in Python, which integrates NumPy, SciPy, Matplotlib and IPython (`www.spyder-ide.org/`).

The short-code examples given in this book be in the form of an interactive Python session: commands typed at a prompt (indicated by `In [x]:`) will produce the indicated output (usually preceded by the prompt `Out [x]:`). It should be possible to duplicate these commands in a Jupyter Notebook or IPython session (e.g., within the interactive programming environments provided by the Anaconda distribution).

2

Basic Python Usage

2.1 Python as a Calculator

Perhaps the simplest use of Python is as a calculator: Numbers can be added, multiplied, subtracted, etc. There are three core *types* of number in Python: Integers (whole numbers) are represented by the type int, real numbers by type float,[1] and complex numbers (which have a real and an imaginary part) by the type complex. In an interactive Python session (e.g., a Jupyter Notebook), numbers of these types can be combined using *operators* in a single *expression* which is evaluated and the result returned to an output prompt. For example:

```
In [x]: 1 + 2
Out[x]: 3

In [x]: 10 / 4
Out[x]: 2.5

In [x]: 911 * 2356
Out[x]: 2146316
```

To make code more understandable, it is sometimes helpful to add *comments*: Everything on a single line following a # character is ignored by the Python interpreter and so can be used to write human-readable remarks providing explanation and context to the code. For example,

```
In [x]: # Molar enthalpy of fusion of ice: convert from kJ.mol-1 to J.mol-1.
In [x]: 6.01 * 1000
Out[x]: 6010.0

In [x]: 6.518 / 1013.25 * 760   # atmospheric pressure on Mars, in Torr
Out[x]: 4.888902047865778
```

---

[1] The name float refers to *floating-point numbers*, the approximate representation of real numbers used by Python (and most other modern computer languages).

The basic algebraic operators are listed in Table 2.1. In using them, it is important
to pay attention to their *precedence*: the order in which they are interpreted in an
expression (see Table 2.2). For example:

```
In [x]: 1 + 3 * 4
Out[x]: 13
```

Here, 3 * 4 is evaluated first, since the multiplication operator, *, has the higher
precedence out of + and *. The result, 12, is then added to 1. Where operators have
equal precedence, the parts of an expression are generally evaluated left-to-right.[2]
Precedence can be overridden using parentheses ("round brackets"):

```
In [x]: 6 / 3 ** 2          # the same as 6 / 9
Out[x]: 0.6666666666666666

In [x]: (6 / 3) ** 2        # the same as 2 ** 2
Out[x]: 4.0
```

Note that these expressions have resulted in a floating point number, even though
we are operating on integers. This is because the division operator, /, always returns a float, even when its result is a whole number. There is a separate integer
division operator, //, which returns the *quotient* of the division ("how many times
does the second number go into the first"); the related *modulus* operator, %, returns
the *remainder*:

```
In [x]: 7 / 3
Out[x]: 2.3333333333333335

In [x]: 7 // 3
Out[x]: 2

In [x]: 7 % 3
Out[x]: 1
```

Something interesting has happened with the expression 7 / 3: The exact value,
$2\frac{1}{3}$ cannot be represented in the way that Python stores real numbers (floating-point
numbers), which has a finite precision (of about 1 in 10^{16}): The nearest float value
to the answer that can be represented is returned.

In addition to the basic algebraic operators of Table 2.1, there are many mathematical functions and the constants π and e provided by Python's math library.
This is a built-in *module* that is provided with every Python installation (no extra
packages need to be installed to use it) but must be imported with the command:

```
import math
```

---

[2] The exception is exponentiation, which is evaluated right-to-left ("top-down").

Table 2.1   Basic Python
arithmetic operators

+	Addition
–	Subtraction
*	Multiplication
/	Floating-point division
//	Integer division
%	Modulus (remainder)
**	Exponentiation

Table 2.2   Python arithmetic operator precedence

**	(highest precedence)
*, /, //, %	
+, –	(lowest precedence)

The functions, some of which are listed in Table 2.3 can then be used by prefixing their name with `math.` – for example:

```
In [x]: import math
In [x]: math.sin(math.pi / 4)
Out[x]: 0.7071067811865475
```

This is an example of a *function call:*[3] The function `math.sin` is passed an *argument* (here, the number $\pi/4$) in parentheses, and returns the result of its calculation.

The NumPy package (see Chapter 9), which is not built in to Python (it needs to be installed separately), provides all of the functionality of `math` and more, so we shall mostly use it instead of `math`. It is usual to import NumPy with the alias `np`, as in the following example:

```
In [x]: import numpy as np
In [x]: 1 / np.sqrt(2)
Out[x]: 0.7071067811865475
```

Although the trigonometric functions in `math` and NumPy use radians rather than degrees, there are a couple of convenience methods for converting between the two:

```
In  [x]: np.degrees(np.pi / 2)
Out [x]: 90.0
```

```
In  [x]: np.sin(np.radians(30))
Out [x]: 0.49999999999999994
```

---

[3] This book will use the terms *function* and *method* interchangeably: a method is a function that "belongs to" an object, but in Python, everything is an object.

Table 2.3   Some functions and constants provided by the `math`
module. *Angles are assumed to be in radians*

`math.pi`	π
`math.e`	e
`math.sqrt(x)`	$\sqrt{x}$
`math.exp(x)`	e^x
`math.log(x)`	$\ln x$
`math.log10(x)`	$\log_{10} x$
`math.sin(x)`	$\sin(x)$
`math.cos(x)`	$\cos(x)$
`math.tan(x)`	$\tan(x)$
`math.asin(x)`	$\arcsin(x)$
`math.acos(x)`	$\arccos(x)$
`math.atan(x)`	$\arctan(x)$
`math.hypot(x, y)`	The Euclidean norm, $\sqrt{x^2 + y^2}$
`math.comb(n, r)`	The binomial coefficient, $\binom{n}{r} \equiv {}^nC_r$
`math.degrees(x)`	Convert x from radians to degrees
`math.radians(x)`	Convert x from degrees to radians

(Note again the finite precision here: The exact answer is 0.5.)

The `math.log` and `np.log` functions return the *natural* logarithm (of base e);
there are separate `math.log10` and `np.log10` variants:

```
In [x]: np.log(10)
Out[x]: 2.302585092994046
```

```
In [x]: 1 / np.log10(np.e)
Out[x]: 2.302585092994046
```

There are a couple of useful *built-in* functions (i.e., those that do require a package
such as `math` or NumPy to be imported). `abs` returns the absolute value of its
argument:

```
In [x]: abs(-4)
Out[x]: 4
```

and `round` rounds a number to a given precision in decimal digits (or to the nearest
integer if no precision is specified):

```
In [x]: round(3.14159265, 4)
Out[x]: 3.1416
```

```
In [x]: round(3.14159265)
Out[x]: 3
```

2.2 Defining Numbers

Unlike in some languages, Python does not require the user to *declare* a type for a number before its use. Numbers that look to the interpreter like integers will be treated as `int` objects; those that look like real numbers will become `float` objects. However, these data types are dimensionless numbers: If they represent physical quantities, the programmer is responsible for keeping track of any units.

Integers in Python can be as large as the computer's memory will allow. For defining very large integers, it can be convenient separate groups of digits with the underscore character, '_'.

```
In [x]: # Avogadro constant (mol-1): exact value by definition.
In [x]: 602_214_076_000_000_000_000_000
Out[x]: 602214076000000000000000
```

Floating point numbers can be written with a decimal point '.', again with optional digit grouping for clarity:

```
In [x]: # Gas constant (J.K-1.mol-1): exact value by definition.
In [x]: 8.31_446_261_815_324
Out[x]: 8.31446261815324
```

or in scientific notation, with the character e (or E) separating the mantissa (significant digits) and the exponent:

```
In [x]: # Boltzmann constant (J.K-1): exact value by definition.
In [x]: 1.380649e-23
Out[x]: 1.380649e-23
```

Complex numbers can be written as the sum of a real and an imaginary part, the latter indicated with a suffixed j:

```
In [x]: 1 + 4j
Out[x]: (1+4j)
```

or by explicitly passing a pair of values to `complex`:

```
In [x]: complex(-2, 3)
Out[x]: (-2+3j)
```

The real and imaginary parts are represented by floating point numbers and can be obtained separately using the `real` and `imag` attributes. The abs built-in returns the magnitude of a complex number:

```
In [x]: (3 + 4j).real
Out[x]: 3.0

In [x]: (3 + 4j).imag
Out[x]: 4.0
```

```
In [x]: abs(3 + 4j)
Out[x]: 5.0
```

2.3 Variables

It is very common to want to store a number in a program so that it can be used repeatedly and referred to by a convenient name: This is the purpose of a *variable*. In Python, a variable can be thought of as a label that is attached to an object (here, an int or a float number). There are some rules about what variables can be called:

- A variable name can contain only letters, digits and the underscore character (often used to indicate a subscript).
- A variable name cannot start with a digit.
- A variable cannot have the same name as any of a set of 30 or so *reserved keywords* (see Table 2.4).

Most modern code editors use syntax highlighting that will indicate when a reserved keyword is used. Note the difference between a valid assignment

```
In [x]: # Avogadro constant (mol-1): exact value by definition.
In [x]: N_A = 602_214_076_000_000_000_000_000
```

and one that fails because of an invalid variable name:

```
In [x]: import = 0
```

```
  File "<ipython-input-84-eb07b9ecdbca>", line 1
    import = 0
           ^
SyntaxError: invalid syntax
```

The attempt to assign 0 to a variable named import fails (a SyntaxError is "raised") because import is a reserved keyword (it forms part of the syntax of Python itself where it is used to import modules, as we have seen). In practice, reserved keywords make unlikely variable names, with the exception of lambda (see Section 6.5 for its meaning), which could be chosen to represent a wavelength, for example; consider using lam in this case. Table 2.4 also contains the three special, unchangeable keywords: True and False, which represent the corresponding Boolean concepts (see Section 5.1), and None, which is used to represent an empty or absent value.

Table 2.4   Python 3 reserved keywords

and	as	assert	async	await	break
class	continue	def	del	elif	else
except	finally	for	from	global	if
import	in	is	lambda	nonlocal	not
or	pass	raise	return	try	while
with	yield	False	True	None	

Well-chosen variable names can make Python code very clear and expressive:

```
In [x]: # Boltzmann constant (J.K-1): exact value by definition.
In [x]: k_B = 1.380649e-23

In [x]: R = N_A * k_B       # the gas constant (J.K-1.mol-1)
```

In the last statement, the right-hand side of the = sign, the expression N_A * k_B, is evaluated first and the variable name R is assigned ("bound") to the result of this calculation.

It is also possible to modify the value associated with a variable name:

```
In [x]: n = 1000
In [x]: n = n + 1
```

It is important to realize that the statement n = n + 1 does not represent some (impossible) mathematical equation to be solved for the number n but is rather an instruction to take the value of n, add one to it, and then reassign ("rebind") the name n to the result. The original value of 1,000 is then "forgotten": The computer memory that was used for its storage is freed and can be used for something else. Expressions like this are so common in programming that Python provides a shortcut (so-called "augmented assignment"):

```
In [x]: n = 1000
In [x]: n += 1
```

achieves the same result. There is a similar syntax for other operators: -=, *=, etc.

Another useful shortcut is the use of comma-separated values (actually, *tuples –* see Section 7.2) to assign several variables at once, for example:

```
In [x]: a, b, c = 42, -1, 0.5
```

In this book, we will try to give objects meaningful variable names: This is a good way to make code "self-documenting" and minimizes the use of explanatory comments that otherwise might be hard to maintain. That said, it is generally understood that the variables i, j and k are used as integer counters or indexes and that _ (the underscore character) is used when the program requires an object to be assigned a variable name but that object is not subsequently used.

2.4 Limitations and Pitfalls

Division by zero is always a hazard with mathematical operations. An Exception (Python error) is "raised" in this case, and execution of the code is halted (see also Section 5.5). The exact form of the error message produced may vary, depending on which calculation raised the Exception:

```
In [x]: 1 / 0
- - - - - - - - - - - - - - - - - - - - - - - - - - - - - - - - - - - - - - - - - - - - - - - - - - - - - - -

ZeroDivisionError                       Traceback (most recent call last)

<ipython-input-112-bc757c3fda29> in <module>
----> 1 1 / 0


ZeroDivisionError: division by zero
```

NumPy is a bit more tolerant of division by zero, and issues only a warning:

```
In [x]: x = 0
In [x]: y = np.sin(x) / x
<ipython-input-116-86b0517f1e0c>:2: RuntimeWarning: invalid value encountered in
double_scalars
  y = np.sin(x) / x
```

```
In [x]: y
Out[x]: nan
```

In this case, y has been assigned to a special floating-point value nan, which stands for "Not a Number," signifying that the result of a calculation is undefined. What the user does with this information is up to them, but be aware that there is no way back from a NaN: Any operation performed on this value simply returns another NaN and nothing is equal to NaN (it isn't even equal to itself).

Some behavior, mostly that resulting from floating-point arithmetic, can be surprising to new users:

```
In [x]: np.tan(np.pi/2)
Out[x]: 1.633123935319537e+16
```

The mathematically correct value of $\tan \frac{\pi}{2}$ is $+\infty$, but because np.pi does not represent π exactly, the value returned is just extremely large rather than infinite.

Floating-point numbers, because they are stored in a fixed amount of memory (8 bytes for so-called *double-precision floating point*), do have a maximum (and minimum) magnitude. Trying to evaluate an expression that generates a number too large to be represented (about 1.8×10^{308}) results in "overflow":

```
In [x]: np.exp(1000)
<ipython-input-120-47a6eab891c2>:1: RuntimeWarning: overflow encountered in exp
  np.exp(1000)

inf
```

Here, the overflow condition has returned the special value inf, which stands for infinity (even though e^{1000} is not infinite, just extremely large).

Underflow can also occur when a number has an absolute value too small to be represented in double-precision floating point form (less than about 2.2×10^{-308}): In this case, you generally won't be warned about it:

```
In [x]: np.exp(-1000)
Out[x]: 0.0
```

2.5 Examples

E2.1

Question

How many molecules of water are there in a 250 mL glass of water?

Take the density, $\rho(H_2O(l))=1$ g cm^{-3} and the molar mass, $M(H_2O)=18$ g mol^{-1}.

Solution

1 mL is the same volume as 1 cm^3 so we might as well keep the unit. We will define some variables first:

```
# Avogadro constant, in mol-1 to 4 s.f.
N_A = 6.022e23

# The volume of the water being considered, in cm3.
V = 250
# The density of water, in g.cm-3.
rho = 1
# The molar mass of H20, in g.mol-1.
M_H20 = 18
```

The water in our glass weighs $m = \rho V$ and contains $n = m/M(H_2O)$ moles of water.

```
# Mass of water, in g.
m = rho * V
```

```
# Amount of water, in mol.
n = m / M_H2O
n
```

```
13.88888888888889
```

The number of water molecules is then $N = nN_A$:

```
n * N_A
```

```
8.363888888888889e+24
```

That is, $N = 8.364 \times 10^{24}$ molecules.

E2.2

Question

The speed of sound in a gas of molar mass M at temperature T is

$$c = \sqrt{\frac{\gamma RT}{M}},$$

where R is the gas constant and for air the *adiabatic index*, $\gamma = \frac{7}{5}$.

Estimate the speed of sound in air at (a) $25\,^\circ$C and (b) $-20\,^\circ$C. Take $M = 29$ g mol^{-1}.

Solution

We have two temperatures to calculate for, so to avoid repeating ourselves, first define a factor, f:

$$c = f\sqrt{T}, \text{ where } f = \sqrt{\frac{\gamma R}{M}}.$$

If we stick to SI units (note that this means expressing M in kg mol^{-1}), then we should expect c to come out in units of m s^{-1}. Explicitly:

$$[c] = \sqrt{\frac{[\cdot][\text{J K}^{-1}\,\text{mol}^{-1}][\text{K}]}{[\text{kg mol}^{-1}]}} = \sqrt{\frac{[\text{J}]}{[\text{kg}]}} = \sqrt{\frac{[\text{kg m}^2\,\text{s}^{-2}]}{[\text{kg}]}} = \text{ms}^{-1}.$$

```
import numpy as np
# The gas constant in J.K-1.mol-1 (4 s.f.).
R = 8.314

# Mean molar mass of air, in kg.mol-1.
```

```
M = 29 / 1000
# Ratio of the heat capacities C_p / C_V (adiabatic index) for a diatomic gas.
gamma = 7 / 5

# Our factor, f, in m.s-1.K-1/2.
f = np.sqrt(gamma * R / M)
```

```
# Convert the first temperature from degC to K.
T = 25 + 273
f * np.sqrt(T)
```

```
345.84234000181505
```

```
# Convert the second temperature from degC to K.
T = -20 + 273
f * np.sqrt(-20 + 273)
```

```
318.66200881509076
```

That is, at $25\,°C$, the speed of sound is $346\ \mathrm{ms}^{-1}$, whereas at $-20\,°C$, it is $319\ \mathrm{ms}^{-1}$.

E2.3

Question

Acetic acid, CH_3CO_2H, is a weak acid with $pK_a = 4.756$. What is the pH of a 0.1 M solution of acetic acid in water?

Solution

Representing acetic acid by HA, the dissociation equilibrium is:

$$\mathrm{HA} \quad \rightleftharpoons \quad \mathrm{H^+} \quad + \quad \mathrm{A^-}$$
$$c - x \qquad\qquad x \qquad\quad x$$

That is, in solution x mol of the acid dissociates, producing x mol of hydrogen ions, the same amount of conjugate base, $\mathrm{A^-}$, and leaving $c - x$ mol of undissociated HA. At equilibrium,

$$K_a = \frac{\{\mathrm{H^+}\}\{\mathrm{A^-}\}}{\{\mathrm{HA}\}},$$

where $\{X\}$ is the *activity* of component X. For this weakly dissociating acid, we may approximate $\{X\} = [X]/c^{\ominus}$ for all species, where $c^{\ominus} = 1\ \mathrm{mol\ dm^{-3}}$ is introduced to ensure that K_a is dimensionless. Therefore,

$$K_a \approx \frac{[H^+][A^-]}{[HA]c^{\ominus}} = \frac{x^2}{(c-x)c^{\ominus}}.$$

By definition, $pK_a = -\log(K_a)$; for a *weak* acid like acetic acid, K_a is small:

```
# The pKa of acetic acid, log10 of the acid dissociation constant.
pKa = 4.756
Ka = 10**-4.756
Ka
```

```
1.7538805018417602e-05
```

That is, $K_a = 1.75 \times 10^{-5}$ and the position of the equilibrium favors the undissociated acid, HA. Nonetheless, some of the acid will dissociate, and the concentration of the resulting hydrogen ions determines the pH of the solution. Since $c \gg x$, it makes sense to approximate $c - x \approx c$ so as to avoid having to solve a quadratic equation. We get:

$$K_a \approx \frac{x^2}{cc^{\ominus}} \Rightarrow x \approx \sqrt{K_a cc^{\ominus}}$$

```
import numpy as np
# "Standard" amount concentration, 1 M = 1 mol.dm-3.
c_std = 1

# The concentration of the acid.
c = 0.1

# The concentration of hydrogen ions at equilibrium.
x = np.sqrt(Ka * c * c_std)
x
```

```
0.0013243415351946643
```

Thus, $[H^+] = 1.3 \times 10^{-3}$ M. Note that it seems justified to have taken $c \gg x$ (we have ignored the even smaller concentration of H^+ due to autoionization of the water solvent).

```
pH = -np.log10(x)
pH
```

```
2.878
```

The pH of 0.1 M acetic acid is therefore about 2.9.

If we felt we absolutely had to solve the equation for x as a quadratic, it is easily done. Rearranging and substituting $K_c = K_a c^{\ominus}$ for convenience yields:

$$x^2 + K_c x - K_c c = 0,$$

for which the only positive root is

$$x_+ = -\frac{K_c}{2} + \frac{1}{2}\sqrt{K_c^2 + 4K_c c}$$

```
Kc = Ka * c_std
x = (-Kc + np.sqrt(Kc**2 + 4*Kc*c))/2
x
```

```
0.0013156011665771512
```

which implies $[H^+] = 1.3 \times 10^{-3}$ M as before. The accurate value for the pH is:

```
pH = -np.log10(x)
pH
```

```
2.8808757500892046
```

which is the same, to three significant figures, as the approximate value.

E2.4

Question

The standard molar enthalpy of formation of gaseous ammonia at 298 K is $\Delta_f H_m^\ominus = -45.92$ kJ mol^{-1}.

The standard molar entropies, also at 298 K, of $H_2(g)$, $N_2(g)$ and $NH_3(g)$ are given below:

$$S_m^\ominus(H_2(g)) = 130.68 \text{ J K}^{-1} \text{ mol}^{-1}$$
$$S_m^\ominus(N_2(g)) = 191.61 \text{ J K}^{-1} \text{ mol}^{-1}$$
$$S_m^\ominus(NH_3(g)) = 192.77 \text{ J K}^{-1} \text{ mol}^{-1}.$$

What is the equilibrium constant, K, for the following equilibrium at 298 K?

$$N_2(g) + 3H_2(g) \rightleftharpoons 2NH_3(g)$$

Solution

First, assign some variable names to the necessary constants and provided thermodynamic quantities:

```
import numpy as np
# The gas constant in J.K-1.mol-1 (4 s.f.).
R = 8.314
```

```
# Standard molar enthalpy of formation of NH3(g) in kJ.mol-1.
DfH_NH3 = -45.92

# Standard molar entropies for H2(g), N2(g) and NH3(g) in J.K-1.mol-1.
Sm_H2 = 130.68
Sm_N2 = 191.61
Sm_NH3 = 192.77

# The temperature considered for the equilibrium.
T = 298
```

Next determine the standard enthalpies and entropies of the reaction. The left-hand side contains only elements in their reference state ($H_2(g)$ and $N_2(g)$) and so their enthalpies of formation are 0. Therefore,

$$\Delta_r H^\ominus = 2\Delta_f H_m^\ominus(NH_3(g)).$$

Entropy is a *state function*, so $\Delta_r S^\ominus$ can be obtained from an extension of Hess's Law:

$$\Delta_r S^\ominus = \sum_{products,P} S^\ominus(P) - \sum_{reactants,R} S^\ominus(R) \tag{2.1}$$

$$= 2S_m^\ominus(NH_3(g)) - 3S_m^\ominus(H_2(g)) - S_m^\ominus(N_2(g)) \tag{2.2}$$

```
# Standard enthalpy of reaction in kJ.mol-1
DrH = 2 * DfH_NH3
# Standard entropy of reaction in J.K-1.mol-1
DrS = 2 * Sm_NH3 - 3 * Sm_H2 - Sm_N2
```

The reaction is exothermic (favoring the products) but is associated with a decrease in entropy of the system (favoring the reactants):

```
DrH, DrS
```

```
(-91.84, -198.11)
```

The position of equilibrium is determined by the standard Gibbs free energy of reaction,

$$\Delta_r G^\ominus = \Delta_r H^\ominus - T\Delta_r S^\ominus.$$

At 298 K, we have:

```
DrG = DrH - T * DrS / 1000
```

Note that we need to convert the units of $\Delta_r S^{\ominus}$ to $kJ\,K^{-1}\,mol^{-1}$ by dividing by 1,000. Python has no conception of units: As far as the interpreter is concerned, the values are pure numbers and we have to keep track of the units ourselves if we are to get meaningful results.

```
DrG
```

```
-32.803219999999996
```

$\Delta_r G^{\ominus}$ is found to be $-32.8\,kJ\,K^{-1}\,mol^{-1}$, implying that products are favored overall.

Finally, the equilibrium constant is obtained by rearranging $\Delta_r G^{\ominus} = -RT\ln K$ to give $K = e^{-\Delta_r G^{\ominus}/RT}$.

```
# NB for this calculation we need the numerator in J.mol-1
K = np.exp(-DrG * 1000 / R / T)
```

There is a pitfall to avoid in this calculation:

```
K = np.exp(-DrG * 1000 / R * T)        # wrong!
```

would be wrong (the division is performed first, and then the quantity $\Delta_r G^{\ominus}/R$ *multiplied by* T. If you're a fan of parentheses, an alternative correct expression is:

```
K = np.exp(-DrG * 1000 / (R * T))
```

The equilibrium constant is:

```
K
```

```
562455.4959880211
```

As expected, $K \gg 1$.

2.6 Exercises

P2.1 The rate of the reaction between H_2 and F_2 to form HF increases by a factor of 10 when the temperature is increased from 25°C to 47°C. What is the reaction activation energy? Assume the Arrhenius equation applies.

P2.2 Body fat (triglyceride) has the average chemical formula $C_{55}H_{104}O_6$. In the absence of other mechanisms (such as ketosis), its metabolism is essentially a low-temperature combustion to form carbon dioxide and water.

Calculate the mass of CO_2 and H_2O produced when 1 kg of fat is "burned off." Take the molar masses to be $M(C) = 12$ g mol^{-1}, $M(H) = 1$ g mol^{-1} and $M(O) = 16$ g mol^{-1}.

What percentage of the original mass of fat is exhaled as CO_2?

P2.3 What is the boiling point of water on the summit of Mt Everest (8,849 m)? Assume that the ambient air pressure, p, decreases with altitude, z, according to $p = p_0 \exp(-z/H)$, where $p_0 = 1$ atm and take the scale height, H to be 8 km. The molar enthalpy of vaporization of water is $\Delta_{vap}H_m = 44$ kJ mol^{-1}.

The Clausius–Clapeyron equation is:

$$\frac{d \ln p}{dT} = \frac{\Delta_{vap}H_m}{RT^2}.$$

3

Strings

3.1 Defining Strings

A Python string (of type `str`) is a sequence of characters, usually representing some textual data; for example, a message that can be printed on the screen. Strings are defined by using either single or double quotes, and can be assigned to variable names, as with numbers:

```
In [x]: 'sodium'
Out[x]: 'sodium'

In [x]: gas = "Carbon dioxide"
```

Strings can be concatenated with the + operator and repeated with the * operator:

```
In [x]: 'CH3' + 'CH2'*3 + 'CH3'
Out[x]: 'CH3CH2CH2CH2CH3'

In [x]: prefix = 'trans-'
In [x]: prefix + 'but-2-ene'
Out[x]: 'trans-but-2-ene'
```

A string between quotes is called a string *literal*; string literals defined next to each other are automatically concatenated (no + required). Note that no space is added into the concatenated string:

```
In [x]: 'Na' 'Cl'
Out[x]: 'NaCl'
```

This can be useful when generating long strings, which can be broken up into parts on separate lines, provided the whole expression is contained in parentheses:

```
In [x]: quote = ('For me chemistry represented an indefinite cloud of future'
                 ' potentialities which enveloped my life to come in black'
                 ' volutes torn by fiery flashes')
```

Table 3.1   Common Python escape sequences

Escape sequence	Meaning
\'	Single quote (')
\"	Double quote (")
\n	Linefeed (LF)
\r	Carriage return (CR)
\t	Horizontal tab
\\	The backslash character itself
\u, \U, \N{}	Unicode character identified by its code point

The line breaks are not included in the string: to represent a new line and other special characters such as tabs, so-called *escape characters* are used, prefixed by a backslash ('\', see Table 3.1). They won't be resolved in the string literal echoed back by the Python interactive shell or Jupyter output cell, however. Instead, pass them to the print function. For example:

```
In [x]: 'Li\tNa\tK\tRb\n3\t11\t19\t37'
Out[x]: 'Li\tNa\tK\tRb\n3\t11\t19\t37'

In [x]: print('Li\tNa\tK\tRb\n3\t11\t19\t37')
Li      Na      K       Rb
3       11      19      37
```

Here, the escape character \t has been printed as a tab and \n ends a line of output and starts a new one.

There are occasions (e.g., when defining strings of LaTeX source, see Chapter 14) when one does not want a backslash to indicate an escape character; in this case either escape the backslash itself ('\\') or define a *raw string* (r'...') as follows:

```
In [x]: print('\tan(\pi)')
        an(\pi)              # Oops: we didn't want \t to be turned into a tab

In [x]: print('\\tan(\pi)')         # Escape the backslash
\tan(\pi)

In [x]: print(r'\tan(\pi)')         # Raw string
\tan(\pi)
```

The print function is extremely powerful and can take a comma-separated sequence of objects (including numbers, which it converts to strings) and output them. By default, it separates them by a single space, and provides a single newline at the end of its output.

```
In [x]: print('Standard atomic weight of Na:', 22.99)
Standard atomic weight of Na: 22.99

In [x]: print('2 + 3 =', 2 + 3)
2 + 3 = 5
```

The difference between \n and \r is that the latter returns the print "cursor" to the start of the line but does not create a new line, so subsequent output overwrites whatever was there before:

```
In [x]: print("Fluorine\rChl")
Chlorine
```

Unicode characters can be identified by their *code points*: each character in almost all the world's major written languages is assigned an integer within the Unicode standard, and this integer code can be used in Python strings using the \u escape code:[1]

```
In [x]: print("\u212b")   # The Swedish Å, used to represent the angstrom
Å
```

Some common characters have specific names that can be referred to using the N escape character:

```
In [x]: print("\N{GREEK CAPITAL LETTER DELTA}")
Δ
```

3.2 String Indexing and Slicing

Individual characters can be accessed using the indexing notation s[i], where the index i is the position of the character in the string, *starting at zero*. For example,

```
In [x]: s = 'Plutonium'
In [x]: s[0]
Out[x]: 'P'

In [x]: s[4]
Out[x]: 'o'
```

If the index i is negative, it is taken to be counted backward from the end of the string (the last character is in "position" −1):

```
In [x]: s[-1]
Out[x]: 'm'
```

[1] For more information on Unicode, see the official site, https://unicode.org/. To search for the code point representing a specific character, there is a useful online service at https://codepoints.net/.

```
In [x]: s[-3]
Out[x]: 'i'
```

Character	P	l	u	t	o	n	i	u	m
Index	0	1	2	3	4	5	6	7	8
Index	−9	−8	−7	−6	−5	−4	−3	−2	−1

A *substring* is obtained by *slicing* the string with the notation s[i:j], where i is the index of the first character to use, and j is *one after* the last character to take. That is, the substring includes the starting character but excludes the ending character:

```
In [x]: s[2:4]
Out[x]: 'ut'       # Substring is s[2] + s[3], the third and fourth characters
```

This takes a bit of getting used to, but it's actually quite useful: The returned string has j - i characters in it, and sequential slices use indexes that start off where the previous one left off:

```
In [x]: s[0:4] + s[4:7] + s[7:9]
Out[x]: 'Plutonium'
```

Strings can be sliced backward or stridden differently with a third value: s[i:j:stride]; where any of i, j, stride are omitted, it is assumed that, respectively, the start of the string (i=0), the end of the string and *stride*=1 (consecutive characters) are intended:

```
In [x]: s[:5]           # The first five characters
Out[x]: 'Pluto'

In [x]: s[1:6:2]        # s[1], s[3], s[5]
Out[x]: 'ltn'

In [x]: s[::-1]
Out[x]: 'muinotulP'     # The whole string, reversed
```

Strings in Python are *immutable*; they cannot be changed in place:

```
In [x]: s = 'alkane'
Out[x]: s[3] = 'e'
-------------------------------------------------------------------
TypeError                                 Traceback (most recent call last)
<ipython-input-65-44f92721ca9f> in <module>
----> 1 s[3] = 'e'

TypeError: 'str' object does not support item assignment
```

Instead, a new string can be created from slices of the old one:

```
In [x]: s[:3] + 'e' + s[4:]        # 'alk' + 'e' + 'ne'
Out[x]: 'alkene'
```

The new string can be assigned to a variable name, of course, including same one, if we want:

```
In [x]: s = s[:3] + 'y' + s[4:]        # 'alk' + 'y' + 'ne'
In [x]: s
Out[x]: 'alkyne'
```

3.3 String Methods

Python comes with a large number of methods for manipulating strings. A full list is given in the documentation,[2] but the most useful ones are demonstrated here:

```
In [x]: f = 'CH3CH2CH2CH3'
In [x]: len(f)                 # sequence length, a built-in function
Out[x]: 12

In [x]: f.count('CH2')      # number of substrings
Out[x]: 2

In [x]: f.lower()           # also f.upper()
Out[x]: 'ch3ch2ch2ch3'

In [x]: f.index('CH2')       # first index of substring
Out[x]: 3

In [x]: # remove any of the characters 'C', 'H', '3' from both ends of f
In [x]: f.strip('CH3')
Out[x]: '2CH2'

In [x]: f.removeprefix('CH3')   # also f.removesuffix(); Python 3.9+ only
Out[x]: 'CH2CH2CH3'

In [x]: f.replace('CH3', 'NH2')
Out[x]: 'NH2CH2CH2NH2'
```

Again, f is immutable and left unchanged by these operations: They return a new Python object. This means that they can be chained together, for example:

```
In [x]: f.removeprefix('CH3').removesuffix('CH3').lower()
Out[x]: 'ch2ch2'
```

See also Section 4.4 for information on the string methods split and join.

---

[2] https://docs.python.org/3/library/stdtypes.html#string-methods.

3.4 String Formatting

When a number is printed to the screen, Python converts it to a string and does its best to choose an appropriate representation: integers are always output with all their digits;[3] for very large or very small floating point numbers, scientific notation is used:

```
In [x]: N_A = 602_214_076_000_000_000_000_000
In [x]: print(N_A)
602214076000000000000000

In [x]: print(float(N_A))
6.02214076e+23
```

The way that numbers are formatted can be controlled using the string `format` method. In its simplest use, values are simply interpolated into a string template at locations indicated by braces (`'{}'`):

```
In [x]: c = 299792458
In [x]: units = 'm.s-1'
In [x]: 'The speed of light is c = {} {}'.format(c, units)
Out[x]: 'The speed of light is c = 299792458 m.s-1'
```

The arguments passed to the `format` method can also be referred to by name or (zero-based) index inside the braces:

```
In [x]: '{0} = {1} {cgs_units}'.format('c', c * 100, cgs_units='cm.s-1')
Out[x]: 'c = 29979245800 cm.s-1'
```

There is a special syntax for refining the formatting of these interpolated strings: the field width, padding number of decimal places and so on. Starting with integers, the width of the space allocated for the number, w, is specified with `:wd` as shown here:

```
In [x]: '{:6d}'.format(42)
Out[x]: '    42'
```

The number 42 is output as a string of six characters, padded on the left with spaces.

Floating-point numbers can be given a precision, p, as well as a width: `:w.pf` and `:w.pe` for positional and scientific notation respectively:

```
In [x]: k_B = 1.380649e-23
In [x]: '{:12.3e}'.format(k_B) # 12 characters, 3 decimal places
Out[x]: '   1.381e-23'

In [x]: '{:.3e}'.format(k_B)    # as many characters as needed, 3 decimal places
```

---

[3] In recent versions of Python, there is actually a default limit of 4300 digits for printing large integers because of a perceived security vulnerability: Converting a huge integer into a string can consume considerable memory and processor resources.

```
Out[x]: '1.381e-23'

In [x]: '{:.28f}'.format(k_B)   # 28 characters, positional notation
Out[x]: '0.00000000000000000000000138065'
```

Note that if the width is omitted, Python makes the string as long as it needs to be
to provide the precision requested.

There are a couple of other useful types of format specifier: :w.pg shifts between
positional and scientific notation depending on the magnitude of the number and the
precision requested (according to some reasonably complicated rules.) Use :ws for
interpolating strings into other strings (which is sometimes necessary for aligning
output).

Since Python version 3.6 it has been possible to interpolate the values of objects
directly, using *f-strings*. The variable name of the object is placed before the colon
in a string defined within quotes preceded by the letter f as follows:[4]

```
In [x]: f"Boltzmann's constant is approximately {k_B:.3e} J.K-1"
Out[x]: "Boltzmann's constant is approximately 1.381e-23 J.K-1"
```

It is even possible to carry out calculations within the interpolation braces: that
is, the interpolated object can derive from an *expression*. To keep code clear, it is
better not to include complex expressions in an f-string, but it is common to carry
out simple unit conversions if necessary:

```
In [x]: print(f'The speed of light is approximately {c/1000:.3f} km.s-1')
The speed of light is approximately 299792.458 km.s-1
```

3.5 Examples

E3.1

Question

Produce a formatted string reporting the mass of a linear alkane, C_nH_{2n+2}, given
n. Take the molar masses of C and H to be 12.0107 g mol^{-1} and 1.00784 g mol^{-1},
respectively.

Solution

```
# Molar masses of carbon, hydrogen atoms (g.mol-1).
mC, mH = 12.0107, 1.00784
# String formatting template.
fmt = 'Mass of {:5s}: {:6.2f} g.mol-1'
```

---

[4] Note that in this example, we have chosen to delimit the string using double quotes, ", because we want to use
a single quote, ', inside the string as an apostrophe.

```
n = 2
nH = 2 * n + 2
formula = 'C{}H{}'.format(n, nH)                    # or f'C{n}H{nH}'
print(fmt.format(formula, n * mC + nH * mH))


n = 8
nH = 2 * n + 2
formula = 'C{}H{}'.format(n, nH)
print(fmt.format(formula, n * mC + nH * mH))
```

```
Mass of C2H6 :   30.07 g.mol-1
Mass of C8H18: 114.23 g.mol-1
```

E3.2

Question

Produce a nicely formatted list of the values of the physical constants, h, c, k_B, R and N_A, to four significant figures, with their units.

Solution

First define variables for the values and units of the physical constants.

```
h, h_units = 6.62607015e-34, 'J.s'
c, c_units = 299792458, 'm.s-1'
kB, kB_units = 1.380649e-23, 'J.K-1'
R, R_units = 8.314462618, 'J.K.mol-1'
N_A, N_A_units = 6.02214076e+23, 'mol-1'
```

We can define our list of constants in a single string by concatenation. It is possible to split the assignment across multiple lines of input, but the right-hand side must be wrapped in parentheses for Python to interpret it as a single expression.

```
s = (f'h    = {h:9.3e} {h_units}\n'
     f'c    = {c:9.3e} {c_units}\n'
     f'kB   = {kB:9.3e} {kB_units}\n'
     f'R    = {R:9.3f} {R_units}\n'
     f'N_A = {N_A:9.3e} {N_A_units}')
```

```
print(s)
```

```
h   = 6.626e-34 J.s
c   = 2.998e+08 m.s-1
kB  = 1.381e-23 J.K-1
R   =     8.314 J.K.mol-1
N_A = 6.022e+23 mol-1
```

3.6 Exercise

P3.1 The following variables define some thermodynamic properties of CO_2 and H_2O:

```
# Triple point of CO2 (K, Pa).
T3_CO2, p3_CO2 = 216.58, 5.185e5
# Enthalpy of fusion of CO2 (kJ.mol-1).
DfusH_CO2 = 9.019
# Entropy of fusion of CO2 (J.K-1.mol-1).
DfusS_CO2 = 40
# Enthalpy of vaporization of CO2 (kJ.mol-1).
DvapH_CO2 = 15.326
# Entropy of vaporization of CO2 (J.K-1.mol-1).
DvapS_CO2 = 70.8

# Triple point of H2O (K, Pa).
T3_H2O, p3_H2O = 273.16, 611.73
# Enthalpy of fusion of H2O (kJ.mol-1).
DfusH_H2O = 6.01
# Entropy of fusion of H2O (J.K-1.mol-1).
DfusS_H2O = 22.0
# Enthalpy of vaporization of H2O (kJ.mol-1).
DvapH_H2O = 40.68
# Entropy of vaporization of H2O (J.K-1.mol-1).
DvapS_H2O = 118.89
```

Use a series of **print** statements with f-strings to produce the following formatted table:

```
                            CO2         H2O
---------------------------------------------
p3     /Pa              518500       611.73
T3     /K               216.58       273.16
DfusH /kJ.mol-1          9.019        6.010
DfusS /J.K-1.mol-1        40.0         22.0
DvapH /kJ.mol-1         15.326       40.680
DvapS /J.K-1.mol-1        70.8        118.9
```

4

Lists and Loops

4.1 Definitions, Syntax and Usage

A string is an example of a *sequence*: a collection of objects (here, characters) that can be taken one at a time (i.e., iterated over). The syntax for retrieving objects from a sequence is:

```
for object in sequence:
    # Do something with object
    ...
```

This is a **for** loop: The block of code *indented by four spaces* is executed once for each object in the sequence. For example,

```
In [x]: name = 'tin'
   ...: for letter in name:
   ...:     print(letter)
   ...:
t
i
n
```

Another important Python sequence is the `list`: an ordered collection of objects that can be of any type. Lists are defined as a comma-separated sequence of objects between square brackets:

```
In [x]: mylist = [4, 'Sn', 'sodium', 3.14159]
```

Lists can be indexed, sliced and iterated over just like strings:

```
In [x]: print(mylist[2])        # The third item in mylist
sodium

In [x]: print(mylist[2::])      # All items from the third onward
['sodium', 3.14159]

In [x]: for item in mylist:     # Loop over all items in mylist
```

```
   ...:        print(item)
   ...:
4
Sn
sodium
3.14159
```

Unlike strings, `lists` are *mutable*: They can be changed in place in various ways, for example by direct assignment:

```
In [x]: mylist[2] = 'potassium'
In [x]: mylist
Out[x]: [4, 'Sn', 'potassium', 3.14159]
```

There are a couple of other important methods for manipulating lists: `append` adds an item onto the end of the `list` (growing its length by one):

```
In [x]: mylist.append(-42)

In [x]: mylist
Out[x]: [4, 'Sn', 'sodium', 3.14159, -42]
```

`extend` can be used to concatenate two lists:

```
In [x]: list1 = ['H', 'He', 'Li']

In [x]: list2 = ['Be', 'B', 'C']

In [x]: list1.extend(list2)

In [x]: list1
Out[x]: ['H', 'He', 'Li', 'Be', 'B', 'C']
```

New `lists` can be created by passing any iterable object into the `list()` constructor:

```
In [x]: list('helium')
Out[x]: ['h', 'e', 'l', 'i', 'u', 'm']
```

Individual numbers are not iterable, so `list(3.14159)` fails:

```
In [x]: list(3.14159)
---------------------------------------------------------------
TypeError                         Traceback (most recent call last)
<ipython-input-32-8870af92a665> in <module>
----> 1 list(3.14159)

TypeError: 'float' object is not iterable
```

However, `str(3.14159)`, the string representation of this number, "`3.14159`", is iterable and so we could create a list of its characters:

```
In [x]: list(str(3.14159))
Out[x]: ['3', '.', '1', '4', '1', '5', '9']
```

4.2 range and enumerate

To generate a sequence of regularly spaced integers (an arithmetic progression), there is a built-in function called range. With a single argument, range(n) generates the integers 0, 1, 2, ... *n-1*:

```
In [x]: for i in range(4):
   ...:      print(i, i**2, i**3)
   ...:
   ...:
0 0 0
1 1 1
2 4 8
3 9 27
```

Note that this sequence follows the general rules of starting at 0 and finishing at *n-1*. This means that it is suitable for generating indexes into other sequences such as lists and strings.

Recall the example of iterating over the characters in a string:

```
In [x]: for letter in 'tin':
   ...:      print(letter)
   ...:
t
i
n
```

The example here demonstrates how expressive Python is as a language. In other languages, it is necessary to keep track of an index into the sequence and use it to retrieve items one at a time, something that could be achieved with the range function:

```
In [x]: name = 'tin'
In [x]: for i in range(len(name)):    # that is, i = 0, 1, 2, 3
   ...:      print(name[i])
   ...:
   ...:
t
i
n
```

This use of range is not encouraged in Python – the simpler and clearer syntax for letter in *name*: is easier to understand and maintain. If, for some reason, you *do* want the index of each item as well as the item itself, there is a further built-in function, `enumerate`: this returns both, which can be unpacked into two variables in the loop:

```
In [x]: for i, letter in enumerate('zinc'):
   ...:         print(i, ':', letter)
   ...:
0 : z
1 : i
2 : n
3 : c
```

4.3 Creating lists

Since a `list` can be grown in-place, one way to create one is to simply append its elements one by one:

```
In [x]: my_list = []
In [x]: for i in range(5):
   ...:         my_list.append((i + 1)**2)

In [x]: my_list
Out[x]: [1, 4, 9, 16, 25]
```

However, there is a more concise syntax which is commonly used in this case: *list comprehension*:

```
In [x]: my_list = [(i + 1)**2 for i in range(5)]

In [x]: my_list
Out[x]: [1, 4, 9, 16, 25]
```

Quite complex expressions can be used in a list comprehension, but it is best to ensure that the code is still easily understandable. Fortunately, Python is expressive enough that multiple functions can be chained without losing readability. For example, to calculate $\sin\theta$ for some angles, θ:

```
In [x]: import math
In [x]: angles = [0, 30, 60, 90, 120, 150, 180]
In [x]: sines = [round(math.sin(math.radians(theta)), 3) for theta in angles]
In [x]: sines
Out[x]: [0.0, 0.5, 0.866, 1.0, 0.866, 0.5, 0.0]
```

4.4 split and join

These two useful methods act on strings. To split a string into a `list` of substrings on a given separator, call `split()`:

```
In [x]: 'a,b,c'.split(',')
Out[x]: ['a', 'b', 'c']

In [x]: 'a::b::c'.split(':')
Out[x]: ['a', '', 'b', '', 'c']

In [x]: 'a::b::c'.split('::')
Out[x]: ['a', 'b', 'c']
```

If you wish to split on *any amount* of whitespace (spaces, tabs), there is no need to provide an argument to `split()`:

```
In [x]: 'a  b    c'.split()
Out[x]: ['a', 'b', 'c']
```

The `join` method does the reverse: It concatenates a sequence of strings, using the specified separator:

```
In [x]: '-'.join(['a', 'b', 'c'])
'a-b-c'
```

The sequence doesn't have to be a `list`: A string is a sequence of characters (which are themselves strings), so the following also works:

```
In [x]: '-'.join('abc')
'a-b-c'
```

4.5 zip

Sometimes it is necessary to iterate over two or more sequences at the same time. This is possible by directly indexing into them, for example with:

```
In [x]: list1 = [1, 2, 3]
In [x]: list2 = ['a', 'b', 'c']
In [x]: for i in range(len(list1)):
   ...:         print(list1[i], list2[i])
   ...:
1 a
2 b
3 c
```

The `zip` built-in allows iteration without explicit indexing: When looped over, it emits the matched items from each sequence in turn:

```
In [x]: for number, letter in zip(list1, list2):
   ...:         print(number, letter)
   ...:
```

```
1 a
2 b
3 c
```

If the sequences have different lengths, `zip` will return matched elements until the shortest sequence is exhausted:

```
In [x]: seq1 = 'abcdefghijk'
In [x]: seq2 = range(1000)
In [x]: seq3 = ['Be', 'Mg', 'Ca', 'Sr']
In [x]: for letter, number, symbol in zip(seq1, seq2, seq3):
   ...:        print(letter, number, symbol)
   ...:
   ...:
a 0 Be
b 1 Mg
c 2 Ca
d 3 Sr
```

4.6 Examples

E4.1

Question

Produce a table of the first 10 element symbols and their corresponding atomic numbers, Z.

Solution

Since Python indexes start at 0, we have to add 1 to the index into an ordered list of element symbols to get Z:

```
symbols = ['H', 'He', 'Li', 'Be', 'B', 'C', 'N', 'O', 'F', 'Ne']
print('Element | Z')
print('------------')
for Z, symbol in enumerate(symbols):
    print('    {:2s} | {:2d}'.format(symbol, Z+1))
```

```
Element | Z
------------
     H |  1
    He |  2
    Li |  3
    Be |  4
     B |  5
     C |  6
     N |  7
     O |  8
     F |  9
    Ne | 10
```

Alternatively, **enumerate** takes an optional argument, **start**, which sets the first integer to be returned as the count:

```
for Z, symbol in enumerate(symbols, start=1):
    print('     {:2s} | {:2d}'.format(symbol, Z))
```

```
    H  |   1
    He |   2
    Li |   3
    Be |   4
    B  |   5
    C  |   6
    N  |   7
    O  |   8
    F  |   9
    Ne |  10
```

E4.2

Question

Calculate the mean molar mass of dry air, given the composition and masses listed in Table 4.1:

Solution

```
# Constituents of dry air: molecule formula, percentage composition by volume
# and molar mass (g.mol-1).
molecule = ['N2', 'O2', 'Ar', 'CO2']
mass = [28.0134, 31.9898, 39.948, 44.0095]
composition = [78.084, 20.946, 0.9290, 0.041]
```

We now have to iterate over two lists at the same time, to calculate the mean:

$$\bar{M}_{air} = \sum_i x_i M_i,$$

where x_i is the fractional amount by volume of the ith constituent of air, with molar mass M_i.

Table 4.1   The composition of dry air

Molecule	Mass /Da	Composition /%
N_2	28.0134	78.084
O_2	31.9898	20.946
Ar	39.948	0.9290
CO_2	44.0095	0.041

```
mean_air_mass = 0
for m, x_perc in zip(mass, composition):
    mean_air_mass += m * x_perc / 100

print('Mean mass of dry air: {:.2f} g.mol-1'.format(mean_air_mass))
```

```
Mean mass of dry air: 28.96 g.mol-1
```

E4.3

Question

A balloon, of mass 0.8 g when uninflated, is filled with helium to a volume of 2 L
at 1 atm ambient pressure. What is the net force on the balloon at the following
temperatures: $-20, 0, 25, 40°C$?

Take $M(He) = 4$ g mol^{-1}, the mean molar mass of air, $\bar{M}_{air} = 29$ g mol^{-1} and
gravitational acceleration, $g = 9.8$ kg m s^{-2}.

Solution

Treating the helium as an ideal gas, its amount is $n = pV/RT$ where the pressure,
p, can be taken to be the same as ambient pressure (ignoring the effects of the
curvature of the inflated balloon's surface).

There are two forces on the balloon: its weight, $F_g = -mg$, where $m = m_{balloon} +
m_{He}$, and the buoyant force due to the air it displaces, $F_b = m_{air}g$. The net force is
then $F = F_g + F_b$ where a negative value of F indicates the balloon will sink (force
directed down), and a positive value that it will rise.

First, define some variables for the constants and parameters in this problem:

```
# Gas constant, J.K-1.mol-1) and gravitational acceleration (kg.m.s-2).
R = 8.314
g = 9.8
# Pressure, Pa
p = 101325
# Inflated volume, m3.
V = 2.e-3
# Mass of uninflated balloon (kg).
m_balloon = 0.8 / 1000
# Molar mass of helium and mean molar mass of air (g.mol-1).
M_He = 4
M_air = 29
```

Now the calculation for the net force can be carried out inside a loop over the
different temperatures:

```
# Temperatures, degC
temperatures = [-20, 0, 25, 40]
```

```
for T in temperatures:
    # Amount (mol) and mass (kg) of He in the inflated balloon.
    n_He = p * V / R / (T + 273)
    m_He = M_He / 1000 * n_He
    # The same amount of air is displaced, but it weighs more.
    m_air = M_air / 1000 * n_He
    # Net force.
    F = (m_air - m_He - m_balloon) * g
    print('{:3d} deg C   {:.4f} N'.format(T, F))
```

```
-20 deg C    0.0158 N
  0 deg C    0.0140 N
 25 deg C    0.0122 N
 40 deg C    0.0112 N
```

Loops can be nested. For example, we can repeat the calculation for the gases Ne and Ar as well.

```
# Gases and their molar masses
gases = ['He', 'Ne', 'Ar']
gas_masses = [4, 20, 40]
# Temperatures, degC
temperatures = [-20, 0, 25, 40]

for T in temperatures:
    # Amount (mol) and mass (kg) of gas in the inflated balloon.
    n = p * V / R / (T + 273)
    # The mass of displaced air.
    m_air = M_air / 1000 * n
    for gas, M_gas in zip(gases, gas_masses):
        m_gas = M_gas / 1000 * n
        # Net force.
        F = (m_air - m_gas - m_balloon) * g
        print('{} at {:3d} deg C: Force = {:8.5f} N'.format(gas, T, F))
```

```
He at -20 deg C: Force =   0.01576 N
Ne at -20 deg C: Force =   0.00066 N
Ar at -20 deg C: Force =  -0.01823 N
He at   0 deg C: Force =   0.01403 N
Ne at   0 deg C: Force =   0.00003 N
Ar at   0 deg C: Force =  -0.01746 N
He at  25 deg C: Force =   0.01220 N
Ne at  25 deg C: Force =  -0.00063 N
Ar at  25 deg C: Force =  -0.01666 N
He at  40 deg C: Force =   0.01124 N
Ne at  40 deg C: Force =  -0.00097 N
Ar at  40 deg C: Force =  -0.01623 N
```

With a bit of thought, we can use `print` to generate a table of the results.

```
print('Net force on balloon (in newtons; +ve is up)')
print('    T       He       Ne       Ar')
for T in temperatures:
```

```
# Amount (mol) and mass (kg) of gas in the inflated balloon.
n = p * V / R / (T + 273)
# The mass of displaced air.
m_air = M_air / 1000 * n
print('{:3d} °C'.format(T), end='')
for gas, M_gas in zip(gases, gas_masses):
    m_gas = M_gas / 1000 * n
    # Net force.
    F = (m_air - m_gas - m_balloon) * g
    print('{:10.5f}'.format(F), end='')
print()
```

```
Net force on balloon (in newtons; +ve is up)
    T          He          Ne          Ar
-20 °C    0.01576    0.00066    -0.01823
  0 °C    0.01403    0.00003    -0.01746
 25 °C    0.01220   -0.00063    -0.01666
 40 °C    0.01124   -0.00097    -0.01623
```

Here, we have used end='' to suppress the default behavior of print to add a new line at the end of its output. The command print() on its own simply outputs a single new line.

5

Comparisons and Flow Control

5.1 Comparisons and Logic

So far, the code we have written is executed in a pre-determined order, one statement after another (or repeated a fixed number of times in the case of a for loop).

For many programs, however, it is important that they can *branch*: Execute different code statements according to one or more conditions. In Python, this is achieved using the if ... elif ... else syntax in which comparisons between objects are evaluated as True or False. A list of useful comparison operators is given in Table 5.1. Note that to compare two objects (e.g., numbers) for equality, the operator is ==: two equals signs, not to be confused with the single equals, =, used for assignment.

```
In [x]: a = 4

In [x]: a == 2 + 2
Out[x]: True

In [x]: a >= 5
Out[x]: False
```

Comparisons are combined with the logic operator keywords and, not and or:

```
In [x]: a, b, c = 1, 2, 3

In [x]: a > b or c > b
Out[x]: True

In [x]: a > b and c > b
Out[x]: False

In [x]: not a > b
Out[x]: True
```

Table 5.1   Python comparison
operators

==	Equal to
!=	Not equal to
>	Greater than
<	Less than
>=	Greater than or equal to
<=	Less than or equal to

There is a precedence issue to consider. The rules are that arithmetic operations (+, *, etc.) always have higher precedence (are evaluated before) comparisons (==, <, etc.), and comparisons always have higher precedence than logical operators. Within the logical operators, the precedence is not (highest precedence) to or (lowest precedence). As with arithmetic, expressions in parentheses have the highest precedence of all. Therefore:

```
In [x]: not a == 1 and b > c       # (not True) and False => False
Out[x]: False

In [x]: not (a == 1 and b > c)     # not (True and False) => (not False) => True
Out[x]: True
```

There are some other useful "membership operators": in and not in test for membership and nonmembership of a sequence, respectively:

```
In [x]: metals = 'Sc Ti V Cr Mn Fe Co Ni Cu Zn'.split()

In [x]: metals
Out[x]: ['Sc', 'Ti', 'V', 'Cr', 'Mn', 'Fe', 'Co', 'Ni', 'Cu', 'Zn']

In [x]: 'Fe' in metals
Out[x]: True

In [x]: 'V' not in metals
Out[x]: False
```

Two more built-in functions are useful for comparisons involving all the items in a sequence. any returns True if any of the items is equivalent to True; all returns True if all of them are:

```
In [x]: all(['a', True, 2])
Out[x]: True

In [x]: any([0, [], '', False])
Out[x]: False
```

As explained below, empty container objects and the number zero are considered equivalent to False; other objects values are equivalent to True.

5.2 if ... elif ... else

Code within an if ... code block (indented by 4 spaces, as for loops) is only executed if the expression following the if keyword evaluates to True:

```
In [x]: a = 3.14159
In [x]: if a > 3:
   ...:        print('a is greater than 3')
   ...:
a is greater than 3

In [x]: if a > 4:
   ...:        print('a is greater than 4')
   ...:
   ...:
```

The else keyword allows an alternative path in the event that the expression is not True:

```
In [x]: if a < 3:
   ...:        print('a is less than 3')
   ...: else:
   ...:        print('a is greater than 3')
   ...:
a is greater than 3
```

Finally, elif allows several conditions to be checked: The code block following the first True comparison is executed:

```
In [x]: if a < 3:
   ...:        print('a is less than 3')
   ...: elif a < 4:
   ...:        print('a is between 3 and 4')
   ...: else:
   ...:        print('a is greater than 4')
   ...:
a is between 3 and 4
```

True and False are Python's built-in constant boolean values (of type bool). They cannot be altered or assigned as variable names. For the purpose of comparisons, Python also has the concept of *truthiness*: Non-boolean objects are treated as True or False according to their value. In particular, a float or int with the value 0 is equivalent to False (all other values are True); the empty string (' ') and list ([]) are False (any string or list with contents is equivalent to True). This allows syntax such as:

```
In [x]: b = 10

In [x]: if b % 2:
   ...:         print('b is odd')
   ...: else:
   ...:         print('b is even')
   ...:
b is even
```

Here, if b is odd, b % 2 is 1 (i.e., b mod $2 = 1$, the remainder of $b/2$) which is equal to True; if b is even, then b % 2 is 0, which is equal to False.

5.3 while loops

An alternative way to create a loop is with the while keyword: whereas a for loop is executed for a fixed number of iterations, a while loop will execute its code block only and for as long as a condition holds. For example:

```
In [x]: i = 0

In [x]: while i < 3:
   ...:         i += 1
   ...:         print(i)
   ...:
1
2
3
```

The counter i is initialized to 0, which is certainly less than 3, so the body of the while loop executes. First i increments, and then its value (1) is printed to the screen. The execution then returns to the while loop test. This repeats until i reaches the value 3: at this point the value is printed, but now the while loop test, i < 3 returns False and the code execution resumes after the end of the loop.

5.4 More Control Flow: break, continue and pass

These three keywords allow further control over the execution of loops. When a break statement is reached, the loop ends immediately:

```
In [x]: for i in range(10):
   ...:         print(i)
   ...:         if i == 3:
   ...:                 break
   ...:
0
1
```

```
2
3
```

The range(10) object will generate the numbers 0, 1, 2, ..., 9 and feed them to
i in turn, but when i reaches 3, the loop is terminated. A common use of break is
to escape from an infinite loop:

```
In [x]: n = 0
   ...: while True:
   ...:     n += 1
   ...:     if not (n % 12 or n % 42):
   ...:         break
   ...: print(n)
84
```

The above code prints the smallest number that is divisible by both 12 and 42.

When a continue statement is encountered the current iteration of the loop is
abandoned and the next iteration (if any) is started. Like break, the continue
keyword can be used inside a for loop as well as a while loop:

```
In [x]: for n in range(10):
   ...:     if n % 3:
   ...:         continue
   ...:     print(n)
0
3
6
9
```

Here, the print statement is only executed if n % 3 is 0 (equivalent to False),
that is if 3 divides n. Otherwise, n % 3 is equivalent to True and the continue
statement is executed, immediately returning the code flow to the top of the for
loop, which increments n.

5.5 Exceptions

Error conditions encountered in Python code are reported as *Exceptions*, usually
with a *traceback* report which can help track down the cause of the error. For
example,

```
In [x]: print(float("2.3D-2"))
---------------------------------------------------------------------
ValueError                               Traceback (most recent call last)
Input In [x], in <module>
----> 1 print(float("2.3D-2"))

ValueError: could not convert string to float: '2.3D-2'
```

Table 5.2   Common Python exceptions

Exception	Cause and description
IndexError	Indexing a sequence (such as a list or string) with a subscript that is out of range.
KeyError	Indexing a dictionary (see Section 7.4) with a key that does not exist in that dictionary.
NameError	Referencing a variable name that has not been defined.
TypeError	Attempting to use an object of an inappropriate type as an argument to a built-in operation or function.
ValueError	Attempting to use an object of the correct type but with an incompatible value as an argument to a built-in operation or function.
ZeroDivisionError	Attempting to divide by zero (either explicitly (using "/" or "//") or as part of a modulo operation "%").
SyntaxError	Using incorrect Python syntax, for example trying to assign a reserved keyword as a variable name, omitting a closing parenthesis or using a single equals sign, = for a comparison instead of ==.

The Exception reported here is a ValueError: the object passed to the float function was a string, "2.3D-2", which is the correct type, but its value contains an invalid character ('D'), and so float was unable to turn it into a floating-point number.

Some common Exceptions are listed in Table 5.2.

Idiomatic Python embraces the philosophy summed up by the abbreviation EAFP: *it is Easier to Ask Forgiveness than to seek Permission*: rather than trying to avoid Exceptions, it is common to run the code anyway, but within a try block, followed by an except block that handles any anticipated error.[1] For example:

```
In [x]: x = 0

In [x]: try:
   ...:       print(4 / x)
   ...: except ZeroDivisionError:
   ...:       print("4 / 0 is not defined!")

4 / 0 is not defined!
```

Code execution then continues as normal after this try ... except block whether the value of 4/x or the warning message was printed.

---

[1] *Unanticipated* errors are expected to stop the code execution and be reported to the user.

5.6 Examples

E5.1

Question

Given the following data on the standard melting and boiling points of various metals, determine their state (solid, liquid or gas) at 500°C.

```
metals = ['Fe', 'Al', 'Hg', 'Ti', 'Pb']
# Standard phase change temperatures for each metal, in K.
melting_points = [1811, 933, 234, 1941, 601]
boiling_points = [3134, 2743, 630, 3560, 2022]
```

Solution

First convert the reference temperature, T, to K, then loop over the three lists simultaneously. For each metal, compare T to the boiling and melting point.

```
# Temperature (K)
T = 500 + 273
for metal, Tm, Tb in zip(metals, melting_points, boiling_points):
    if T > Tb:
        print(f'{metal} is a gas at {T} K')
    elif T > Tm:
        print(f'{metal} is a liquid at {T} K')
    else:
        print(f'{metal} is a solid at {T} K')
```

```
Fe is a solid at 773 K
Al is a solid at 773 K
Hg is a gas at 773 K
Ti is a solid at 773 K
Pb is a liquid at 773 K
```

To produce a list of boolean values, True or False, denoting whether T is less than the boiling point, we could use the append method:

```
is_gas = []
for Tb in boiling_points:
    if T > Tb:
        is_gas.append(True)
    else:
        is_gas.append(False)
is_gas
```

```
[False, False, True, False, False]
```

We can, however, do away with the if ... else block altogether: Since we are appending the result of the comparison T < Tb (True or False), a nicer way would be:

```
is_gas = []
for Tb in boiling_points:
    is_gas.append(T > Tb)
is_gas
```

```
[False, False, True, False, False]
```

Even better, we can create this list without having to use the **append** method: use a list comprehension (see Section 4.3) to generate the **list**:

```
is_gas = [T > Tb for Tb in boiling_points]
is_gas
```

```
[False, False, True, False, False]
```

The built-in functions **any** and **all** can be used to establish, respectively, if any and all of the metals are gases at the temperature T:

```
any(is_gas), all(is_gas)
```

```
(True, False)
```

E5.2

The Iterative Weak Acid Approximation

Example E2.3 demonstrated a calculation of the pH of a weak acid which involved calculating the hydrogen ion concentration, $x = [H^+]$ from the acid dissociation constant,

$$K_a \approx \frac{[H^+][A^-]}{[HA]c^\ominus} = \frac{x^2}{(c-x)c^\ominus}.$$

Since for a weak acid $x \ll c$, an estimate of x could be obtained by setting $c - x \approx c$ in this equation, to give $x \approx \sqrt{K_a c c^\ominus}$.

The *iterative weak acid* approximation improves on this estimate with the following iterated calculation:

$$x^{(n+1)} = \sqrt{K_a c^\ominus (c - x^{(n)})}.$$

Even starting with $x^{(0)} = 0$, this process will rapidly converge on the correct value of x.

Returning to 0.1 M acetic acid ($K_a = 1.754 \times 10^{-5}$):

```
import numpy as np

# The acid dissociation constant of acetic acid.
Ka = 1.754e-5

# "Standard" amount concentration, 1 M = 1 mol.dm-3.
c_std = 1

# The concentration of the acid.
c = 0.1

# Iterate until x changes by less than this amount.
TOL = 1.e-5

x_old = 0
while True:
    x_new = np.sqrt(Ka * c_std * (c - x_old))
    if abs(x_new - x_old) < TOL:
        break
    x_old = x_new
print(x_new)
print('pH = {:.3f}'.format(-np.log10(x_new)))
```

```
0.0013155874194253987
pH = 2.881
```

This is the same value (to 4 s.f.) as that obtained by solving the quadratic equation in the Example E2.3.

E5.3

Question

Determine the molar volume, V_m, of hexane vapor at $T = 25\,°C$ and $p = 10\,kPa$.

Use the van der Waals equation of state with parameters $a = 2.471\ m^6\ bar\ mol^{-2}$ and $b = 1.735 \times 10^{-4}\ m^3\ mol^{-1}$.

Solution

The van der Waals equation of state is

$$p = \frac{RT}{V_m - b} - \frac{a}{V_m^2}$$

Our first thought might be to rearrange this equation for V_m:

$$pV_m^3 - (bp + RT)V_m^2 + aV_m - ab = 0$$

Unfortunately, this is a cubic equation in V_m, and there is no easy way to find the (relevant) root directly (although there are formulas that can be used, they are messy and we would need to exclude the imaginary roots).

However, there is a numerical approach, based on iteration. A different rearrangement gives:

$$V_m = b + \frac{RT}{p + \frac{a}{V_m^2}}$$

where the intermolecular interaction term, $\frac{a}{V_m^2} \ll p$. If we start from an initial guess, $V_m^{(0)}$, we can expect to get progressively better approximations to the true V_m from the iterative process:

$$V_m^{(n+1)} = b + \frac{RT}{p + \frac{a}{V_m^{(n)2}}}.$$

A sensible initial guess would be $V_m^{(0)} = \frac{RT}{p}$, the molar volume predicted by the ideal gas equation of state.

First, define the constants and parameters we need:

```
import math
# The gas constant, in J.K-1.mol-1.
R = 8.314
# Van der Waals parameters for hexane: m6.Pa.mol-2 and m3.mol-1, respectively.
a, b = 2.471, 1.735e-4
# The temperature (K) and pressure (Pa) for the problem.
T, p = 25 + 273.15, 10 * 1000
```

The ideal gas molar volume, in m^3, is:

```
Vm = R * T / p
Vm
```

```
0.0024788191
```

We will iterate until the value of V_m changes by less than some small amount, TOL.

```
TOL = 1.e-12
Vold = R * T / p
while True:
    Vm = b + R * T / (p + a / Vold**2)
    print(Vm)
    if abs(Vm - Vold) < TOL:
        break
    Vold = Vm
print()
print('Vm = {:.4f} L.mol-1'.format(Vm*1000))
```

```
0.000233644290191 25273
0.000174047502994 08598
0.0001738038466444478
0.000173802996608 70434
0.000173802993645 29655
0.000173802993634 96549

Vm = 0.1738 L.mol-1
```

Finally, we should check that this procedure converged on the correct answer by feeding this value for V_m back into the van der Waals equation of state:

```
R * T / (Vm - b) - a / Vm**2
```

```
9999.990276411176
```

This is close to the expected value of $p = 10$ kPa.

E5.4

Question

Given a string representing the electronic configuration of an atom in the form `'1s2.2s2.2p4'`, write a program to verify that the configuration is valid. The following tests should be made:

1. The orbitals should exist (`'1s2.2d3'` is invalid);
2. Orbitals must not be duplicated (`'1s2.2s1.2s1'` is invalid); and
3. Orbitals must not contain more than the maximum number of electrons allowed for them.

Solution

An atomic orbital configuration is written nl^N where $n = 1, 2, 3, \ldots$ is the principal quantum number, $l = 0, 1, \ldots, n-1$ is the orbital angular momentum quantum number written as a letter (s, p, d, etc.), and N is the occupancy of the orbital. The spatial degeneracy of the orbital is $2l + 1$ (the distinct values of the magnetic quantum number $m_l = -l, -l+1, \ldots, l-1, l$), and each electron's spin projection quantum number can be $m_s = \pm\frac{1}{2}$, so the maximum value of N is $2(2l+1)$.

We will build up the code step-by-step, and test it with a (very) invalid configuration string. The occupied orbitals are separated by the character `'.'`, so they can be turned into a list with the string split method:

```
# A deliberately bad atomic configuration.
s = '1p2.2s2.2p6.2s3.4d10'

s.split('.')
```

```
['1p2', '2s2', '2p6', '2s3', '4d10']
```

Next, parse the occupied orbital strings to extract n, l and N. Since the letter 1 looks similar to the digit 1 in some fonts, we will call its variable name ell; to further avoid confusion we will call the principal quantum number n and the number of electrons in each orbital, N, n_electrons.

```
# Split the string into substrings for each occupied orbital; loop over them
for occ_orbital in s.split('.'):
    # The orbital is the first two characters of the substring.
    # NB this means we can't have n > 9 ...
    orbital = occ_orbital[:2]
    n = int(orbital[0])
    ell_letter = orbital[1]
    # The easiest way of converting ell_letter to integer ell: find its index
    # in an ordered string:
    ell = 'spdf'.index(ell_letter)
    n_electrons = int(occ_orbital[2:])
    print(f'n={n}, l={ell} ("{ell_letter}"), N={n_electrons}')
```

```
n=1, l=1 ("p"), N=2
n=2, l=0 ("s"), N=2
n=2, l=1 ("p"), N=6
n=2, l=0 ("s"), N=3
n=4, l=2 ("d"), N=10
```

Looks good. Now check that $l < n$ and $N \leq 2(2l+1)$ and store the orbital part of the configuration (without its occupancy) so we can check that the same orbital doesn't come up twice:

```
# A deliberately bad atomic configuration.
s = '1p2.2s2.2p6.2s3.4d10'

# Split the string into substrings for each occupied orbital; loop over them.
orbitals = []
for occ_orbital in s.split('.'):
    orbital = occ_orbital[:2]
    n = int(orbital[0])
    ell_letter = orbital[1]
    ell = 'spdf'.index(ell_letter)
    n_electrons = int(occ_orbital[2:])

    if ell > n - 1:
        print(f'No such orbital: {orbital}')
    if n_electrons > 2 * (2 * ell + 1):
        print(f'Too many electrons in orbital: {occ_orbital}')
    if orbital in orbitals:
        print(f'Repeated orbital: {orbital}')
    orbitals.append(orbital)
```

```
No such orbital: 1p
Too many electrons in orbital: 2s3
Repeated orbital: 2s
```

5.7 Exercises

P5.1   The following list contains the element symbols:

```
element_symbols = [
 'H',  'He', 'Li', 'Be',  'B',  'C',  'N',  'O',  'F', 'Ne', 'Na', 'Mg', 'Al',
'Si',  'P',  'S', 'Cl', 'Ar',  'K', 'Ca', 'Sc', 'Ti',  'V', 'Cr', 'Mn', 'Fe',
'Co', 'Ni', 'Cu', 'Zn', 'Ga', 'Ge', 'As', 'Se', 'Br', 'Kr', 'Rb', 'Sr',  'Y',
'Zr', 'Nb', 'Mo', 'Tc', 'Ru', 'Rh', 'Pd', 'Ag', 'Cd', 'In', 'Sn', 'Sb', 'Te',
 'I', 'Xe', 'Cs', 'Ba', 'La', 'Ce', 'Pr', 'Nd', 'Pm', 'Sm', 'Eu', 'Gd', 'Tb',
'Dy', 'Ho', 'Er', 'Tm', 'Yb', 'Lu', 'Hf', 'Ta',  'W', 'Re', 'Os', 'Ir', 'Pt',
'Au', 'Hg', 'Tl', 'Pb', 'Bi', 'Po', 'At', 'Rn', 'Fr', 'Ra', 'Ac', 'Th', 'Pa',
 'U', 'Np', 'Pu', 'Am', 'Cm', 'Bk', 'Cf', 'Es', 'Fm', 'Md', 'No', 'Lr', 'Rf',
'Db', 'Sg', 'Bh', 'Hs', 'Mt', 'Ds', 'Rg', 'Cn', 'Nh', 'Fl', 'Mc', 'Lv', 'Ts',
'Og'
]
```

Determine how many times each letter of the alphabet appears in these element symbols (in either upper or lower case). Which letters do not appear?

6

Functions

Until now, the code we have written has not been very reusable. It is helpful to group code into blocks that can be named and executed repeatedly (perhaps with different parameters). That is, we want to write our own *functions*. This is useful not only to improve portability and to avoid repeating lines of code: It also helps break down complex tasks into simpler ones, which leads to more understandable and maintainable code.

6.1 Defining Functions

A function is defined with a `def` statement which gives it a name and specifies any arguments that it takes in brackets, followed by a colon. The body of the function is indented by 4 spaces. When called, the function's statements are executed; if the `return` keyword is encountered at any point, the specified values are returned to the caller. For example:

```
In [x]: def cube(x):
   ...:     x_cubed = x**3
   ...:     return x_cubed
   ...:

In [x]: cube(3)
Out[x]: 27

In [x]: cube(4)
Out[x]: 64
```

The function here, cube, takes a single argument, x (the number to be cubed), and returns a single value. To return two or more values from a function, separate them by commas. For example, the program below defines a function to return both roots of the quadratic equation $ax^2 + bx + c$ (assuming it *has* two real roots). The appropriate parameters taken by this function are the coefficients a, b and c defining the polynomial:

```
In [x]: import math
   ...:
   ...: def roots(a, b, c):
   ...:     d = b**2 - 4*a*c
   ...:     r1 = (-b + math.sqrt(d)) / 2 / a
   ...:     r2 = (-b - math.sqrt(d)) / 2 / a
   ...:     return r1, r2
   ...:

In [x]: print(roots(1, 3, -10))
(2.0, -5.0)
```

6.2 Keyword and Default Arguments

So far, we have used *positional arguments* in calling functions: The values to be assigned to each function argument is determined by the order in which they are given. Thus, roots(1, 3, -10) results in the function roots being called with the arguments a=1, b=3 and c=-10. It is also possible to pass values to a function in an arbitrary order using *keyword arguments*:

```
In [x]: roots(b=3, c=-10, a=1)
Out[x]: (2.0, -5.0)
```

Positional and keyword arguments can be used in the same function call, but the positional arguments must come first:

```
In [x]: roots(1, c=-10, b=3)
Out[x]: (2.0, -5.0)

In [x]: roots(1, b=3, -10)
  File "<ipython-input-10-b31133d01219>", line 1
    roots(1, b=3, -10)
                    ^
SyntaxError: positional argument follows keyword argument
```

Sometimes it is helpful to make an argument to a function optional; that is, the function code will use a default value for that argument if one is not provided when the function is called. This can be done in the function definition:

```
In [x]: def Vm_ideal(T=298.15, p=101325):
   ...:     R = 8.31446    # gas constant, J.K-1.mol-1
   ...:     return R * T / p
   ...:
```

Now, the function Vm_ideal can be called with no arguments, either of T and p, or both of them:

```
In [x]: Vm_ideal()                      # No arguments: use defaults
Out[x]: 0.024465395993091537

In [x]: Vm_ideal(273.15)                # Use provided T and default p
Out[x]: 0.02241396248704663

In [x]: Vm_ideal(p=1.e5)                # Use default T and provided p
Out[x]: 0.02478956249

In [x]: Vm_ideal(600, 1.e3)             # Use provided values for both p and T
Out[x]: 4.988676000000001
```

Sometimes it is necessary to pass the contents of a `list` to a function as its arguments (rather than pass the `list` as a single object); The "star syntax" unpacks a sequence in this way:

```
In [x]: state = [298.15, 101325]
In [x]: Vm_ideal(*state)                # Unpack state to arguments T, p
Out[x]: 0.024465395993091537
```

6.3 Docstrings

Even the most basic functions should be documented with a *docstring*: this is a string literal placed after the function signature describing what the function does, what values it returns and what arguments it takes. If the string is written with triple quotes ("""...""" or '''...'''), then it may contain line breaks.

```
In [x]: def Vm_ideal(T=298.15, p=101325):
   ...:     """Return the molar volume of an ideal gas, in m3.
   ...:
   ...:     The temperature, T (in K) and pressure, p (in Pa) can be provided.
   ...:
   ...:     """
   ...:
   ...:     R = 8.31446    # gas constant, J.K-1.mol-1
   ...:     return R * T / p
```

To inspect a function's docstring in an interactive Python session (e.g., IPython or a Jupyter notebook), type `help(func)` or `func?`:

```
In [x]: Vm_ideal?
Signature: Vm_ideal(T=298.15, p=101325)
Docstring:
Return the molar volume of an ideal gas, in m3.

The temperature, T (in K) and pressure, p (in Pa) can be provided.
File:       ...
Type:       function
```

This also works for built-in and imported functions:

```
In [x]: math.asin?
Signature: math.asin(x, /)
Docstring:
Return the arc sine (measured in radians) of x.

The result is between -pi/2 and pi/2.
Type:      builtin_function_or_method
```

6.4 Scope

The term *scope* refers to the way variable names are resolved in a Python pro-gram: that is, how the Python interpreter knows which object they refer to. This is relevant, because it is possible to use the same variable name to refer to different objects inside and outside a function. For example:

```
In [x]: x, y = 1, 3

In [x]: def add_y(z):
   ...:     x = y + z
   ...:     return x

In [x]: add_y(x)
Out[x]: 4

In [x]: x
Out[x]: 1
```

It is helpful to start by distinguishing between three types of scope: local scope, global scope and built-in scope.[1] The definition x, y = 1, 3 above creates two variable names in *global scope*: as the name suggests, such variables are accessible from anywhere in your code.

The function definition add_y takes a single argument, which is bound to a new variable, called z. The body of the function has access to this object as well as the objects x and y. However, when the assignment x = y + z is made, it isn't the global x that is updated; a new variable name, also called x, which is local to that function is created: changes to x inside this function do not affect the global x. Therefore, the value returned from the function is 4, but inspecting the value of x after the function has returned gives the value of the (unchanged) global x (1).

There's quite a lot going on here, so it is worth describing it in a bit more detail. Inside the function add_y, there is an assignment statement, x = y + z. Assign-ments in Python are always evaluated right-to-left (the interpreter needs to know

---

[1] There is a further scope, nonlocal or enclosing scope, relevant to nested functions, but we will not describe it here.

what object to make an assignment to before attaching the variable name to it). Therefore, the expression y + z must be evaluated, and hence the value of y must be obtained (z was passed as an argument to the function). The scope rules determine how the value of y is retrieved: First, Python looks in the local scope of the function, but since no variable named y exists in this scope, it doesn't find anything. So next, it looks in the global scope of the code (or interactive Python session) and finds a variable named y attached to the integer 3. So the right-hand side of the statement evaluates to the integer 4, and Python is instructed to assign this object to the variable name x. Since this happens within the function, a *new variable* named x is created in a scope local to that function. It is this x that is referred to in the `return` statement, and hence the object 4 is returned. Outside the function, the only x is the one in global scope, and that hasn't changed.

Of course, we may want to actually increment the integer x in global scope, in which case we can reassign the object returned by add_y to x:

```
In [x]: x = add_y(x)

In [x]: x
Out[x]: 4
```

Functions are not the only place that local scopes exist. For example, list comprehensions also use a local scope:

```
In [x]: x = 1

In [x]: [x**2 for x in range(6)]
Out[x]: [0, 1, 4, 9, 16, 25]

In [x]: x
Out[x]: 1
```

Built-in scope is the place that Python puts the keywords, built-in functions, exceptions and other attributes that make up its core syntax and functionality. It is the last place the interpreter looks when trying to resolve a variable name. For example, if you (unwisely) rebind the name `len` to an object, this name then no longer refers to the built-in `len()` function (for finding the length of sequences):

```
In [x]: len('abc')
Out[x]: 3

In [x]: len = 100          # allowed; not a good idea

In [x]: len('abcdef')
---------------------------------------------------------------
TypeError                                 Traceback (most recent call last)
<ipython-input-3-4f6544af4f86> in <module>
```

```
----> 1 len('abc')

TypeError: 'int' object is not callable
```

In its first use, the name len is only found in the built-in scope as the length function, and that is how it is used. After len is assigned to the integer 100, it exists in global scope and is found there *before* it is looked for in built-in scope, so its use as a function raises an Exception: len('abcdef') attempts to call the integer object 100 with an argument, 'abcdef', which is an error.

6.5 lambda (Anonymous) Functions

There are occasions when it is helpful to define and use a simple function, but without wrapping it in a def block or necessarily giving it a name. In this case, a lambda function can be used. The syntax defining a lambda function is:

```
lambda arguments: expression
```

where *arguments* is a comma-separated sequence of arguments, as for regular functions, and *expression* is a simple Python expression (no loop blocks or conditionals are allowed).

For example, the following are equivalent:

```
In [x]: def adder(a, b):
   ...:        return a + b
   ...:

In [x]: adder(5, 6)
Out[x]: 11

In [x]: adder = lambda a, b: a + b

In [x]: adder(5, 6)
Out[x]: 11
```

In practice, lambda functions are, arguably, rather over-used and usually if a lambda function is bound to a variable name like this (and is, therefore, surely no longer anonymous), regular def functions are to be preferred. However, they are useful in one-off cases where the function object itself is passed to another function. A common example is in sorting a list of tuples. Consider the following list of three tuples, each of which stores an element symbol and its mass number:

```
In [x]: masses = [('H', 1), ('Ar', 40), ('C', 12)]
```

Sorting this list with the sort method will result a list of tuples ordered by the alphabetical order of their first item:

```
In [x]: masses.sort()

In [x]: masses
Out[x]: [('Ar', 40), ('C', 12), ('H', 1)]
```

Suppose we want to sort the list by the *second* item of each tuple, the mass number. The `sort` method takes an optional argument, `key`, which expects to receive a function: this function is passed each item in the list, and the list is sorted by whatever it returns for each item. So to explicitly define a named function:

```
In [x]: def get_mass(e):
   ...:     return e[1]

In [x]: masses.sort(key=get_mass)

In [x]: masses
Out[x]: [('H', 1), ('C', 12), ('Ar', 40)]
```

Note the syntax: `key` must be a function:[2] its single argument is an item from the list (here, called e, but any name will do). This function, named `get_mass`, has the simple job of returning the second item in the `tuple` passed to it. However, since this might be the only place where such a function is used, it is easy to use an anonymous `lambda` function in-line:

```
In [x]: masses = [('H', 1), ('Ar', 40), ('C', 12)]
In [x]: masses.sort(key=lambda e: e[1])

In [x]: masses
Out[x]: [('H', 1), ('C', 12), ('Ar', 40)]
```

Similarly, sorting of strings ignoring any leading whitespace can be achieved by using `strip()` in a `lambda` function:

```
In [x]: sorted([' alpha', 'beta', '   gamma'])
Out[x]: ['   gamma', ' alpha', 'beta']  # the whitespace is included in the sort

In [x]: # Sorting by the strings without their leading space.
In [x]: sorted([' alpha', 'beta', '   gamma'], key=lambda s: s.strip())
Out[x]: [' alpha', 'beta', '   gamma']
```

In fact, in this case, we don't need a `lambda` function since the `strip` string method is available directly from the `str` type object:

```
In [x]: sorted([' alpha', 'beta', '   gamma'], key=str.strip)
Out[x]: [' alpha', 'beta', '   gamma']
```

---

[2] Functions are objects, like everything else in Python; just like other objects, they can be named, renamed and handed to other functions.

`lambda` functions are used extensively to perform small data-cleaning jobs in pandas `DataFrames` (see Chapter 31).

6.6 Examples

E6.1

Question

The Einstein model for the heat capacity of a crystal is

$$C_{V,\mathrm{m}} = 3R \left(\frac{\Theta_E}{T} \right)^2 \frac{\exp\left(\frac{\Theta_E}{T}\right)}{\left[\exp\left(\frac{\Theta_E}{T}\right) - 1 \right]^2},$$

where the *Einstein temperature*, Θ_E, is a constant for a given material.

Write a Python function to calculate $C_{V,\mathrm{m}}$ at 300 K for (a) sodium ($\Theta_E = 192$ K) and (b) diamond ($\Theta_E = 1450$ K).

Solution

The function `CV_Einstein` takes two arguments: `T`, the temperature at which to calculate the heat capacity; and `ThetaE`, the characteristic Einstein temperature for the material.

```python
import numpy as np

def CV_Einstein(T, ThetaE):
    """Return the Einstein molar heat capacity of a material, C_Vm.

    T is the temperature (K) and ThetaE the Einstein temperature for the
    material being considered. C_Vm is returned in J.K-1.mol-1.

    """

    # Gas constant, J.K-1.mol-1
    R = 8.314462618

    x = ThetaE / T
    f = np.exp(x)
    return 3 * R * f * (x / (f - 1))**2
```

```python
T = 300
ThetaE_Na, ThetaE_dia = 192, 1450
CV_Na, CV_dia = CV_Einstein(T, ThetaE_Na), CV_Einstein(T, ThetaE_dia)
print(f'Material    CVm({T} K)')
print('{:7s} {:8.2f} J.K-1.mol-1'.format('Na', CV_Na))
print('{:7s} {:8.2f} J.K-1.mol-1'.format('diamond', CV_dia))
```

```
Material     CVm(300 K)
Na           24.11 J.K-1.mol-1
diamond       4.71 J.K-1.mol-1
```

E6.2

Question

For the following equations of state, define Python functions returning the pressure of a gas, p, given its volume, V, amount, n, and temperature, T.

Ideal gas equation of state:

$$p = \frac{nRT}{V}$$

Van der Waals equation of state:

$$p = \frac{nRT}{V - nb} - \frac{n^2 a}{V^2}$$

Dieterici equation of state:

$$p = \frac{nRT \exp\left(-\frac{na}{RTV}\right)}{V - nb}.$$

Compare the pressure predicted for 1 mol of CO_2 at $T = 273.15$ K confined to a volume of 20 L. Take the van der Waals parameters $a = 3.640$ L^2 bar mol^{-2} and $b = 0.04267$ L mol^{-1}; and Dieterici parameters $a = 4.692$ L^2 bar mol^{-2} and $b = 0.04639$ L mol^{-1}.

Repeat the calculations for an amount of 0.01 mol.

Solution

First, define a Python function for each of the equations of state.

```python
import numpy as np
# Gas constant, J.K-1.mol-1
R = 8.314462618

def p_ideal(n, T, V):
    """Ideal gas equation of state, SI units."""
    return n * R * T / V

def p_vdw(n, T, V, a, b):
    """Van der Waals equation of state, SI units."""
    return n * R * T / (V - n*b) - n**2 * a / V**2

def p_dieterici(n, T, V, a, b):
    """Dieterici equation of state, SI units."""
    return n * R * T * np.exp(-n * a / R / T / V) / (V - n * b)
```

Now call the functions with the appropriate state variables, *n*, *T* and *V* and parameters, *a* and *b*. Take care to convert everything into SI units.

```
# Amount of gas (mol), temperature (K) and volume (m3).
n, T, V = 1, 273.15, 20 * 1.e-3

print(f'n = {n} mol, T = {T} K, V = {V} m3')
# Ideal gas equation of state.
print(f'Ideal:          p = {p_ideal(n, T, V):.3f} Pa')

# Van der Waals equation of state (a in m6.Pa.mol-2, b in m3.mol-1).
a_vdw, b_vdw = 3.640 / 10, 0.04267 / 1.e3
print(f'Van der Waals: p = {p_vdw(n, T, V, a_vdw, b_vdw):.3f} Pa')

# Dieterici equation of state (a in m6.Pa.mol-2, b in m3.mol-1).
a_diet, b_diet = 4.692 / 10, 0.04639 / 1.e3
print(f'Dieterici:      p = {p_dieterici(n, T, V, a_diet, b_diet):.3f} Pa')
```

```
n = 1 mol, T = 273.15 K, V = 0.02 m3
Ideal:          p = 113554.773 Pa
Van der Waals: p = 112887.560 Pa
Dieterici:      p = 112649.100 Pa
```

Now repeat the calculation for 0.01 mol:

```
n = 0.01
print(f'n = {n} mol, T = {T} K, V = {V} m3')
print(f'Ideal:          p = {p_ideal(n, T, V):.3f} Pa')
print(f'Van der Waals: p = {p_vdw(n, T, V, a_vdw, b_vdw):.3f} Pa')
print(f'Dieterici:      p = {p_dieterici(n, T, V, a_diet, b_diet):.3f} Pa')
```

```
n = 0.01 mol, T = 273.15 K, V = 0.02 m3
Ideal:          p = 1135.548 Pa
Van der Waals: p = 1135.481 Pa
Dieterici:      p = 1135.457 Pa
```

At lower pressures (achieved by confining less gas to the same volume at the same temperature as before), there is closer agreement between the equations of state.

7

Data Structures

In addition to the basic data types, `int`, `float`, `bool`, `str` and so on, Python provides a rich variety of compound data types: those used to group together objects. The `list`, introduced in Chapter 4, is perhaps the simplest example of this. `lists` are described in more detail in this chapter, along with some other useful data structures for storing objects and the important distinction between mutable and immutable objects.

7.1 Lists

The basic usage of Python `lists` – creating, indexing, slicing, appending and extending them – has been described in Chapter 4. A `list` is an ordered, mutable, sequence of objects of any type. It is ordered in the sense that its elements can be referred to with a unique integer index, and they have a length that can be obtained using the `len` built-in function:

```
In [x]: # A list with elements indexed at lst[0], lst[1], lst[2], lst[3].
In [x]: lst = ['A', 2.0, 'C', 9]

In [x]: lst[2]   # the third element of the list
Out[x]: 'C'

In [x]: len(lst)
Out[x]: 4
```

`lists` are mutable in the sense that they can be altered in-place (unlike, for example, strings):

```
In [x]: lst[2] = -1

In [x]: lst
Out[x]: ['A', 2.0, -1, 9]
```

This is fine: `lst[2]` was a reference to the string object `'C'` and we simply re-assigned it to the new object, the integer `-1`. Strings themselves, however, are immutable:

```
In [x]: my_str = "hello"

In [x]: my_str
Out[x]: 'hello'

In [x]: my_str[1] = "a"
--------------------------------------------------------------------
TypeError                           Traceback (most recent call last)
Input In [x], in <module>
----> 1 my_str[1] = "a"

TypeError: 'str' object does not support item assignment
```

In this situation, the only solution is to define an entirely new `str` object, for example:

```
In [x]: my_str = my_str[0] + "a" + my_str[2:]

In [x]: my_str
Out[x]: 'hallo'
```

Immutable objects may seem less flexible, but they confer advantages of speed and security, and they can be handed around to different functions and other objects without the need to copy them in case they change in future.

The fact that a mutable object can change after its creation has an important consequence when the same object is assigned to more than one variable name. An example should make this clear:

```
In [x]: lst1 = ['a', 'b', 'c']
In [x]: lst2 = lst1

In [x]: lst1.append('d')        # Append to lst1

In [x]: lst1
Out[x]: ['a', 'b', 'c', 'd']    # As expected

In [x]: lst2
Out[x]: ['a', 'b', 'c', 'd']    # lst2 has changed too!
```

What happened here? To understand this, it is best to think of a variable name such as `lst1` as a label attached to the underlying `list` object: It is not the object itself. In the above code, the `list` object `['a', 'b', 'c']` is created somewhere in memory, and the variable name `lst1` assigned to reference that object. The assignment `lst2 = lst1` then assigns the name `lst2` to point to the same object: that is, `lst1` and `lst2` are references to the same `list`: Accessing the list using either name will change it.

Immutable objects such as strings and numbers can certainly be assigned multiple variable names, but because they cannot be changed at all, there is no surprising behavior. Of course, any given variable name can be assigned to a new object – consider the following:

```
In [x]: n = 1

In [x]: m = n

In [x]: m = m + 1

In [x]: n
Out[x]: 1

In [x]: m
Out[x]: 2
```

Throughout these operations, n remains a reference to the immutable object 1. m is also assigned to this object, but the assignment m = m + 1 creates a new object, 2, and reassigns the name m to it (it doesn't change the 1 that n is pointing to. Python has a built-in method, id() to follow this: Typically, this method will return the address of an object in the computer's memory:

```
In [x]: n = 1

In [x]: id(n)
Out[x]: 4400245040

In [x]: m = n

In [x]: id(m)
Out[x]: 4400245040        # Same object, same id

In [x]: m = m + 1

In [x]: id(m)             # Now m is a new int object with a different id
Out[x]: 4400245072

In [x]: id(n)             # n has not changed
Out[x]: 4400245040
```

If you do want an independent copy of a list, call its copy() method:

```
In [x]: lst1 = ['a', 'b', 'c']

In [x]: lst2 = lst1.copy()     # alternatively, take a complete slice: lst1[:]

In [x]: lst1.append('d')

In [x]: lst1
Out[x]: ['a', 'b', 'c', 'd']

In [x]: lst2                    # lst2 hasn't changed ...
```

```
Out[x]: ['a', 'b', 'c']

In [x]: id(lst1), id(lst2)     # ... because it's a different object from lst1
Out[x]: (4456958976, 4445977280)
```

To obtain the index of an element in a `list`, call its `index` method:

```
In [x]: lst2.index('c')
Out[x]: 2
```

`lists` can be sorted in-place with the `sort` method (which does not return anything):

```
In [x]: masses = [44, 28, 16, 32, 4]

In [x]: masses.sort()

In [x]: masses
Out[x]: [4, 16, 28, 32, 44]
```

Alternatively, the `sorted` built-in returns a new `list` object with the original unchanged:

```
In [x]: masses = [44, 28, 16, 32, 4]

In [x]: sorted(masses)          # returns a new, ordered list
Out[x]: [4, 16, 28, 32, 44]

In [x]: masses                  # not changed
Out[x]: [44, 28, 16, 32, 4]
```

Python refuses to sort a `list` if its elements cannot be sensibly ordered by their values:

```
In [x]: lst1 = [1, 0, "a"]

In [x]: lst1.sort()
---------------------------------------------------------------------------
TypeError                                 Traceback (most recent call last)
Input In [x], in <module>
----> 1 lst1.sort()

TypeError: '<' not supported between instances of 'str' and 'int'
```

Since there is no sensible way to compare the values of string and integer objects, Python raises a `TypeError` exception.

7.2 Tuples

A `tuple` is an *immutable* ordered sequence of objects. Like `lists`, they can contain objects of different types. `tuples` are defined by providing the sequence of

elements between parentheses or by passing a suitable (i.e., iterable) object to the
`tuple` constructor:

```
In [x]: tpl1 = (0, 'a', 2.3)

In [x]: tpl1
Out[x]: (0, 'a', 2.3)

In [x]: tpl2 = tuple('abcdef')

In [x]: tpl2
Out[x]: ('a', 'b', 'c', 'd', 'e', 'f')
```

In fact, in many cases, the parentheses are not necessary and a comma-separated
sequence:

```
In [x]: tpl1 = 0, 'a', 2.3

In [x]: tpl1
Out[x]: (0, 'a', 2.3)
```

To create a tuple with a single element a trailing comma is necessary:

```
In [x]: a = (3)           # the integer 3

In [x]: a
Out[x]: 3

In [x]: a = (3,)        # a tuple of length 1 (a "singleton")

In [x]: a
Out[x]: (3,)
```

As an immutable data structure, assignment to a `tuple` is not allowed and appending and removing elements is not supported:

```
In [x]: tpl2[0] = "z"
---------------------------------------------------------------------
TypeError                                 Traceback (most recent call last)
Input In [x], in <module>
----> 1 tpl2[0] = "z"

TypeError: 'tuple' object does not support item assignment
```

7.3 Sets

A Python `set` is a mutable, unordered collection of objects with no duplicate elements:

```
In [x]: s = set([1, 1, 'a', 'b', 1, 2, 'a', 1.0])

In [x]:
Out[x]: {1, 2, 'a', 'b'}
```

Note that 1 and 1.0 are considered duplicates, even though they are objects with different types (an int and a float) because they are equal. A non-empty set can also be defined by providing the elements between braces:

```
In [x]: s = {1, 1, 'a', 'b', 1, 2, 'a', 1.0}
```

(but s = {} defines a dictionary (see below), not the empty set).
    Since they don't have an inherent order, sets cannot be indexed:

```
In [x]: s[0]
---------------------------------------------------------------
TypeError                               Traceback (most recent call last)
Input In [x], in <module>
----> 1 s[0]

TypeError: 'set' object is not subscriptable
```

Since a set is a mutable object, it has methods for adding and removing elements:

```
In [x]: s.add(3)              # add a single element

In [x]: s
Out[x]: {1, 2, 3, 'a', 'b'}

In [x]: s.update([2, 3, 4])   # add multiple elements from a collection

In [x]: s
Out[x]: {1, 2, 3, 4, 'a', 'b'}

In [x]: s.remove(4)

In [x]: s                     # remove a specific element
Out[x]: {1, 2, 3, 'a', 'b'}

In [x]: s.pop()               # remove and return an arbitrary element
Out[x]: 1

In [x]: s
Out[x]: {2, 3, 'a', 'b'}
```

sets support the following operators: -: difference, ^: symmetric difference, &: intersection, |: union:

```
In [x]: s1 = {"A", "B", "C", "D"}

In [x]: s2 = {"C", "D", "X", "Y", "Z"}

In [x]: s1 - s2          # elements in s1 but not s2
Out[x]: {'A', 'B'}

In [x]: s2 - s1          # elements in s2 but not s1
Out[x]: {'X', 'Y', 'Z'}

In [x]: s1 ^ s2          # symmetric difference: elements not in both sets
Out[x]: {'A', 'B', 'X', 'Y', 'Z'}

In [x]: s1 & s2          # intersection: elements common to both sets
Out[x]: {'C', 'D'}

In [x]: s1 | s2          # union: all unique elements from both sets
Out[x]: {'A', 'B', 'C', 'D', 'X', 'Y', 'Z'}
```

7.4 Dictionaries

A Python dictionary is a type of *associative array*: A mapping of *key* objects to *value* objects. Instead of indexing with an integer, each value is associated with a unique key (which must be an immutable object). In the simplest way to define a dictionary, the key-value pairs are provided as a comma-separated sequence between braces:

```
In [x]: densities = {'Fe': 7.9, 'Al': 2.73, 'Pt': 21.4}        # in g/cm3

In [x]: densities['Pt']
Out[x]: 21.4
```

Attempting to retrieve a value using a key that does not exist in the dictionary raises a KeyError exception:

```
In [x]: densities["Ni"]
---------------------------------------------------------------------------
KeyError                                   Traceback (most recent call last)
Input In [x], in <module>
----> 1 densities["Ni"]

KeyError: 'Ni'
```

The get method allows one to provide a default value in the event that the provided key is not in the dictionary:

```
In [x]: print(densities.get("Ni"))     # return None if Ni not in densities.keys()
None

In [x]: print(densities.get("Ni", "No data available"))
'No data available'
```

Dictionaries are mutable, so existing values can be changed and additional key-value pairs can inserted:

```
In [x]: densities['Fe'] = 7.85
In [x]: densities['Cu'] = 8.94
In [x]: densities.update({'Hg': 13.6, 'Au': 19.3})

In [x]: densities
Out[x]: {'Fe': 7.85, 'Al': 2.73, 'Pt': 21.4, 'Cu': 8.94, 'Hg': 13.6, 'Au': 19.3}
```

Iterating over a dictionary returns its *keys* in the order of their insertion:

```
In [x]: for element in densities:
   ...:         print(element)
Fe
Al
Pt
Cu
Hg
Au
```

(The dictionary keys can be obtained more explicitly from `densities.keys()`.)
To access the values only, use `densities.values()`:

```
In [x]: for density in densities.values():
   ...:         print(density)
7.85
2.73
21.4
8.94
13.6
19.3
```

Often, however, one wants to iterate over both the keys and the values; these are provided by `densities.items()`:

```
In [x]: for element, density in densities.items():
   ...:         print(element, density)
Fe 7.85
Al 2.73
Pt 21.4
Cu 8.94
Hg 13.6
Au 19.3
```

Finally, note that an empty dictionary can be created with a pair of braces:

```
In [x]: my_dict = {}
```

(this is not an empty set!)

7.5 Examples

E7.1

Dictionaries as Simple Databases

One application of a dictionary is as a simple database: Values can be stored in association with keys instead of assigned to individual variable names. This is much easier to maintain and has the advantage that the keys can be arbitrary strings (or other immutable data types) whereas variable names are constrained by the syntax of the Python language (see Section 2.3).

For example, to store some properties of the elements C, Si and Ge, one could construct a nested dictionary in the following way:

```
element_properties = {
    'C': {'mass /u': 12.0, 'Tm /K': 3823, 'Tb /K': 5100,
          'rho /g.cm-3': 3.51, 'IE /eV': 11.26, 'atomic radius /pm': 77.2},
    'Si': {'mass /u': 28.1, 'Tm /K': 1683, 'Tb /K': 2628,
          'rho /g.cm-3': 2.33, 'IE /eV': 8.15, 'atomic radius /pm': 117},
    'Ge': {'mass /u': 72.6, 'Tm /K': 1211, 'Tb /K': 3103,
          'rho /g.cm-3': 5.32, 'IE /eV': 7.90, 'atomic radius /pm': 122.5}
}
```

```
element_properties['C']
```

```
{'mass /u': 12.0,
 'Tm /K': 3823,
 'Tb /K': 5100,
 'rho /g.cm-3': 3.51,
 'IE /eV': 11.26,
 'atomic radius /pm': 77.2}
```

```
IE = element_properties['Ge']['IE /eV']
print(f'The ionization energy of Ge is {IE} eV.')
```

```
The ionization energy of Ge is 7.9 eV.
```

```
print('Densities of Group IV elements (g.cm-3)')
for symbol, properties in element_properties.items():
    density = properties['rho /g.cm-3']
    print(f'{symbol:>2s}: {density:5.2f}')
```

```
Densities of Group IV elements (g.cm-3)
 C:   3.51
Si:   2.33
Ge:   5.32
```

As mutable data structures, dictionaries can be updated. For example, to add an entry corresponding to the melting point in degrees centigrade for each element:

```
for symbol, properties in element_properties.items():
    Tm_K = properties['Tm /K']
    element_properties[symbol]['Tm /degC'] = Tm_K - 273
```

```
print(f"Melting point of silicon: {element_properties['Si']['Tm /degC']} degC")
```

```
Melting point of silicon: 1410 degC
```

Two dictionaries can be merged (since Python 3.9) using the | operator:

```
more_element_properties = {
        'Sn': {'mass /u': 118.7, 'Tm /K': 505, 'Tb /K': 2543,
                'rho /g.cm-3': 7.29, 'IE /eV': 7.34, 'atomic radius /pm': 140.5},
        'Pb': {'mass /u': 207.2, 'Tm /K': 601, 'Tb /K': 2013,
                'rho /g.cm-3': 11.3, 'IE /eV': 7.42, 'atomic radius /pm': 175}
}

# Add the properties of elements Sn and Pb to the dictionary.
element_properties = element_properties | more_element_properties
```

```
print(element_properties.keys())
```

```
dict_keys(['C', 'Si', 'Ge', 'Sn', 'Pb'])
```

Note that, in practice, this sort of data might be better held in a pandas `DataFrame` (see Chapter 31).

E7.2

Question

Write a function to parse a chemical formula in the form `'CH3CH2Cl'` into a list of the element symbols it contains in the order in which they are encountered (i.e., in this case, the list `['C', 'H', 'C', 'H', 'Cl']`). Use the Python `set` built-in to determine the unique element symbols (here, `{'C', 'H', 'Cl'}`) and then to categorize the formula as representing a hydrocarbon (containing only carbon and hydrogen) or not.

Solution

The following function determines the element symbols in a formula string and returns them in a list.

```python
def get_element_symbols(formula):
    """Return all the element symbols in the string formula."""

    n = len(formula)
    i = 0
    symbols = []
    # Iterate over the string, noting the capital letters, which indicate
    # the start of an element symbol.
    while i < n:
        if formula[i].isupper():
            # A new element symbol
            symbols.append(formula[i])
        elif formula[i].islower():
            # If we encounter a lowercase letter, it is the second character
            # of an element symbol.
            symbols[-1] += formula[i]
        i += 1
    return symbols
```

```python
# Test the function.
formula = 'CH3CH2Cl'
symbols = get_element_symbols(formula)
print(formula, ':', symbols)
```

```
CH3CH2Cl : ['C', 'H', 'C', 'H', 'Cl']
```

To determine the unique element symbols, create a `set` from the list:

```python
set(symbols)
```

```
{'C', 'Cl', 'H'}
```

```python
# Another test.
formula = 'CH3CBr2CO2H'
symbols = get_element_symbols(formula)
print(formula, ':', symbols)
print('Unique symbols:', set(symbols))
```

```
CH3CBr2CO2H : ['C', 'H', 'C', 'Br', 'C', 'O', 'H']
Unique symbols: {'C', 'O', 'Br', 'H'}
```

With this approach, we can define a function to determine if a formula represents a hydrocarbon or not:

```
def is_hydrocarbon(formula):
    symbol_set = set(get_element_symbols(formula))
    hydrocarbon_set = {'C', 'H'}
    return hydrocarbon_set == symbol_set
```

```
is_hydrocarbon('CH3CH2Cl')
```

```
False
```

```
is_hydrocarbon('CH3CH3')
```

```
True
```

7.6 Exercises

P7.1 The crystal lattice structures of the first-row transition metal elements are given below. Some elements have different crystal structures under different conditions of temperature and pressure. Use Python sets to group them and determine which metals (a) only exist in face-centered cubic (fcc), body-centered cubic (bcc) or hexagonal close-packed (hcp) structures; (b) exist in two of these structures; (c) do not form an hcp structure.

Do any of them exist in all three structures?

- fcc: Cu, Co, Fe, Mn, Ni, Sc
- bcc: Cr, Fe, Mn, Ti, V
- hcp: Co, Ni, Sc, Ti, Zn

P7.2 Extend the code from Example E7.2 to deduce the stoichiometric formula of an arbitrary chemical formula. For example:

```
formula = 'CClH2CBr2COH'
print(get_stoichiometry(formula))
print(get_stoichiometric_formula(formula))
```

```
{'C': 3, 'Cl': 1, 'H': 3, 'Br': 2, 'O': 1}
C3ClH3Br2O
```

8

File Input/Output

It is often necessary to read data in to a Python program, or to write data produced by it to a file. Python code can interact with an external data file through a `file` object, which is created with the built-in function, `open`.

8.1 Writing Files

To create a `file` object for *writing to*, provide the `open` function with a filename, and set `'w'` as the `mode`. Here, we will assign the variable `f` to the `file` object returned:

```
In [x]: f = open('output_file.txt', 'w')
```

Output from `print` that would have been sent to the screen can then be redirected to this open file object, by specifying `file=f`:

```
In [x]: print('a line of data', file=f)
```

After any amount of output, the file object is closed by calling its `close` function:

```
In [x]: f.close()
```

If you inspect the text file `output_file.txt` you should find it contains a single line reading a `line of data`.

In practice, it is better to use a slightly different syntax for inputting and outputting to files:

```
In [x]: with open('output_file.txt', 'w') as f:
   ...:     print('a line of data', file=f)
   ...:     print('a second line of data', file=f)
   ...:
```

The `with` keyword creates a *context* within which the file object `f` can be used. This context ensures that the file is closed after its indented block of code is executed (no need to call `f.close()`) and handles gracefully any errors that could occur during the file-writing process to minimizing the risk of losing data.

To write to the file only the exact contents of a string (with no additional line-endings, etc.), call `f.write`:

```
In [x]: with open("data.txt", "w") as f:
   ...:        f.write("text1")
   ...:        f.write("text2")
```

This will produce a text file with the contents `text1text2` (with no newline characters or spaces).

8.2 Reading Files

To read a text data file, open it with `'r'` as its mode. This is the default for the `open` function and so does not need to be specified explicitly.

There are several `file` object methods for reading from the open file: `f.read()` (to read the entire file), `f.readline()` (read a single line from the file), or `f.readlines()` (read all of the lines into a `list` of strings).

```
In [x]: with open('output_file.txt') as f:
   ...:        data = f.read()
   ...:

In [x]: data
Out[x]: 'a line of data\na second line of data\n'
```

Note the escaped line-ending characters, `\n`. To see the `data` string as it appears in a text editor, `print` it:

```
In [x]: print(data)
a line of data
a second line of data
```

Alternatively:

```
In [x]: with open('output_file.txt') as f:
   ...:        data = f.readlines()
   ...:

In [x]: data
Out[x]: ['a line of data\n', 'a second line of data\n']
```

Here, `data` is a Python `list` containing the two lines from the file (including their line-endings).

Table 8.1   Python file open modes

Character	Description
'r'	open for reading (the default)
'w'	open for writing, overwriting the file if it exists
'x'	open for writing, failing if the file exists already
'a'	open for writing, appending if the file exists
't'	read/write text data (the default)
'b'	read/write binary data
'+'	open for updating (reading and writing)

Python can also read and write binary files, and append to existing files by passing additional flags to the mode argument (see Table 8.1). For example, open ('file.dat', 'xb') opens a file called 'file.dat' to write binary data, but will fail if the file already exists.

8.3 Character Encoding

Computers represent the human-readable characters of strings as numbers; the mapping which associates each number (code point) to a specific character is called the *character encoding*. Many early computers used the ASCII (American Standard Code for Information Interchange) encoding which represented each of a set of 95 characters as a number between 32 and 126 (e.g., 65 = "A").

Unfortunately, until recently, there was no widely agreed standard for the expanded encoding needed to represent the many characters used in the hundreds of different human languages: different operating systems and platforms used different encodings. If you've ever seen text displayed as "caf¿" or "cafÃ©" instead of "café," the cause is likely a mismatch between the assumed encoding of the string and one actually used by the author.

The encoding used by Python in reading and writing text files is platform-dependent, and files written and read on the same platform should make the round-trip unaltered. However, it might be necessary to explicitly specify the encoding when opening a file if it comes from elsewhere; this can be done with the encoding argument. Since the most widely-adopted character encoding in use today is UTF-8, this is a good place to start:

```
In [x]: with open("data.txt", encoding="utf8") as f:
   ...:     print(f.read())
```

The exception is (of course) Microsoft Windows, which favors the cp1252 encoding. There is some overlap between character encodings (e.g., the first 95 printable characters encoded by UTF8 are the same as those in the ASCII character set), but if reading a file fails or gives nonsensical results for some reason, it will be necessary to deduce (or guess) the encoding somehow.[1]

8.4 Example

E8.1

Question

Write a text file, `atomic_numbers.txt`, listing the symbols for the first ten elements of the periodic table, with their atomic numbers, Z.

Solution

First create an ordered `list` of the element symbols. Each element's atomic number is therefore its index in this `list` plus one (because indexes start at zero).

```
elements = 'H He Be Li B C N O F Ne'
elements = elements.split()
elements
```

```
['H', 'He', 'Be', 'Li', 'B', 'C', 'N', 'O', 'F', 'Ne']
```

To write a text file, create a `file` object, fo, in "write" mode (`'w'`) and iterate over the `elements` `list`. We can use `enumerate` to get the index and element symbol for each iteration:

```
with open('atomic_numbers.txt', 'w') as fo:
    for i, symbol in enumerate(elements):
        print(f'{i+1:2d} {symbol}', file=fo)
```

To read the text file, we could create another `file` object, this time in "read" mode (`'r'`) and read all its lines into a `list` of strings:

```
with open('atomic_numbers.txt', 'r') as fi:
    lines = fi.readlines()
lines
```

```
[' 1 H\n',
 ' 2 He\n',
 ' 3 Be\n',
 ' 4 Li\n',
 ' 5 B\n',
 ' 6 C\n',
 ' 7 N\n',
```

---

[1] A list of available encodings and more information on this topic is available in the Python documentation, at `https://docs.python.org/3/library/codecs.html#standard-encodings`.

```
'  8 O\n',
'  9 F\n',
'10 Ne\n']
```

Each line includes the newline character, '\n', at the end of it. To make the output
a bit tidier, we could iterate over this list and parse its fields individually. There
are two fields per line, separated by whitespace:

```
with open('atomic_numbers.txt', 'r') as fi:
    for line in fi.readlines():
        fields = line.split()
        Z = int(fields[0])
        symbol = fields[1]
        print(f'Z({symbol:2s}) = {Z:2d}')
```

```
Z(H ) =  1
Z(He) =  2
Z(Be) =  3
Z(Li) =  4
Z(B ) =  5
Z(C ) =  6
Z(N ) =  7
Z(O ) =  8
Z(F ) =  9
Z(Ne) = 10
```

8.5 Exercises

P8.1 The file thermo-data.csv, available at https://scipython.com/chem/
xfa/, contains thermodynamic data on several compounds under standard conditions of pressure ($p = 1$ bar) and at $T = 298$ K in the format:

```
Formula,    DfHm /kJ.mol-1,    Sm /J.K-1.mol-1
H2O(g),     -241.83,           188.84
CO2(g),     -393.52,           213.79
LiOH(s),    -484.93,           42.81
Li2CO3(s),  -1216.04,          90.31
```

Read these data into suitable dictionaries of the standard molar enthalpies of formation and molar entropies, and use them to calculate the equilibrium constant for
the following reaction, which forms the basis for the operation of carbon dioxide
removal from the crew cabins of some space craft.

$$2LiOH(s) + CO_2(g) \rightarrow Li_2CO_3(s) + H_2O(g)$$

P8.2   The text file vap-data.txt, which can be downloaded from https://
scipython.com/chem/xfb/, contains data for the boiling points, T_b, and enthalpies of vaporization, $\Delta_{vap}H$, for some liquids:

```
Compound      Tb /K      DvapH /kJ.mol-1
CH3Cl         247        21.5
CH2CHCHCH2    268.6      22.47
C2H5OH        351.39     38.7
H2O           373.15     40.7
CH4           111.7       8.17
HCO2H         373.9      22.69
CH3OH         337.8      35.21
C8H18         398.7      34.41
C6H5CH3       383.8      33.18
```

Read in these data and calculate the entropy of vaporization, $\Delta_{vap}S$, for each substance. Comment on any deviation from *Trouton's rule* (that, for many liquids,
$\Delta_{vap}S \approx 85\,\mathrm{J\,K^{-1}\,mol^{-1}}$). Write another text file with these data, including a column for the entropy of vaporization.

9

Basic NumPy

9.1 Creating NumPy Arrays

Python's `lists` are sequences of items that can be indexed into, iterated over, appended to, etc.:

```
In [x]: lst = [1, 'a', 'pi', -0.25]

In [x]: lst[2]
Out[x]: 'pi'

In [x]: lst.append(99.)

In [x]: for item in lst:
   ...:         print(item)
   ...:
1
a
pi
-0.25
99.0
```

`lists` are convenient for storing relatively small amounts of heterogeneous data, but are not well-suited to larger-scale numerical calculations: Creating and iterating over Python lists can be slow and cumbersome. For example, suppose we want to calculate the function $y = \sin(x)$ for a list of 10 evenly spaced values of x between 0 and 2π (inclusive):

```
In [x]: import math
In [x]: xvals, yvals = [], []
In [x]: n = 10
In [x]: x = 0
In [x]: dx = 2 * math.pi / (n - 1)
In [x]: for i in range(n):
   ...:         x = i * dx
   ...:         xvals.append(x)
   ...:         yvals.append(math.sin(x))
```

```
    ...:

In [x]: xvals
Out[x]:
([0.0,
  0.6981317007977318,
  1.3962634015954636,
  2.0943951023931953,
  2.792526803190927,
  3.490658503988659,
  4.1887902047863905,
  4.886921905584122,
  5.585053606381854,
  6.283185307179586])

In [x]: yvals
Out[x]:
 [0.0,
  0.6427876096865393,
  0.984807753012208,
  0.8660254037844388,
  0.3420201433256689,
  -0.34202014332566866,
  -0.8660254037844384,
  -0.9848077530122081,
  -0.6427876096865396,
  -2.4492935982947064e-16])
```

That's a lot of work: we've got to work out the spacing between the values of x,

$$\Delta x = \frac{2\pi}{n - 1}$$

for n points and slowly iterate using counter, i, that ranges from 0 to $n - 1$ in order to build up the lists `xvals` and `yvals` using `append`.

NumPy provides an efficient way to achieve the same thing without explicit loops:

```
In [x]: import numpy as np
In [x]: x = np.linspace(0, 2*np.pi, 10)

In [x]: y = np.sin(x)

In [x]: x
Out[x]:
array([0.        , 0.6981317 , 1.3962634 , 2.0943951 , 2.7925268 ,
       3.4906585 , 4.1887902 , 4.88692191, 5.58505361, 6.28318531])

In [x]: y
Out[x]:
array([ 0.00000000e+00,  6.42787610e-01,  9.84807753e-01,  8.66025404e-01,
        3.42020143e-01, -3.42020143e-01, -8.66025404e-01, -9.84807753e-01,
       -6.42787610e-01, -2.44929360e-16])
```

The NumPy function linspace acts a bit like the range built-in, but it returns a new object, a NumPy *array*, holding a specified number of evenly spaced values between some start and end point. By default, this end value is included in the sequence, unlike with range. It's most basic call signature is:

np.linspace(*start*, *stop*, *num*)

For example:

```
In [x]: arr = np.linspace(1, 3, 5)

In [x]: arr
Out[x]: array([1. , 1.5, 2. , 2.5, 3. ])
```

When a NumPy function is used with an array, the function is applied to every item in the array automatically, with no need to explicitly loop over it. The same is true for the regular arithmetic operations:

```
In [x]: arr * 2
Out[x]: array([2., 3., 4., 5., 6.])

In [x]: 1 / arr
Out[x]: array([1.        , 0.66666667, 0.5       , 0.4       , 0.33333333])

In [x]: np.log(arr)
Out[x]: array([0.        , 0.40546511, 0.69314718, 0.91629073, 1.09861229])
```

This property of NumPy arrays is called *vectorization*, and because the code that executes the calculations is pre-compiled C code within the NumPy library, it is much faster than implementing the same loops in Python with, for example, a for loop.

Arrays can also be created directly:

```
In [x]: b = np.array([0, -1.5, 2])

In [x]: b
Out[x]: array([ 0. , -1.5,  2. ])
```

but note that they have only one data type (we will be concerned mostly with arrays of integers or arrays of floats). A matrix (in the mathematical sense) is most conveniently represented by a two-dimensional array, which can be created from a list of lists:

```
In [x]: A = np.array( [[1, 2, 0], [1, 3, 4], [-2, 0.5, 1]] )

In [x]: A
Out[x]:
array([[ 1. ,  2. ,  0. ],
       [ 1. ,  3. ,  4. ],
       [-2. ,  0.5,  1. ]])
```

There are some further useful array-creation methods: np.ones creates an array of a given shape filled with the value 1, np.zeros does the same but for the value 0:

```
In [x]: X = np.zeros((2,3))       # two rows, three columns

In [x]: X
Out[x]:
array([[0., 0., 0.],
       [0., 0., 0.]])

In [x]: Y = np.ones((3,2))        # three rows, two columns

In [x]: Y
Out[x]:
array([[1., 1.],
       [1., 1.],
       [1., 1.]])

In [x]: Z = 4 * np.ones((2,2))

In [x]: Z
Out[x]:
array([[4., 4.],
       [4., 4.]])
```

As already mentioned, NumPy arrays have a single data type (dtype), which in the arrays considered so far is double-precision floating point (64 bits, the same as regular Python floats):

```
In [x]: Z.dtype
Out[x]: dtype('float64')
```

If you want a different dtype, you can set it when the array is created:

```
In [x]: Z = 4 * np.ones((2,2), dtype=int)
In [x]: Z
Out[x]:
array([[4, 4],
       [4, 4]])

In [x]: Z.dtype
Out[x]: dtype('int64')
```

Note that, unlike the Python int types, which can hold any size of integer (so long as it fits within the computer's memory), by default NumPy represents signed integers in 64 bits (giving a range of -9223372036854775808 to $+9223372036854$ 775807). No warning or error is issued if a calculation overflows this limit:

```
In [x]: arr = np.array([9223372036854775807])

In [x]: arr
Out[x]: array([9223372036854775807])
```

```
In [x]: arr + 1
Out[x]: array([-9223372036854775808])
```

Because of the way that integers are stored, adding one has unintentionally set the flag indicating a negative number.

np.empty creates a NumPy array of a specified shape but without assigning any particular value to the contents: they are initialized to whatever happens to be in the memory reserved for them at the time (probably garbage, when interpreted as numbers):

```
In [x]: E = np.empty((4, 2))

In [x]: E
Out[x]:
array([[1.49166815e-154, 1.49166815e-154],
       [6.37344683e-322, 0.00000000e+000],
       [0.00000000e+000, 0.00000000e+000],
       [0.00000000e+000, 0.00000000e+000]])
```

Finally, the method np.arange is more closely related to the range built-in than np.linspace: Given values for start, stop and step it returns values, evenly spaced by step, from the interval [start, stop): that is, including start but excluding stop:

```
In [x]: x = np.arange(1, 10, 2)

In [x]: x
Out[x]: array([1, 3, 5, 7, 9])

In [x]: y = np.arange(0, 2, 0.2)

In [x]: y
Out[x]: array([0. , 0.2, 0.4, 0.6, 0.8, 1. , 1.2, 1.4, 1.6, 1.8])
```

Arrays can be conveniently reshaped with reshape:

```
In [x]: z = y.reshape((5, 2))

In [x]: z

Out[x]:
array([[0. , 0.2],
       [0.4, 0.6],
       [0.8, 1. ],
       [1.2, 1.4],
       [1.6, 1.8]])
```

It is important to understand that reshape returns a so-called *view* on the original array: z here does not contain an independent copy of the data in y. Therefore, changing values in z will also affect y: They are just different ways of interpreting the same underlying data in memory. For example:

```
In [x]: y[0] = 99

In [x]: y
Out[x]: array([99. ,  0.2,  0.4,  0.6,  0.8,  1. ,  1.2,  1.4,  1.6,  1.8])

In [x]: z        # z has also changed (it is a view on the same data as y)
Out[x]:
array([[99. ,  0.2],
       [ 0.4,  0.6],
       [ 0.8,  1. ],
       [ 1.2,  1.4],
       [ 1.6,  1.8]])
```

It is common to string multiple array methods together if the intermediate objects are not needed:

```
In [x]: np.arange(0, 24).reshape(2,3,4)    # create a 2 x 3 x 4 array
Out[x]:
array([[[ 0,  1,  2,  3],
        [ 4,  5,  6,  7],
        [ 8,  9, 10, 11]],

       [[12, 13, 14, 15],
        [16, 17, 18, 19],
        [20, 21, 22, 23]]])
```

9.2 Indexing and Slicing NumPy Arrays

Indexing into NumPy arrays is the same as for other sequences in the case of one-dimensional arrays:

```
In [x]: b[1]
Out[x]: -1.5
```

However, whilst nested lists must be indexed with sequences of square brackets:

```
In [x]: lst = [[1, 2, 0], [1, 3, 4], [-2, 0.5, 1]]

In [x]: lst[1][2]
Out[x]: 4
```

The preferred way of indexing into multidimensional NumPy arrays is with comma-separated values:

```
In [x]: A = np.array(lst)
In [x]: A
Out[x]:
```

```
array([[ 1. ,   2. ,   0. ],
       [ 1. ,   3. ,   4. ],
       [-2. ,   0.5,   1. ]])

In [x]: A[1, 2]
Out[x]: 4.0
```

(Note also that NumPy converted the original, integer 4 in to a float: The presence of 0.5 in the initializing statement forced the entire array to become one of floating point values: NumPy arrays only ever have a single type.)

Slicing NumPy arrays also works in an analogous way:

```
In [x]: A[:2]               # The first two rows and all columns
array([[1., 2., 0.],
       [1., 3., 4.]])

In [x]: A[:, :2]            # All rows and the first two columns
Out[x]:
array([[ 1. ,   2. ],
       [ 1. ,   3. ],
       [-2. ,   0.5]])


In [x]: A[1, 1:]           # All elements after the first on the second row
Out[x]: array([3., 4.])

In [x]: A[:-1,:-1]          # All elements apart from the last row and column
Out[x]:
array([[1., 2.],
       [1., 3.]])
```

For efficiency reasons, a slice of a NumPy array is a view on the original array. Therefore, if a slice is assigned to a variable, changing one will change the other:

```
In [x]: A
Out[x]:
array([[ 1. ,   2. ,   0. ],
       [ 1. ,   3. ,   4. ],
       [-2. ,   0.5,   1. ]])

In [x]: b = A[:, 1]         # The second column of A

In [x]: b
Out[x]: array([2. ,  3. ,  0.5])

In [x]: b[1] = -99

In [x]: b                   # b has changed ...
Out[x]: array([  2. ,  -99. ,    0.5])
```

```
In [x]: A                # ... so has A!
Out[x]:
array([[  1. ,    2. ,    0. ],
       [  1. ,  -99. ,    4. ],
       [ -2. ,    0.5,    1. ]])
```

Striding over a NumPy array works in the same way as for regular Python sequences:

```
In [x]: arr
Out[x]: array([ 0,  1,  2,  3,  4,  5,  6,  7,  8,  9, 10])

In [x]: arr[::2]      # every other number
Out[x]: array([ 0,  2,  4,  6,  8, 10])

In [x]: arr[2:7:3]    # start at second element, output every third; stop before 7
Out[x]: array([2, 5])
```

This includes backward slicing with a negative stride:

```
In [x]: arr = np.arange(0, 12).reshape((3,4))

In [x]: arr
Out[x]:
array([[ 0,  1,  2,  3],
       [ 4,  5,  6,  7],
       [ 8,  9, 10, 11]])

In [x]: arr[:,::-1]      # all the rows, but with the columns backwards
Out[x]:
array([[ 3,  2,  1,  0],
       [ 7,  6,  5,  4],
       [11, 10,  9,  8]])
```

9.3 NumPy Array Aggregation

NumPy provides a number of useful functions for statistically summarizing the values in an array, including finding their sum, maximum, minimum, mean, variance and so on. The important ones are listed in Table 9.1. For example:

```
In [x]: arr = np.array([0, 1.2, 2.3, 1.5, 0.6, 1.2, 1.9])

In [x]: np.sum(arr)
Out[x]: 8.7

In [x]: np.mean(arr)
Out[x]: 1.2428571428571427

In [x]: np.var(arr)
Out[x]: 0.5110204081632653

In [x]: np.std(arr)
Out[x]: 0.7148569144683888
```

Table 9.1   Important NumPy data summary methods

min	Minimum
argmin	Index of minimum value in array
max	Maximum
argmax	Index of maximum value in array
sign	Sign of each element (returns -1, 0 or $+1$)
sum	Sum of array elements
prod	Product of array elements
mean	Mean average
average	Weighted average
median	Median average
std	Standard deviation
var	Variance

```
In [x]: np.min(arr)
Out[x]: 0.0

In [x]: np.max(arr)
Out[x]: 2.3
```

Sometimes it is useful to know not the maximum or minimum value itself, but rather the *index* of the maximum or minimum value: this is what np.argmax and np.argmin do:

```
In [x]: np.argmax(arr)
Out[x]: 2                    # arr[2] (=2.3) is the maximum element

In [x]: np.argmin(arr)
Out[x]: 0                    # arr[0] (=0.0) is the minimum element
```

These statistics apply to the whole array. For multidimensional arrays, to apply them to a particular axis (e.g., across all columns), use the axis argument. For example, a two dimensional array has rows (axis=0) and columns (axis=1):

```
In [x]: arr = np.array([[3, 0, -2, 1], [-2, -2, 4, 0], [-4, 1, 1, 1]])
In [x]: arr
Out[x]: array([[ 3,  0, -2,  1],
               [-2, -2,  4,  0],
               [-4,  1,  1,  1]])
```

We can find the global minimum with:

```
In [x]: np.min(arr)
Out[x]: -4
```

To find the minimum value within each column (i.e., down the rows), use (axis=0):

```
In [x]: np.min(arr, axis=0)
Out[x]: array([-4, -2, -2,  0])
```

To find the minimum value within each row (i.e., along the columns), use (`axis=1`):

```
In [x]: np.min(arr, axis=1)
Out[x]: array([-2, -2, -4])
```

9.4 NaN: Not a Number

In the standard floating point description of numbers, there is a special value, "NaN," which stands for "Not a Number," denoting an undefined or invalid value. NumPy's representation of NaN is `numpy.nan`. For example:

```
In [x]: arr = np.arcsin([-0.5, 0, 0.5, np.sqrt(3)/2, 1, 1.5])
<ipython-input-48-4ddf>:1: RuntimeWarning: invalid value encountered in arcsin
  arr = np.arcsin([-0.5, 0, 0.5, np.sqrt(3)/2, 1, 1.5])

In [x]: arr
Out[x]:
array([-0.52359878,  0.        ,  0.52359878,  1.04719755,  1.57079633,
                nan])
```

There is no number, x, for which $\sin(x) = 1.5$ is defined, and so attempting to find the arcsin of 1.5 yields NaN (with a warning).

Dealing with NaN entries in an array can be cumbersome and if an array includes such values, the above functions will also return `numpy.nan`:

```
In [x]: np.max(arr)
Out[x]: nan
```

Luckily, where it is appropriate to simply ignore these invalid array entries there are a family of related methods, `nanmin`, `nanmax`, etc. that do so:

```
In [x]: np.nanmax(arr)         # Maximum valid value, ignoring nans
Out[x]: 1.5707963267948966

In [x]: np.nanmean(arr)        # Mean of all valid values in arr
Out[x]: 0.5235987755982988

In [x]: np.nanargmax(arr)      # Index of the maximum valid value in arr
Out[x]: 4
```

9.5 Boolean Arrays and Indexing

Comparison operations on a NumPy array are vectorized and return an array of `True` and `False` values:

```
In [x]: arr = np.array([0, 1, 2, 4, 2, 1])

In [x]: arr > 1
Out[x]: array([False, False,  True,   True,   True, False])

In [x]: arr == 1
Out[x]: array([False,   True, False, False, False,   True])


In [x]: arr**2 > 10
Out[x]: array([False, False, False,   True, False, False])
```

Boolean arrays can be combined using logic operators, but instead of and, or and
not, the corresponding *bitwise* operators &, | and ~ must be used:

```
In [x]: (arr > 0) & (arr < 4)
Out[x]: array([False,   True,   True, False,   True,   True])

In [x]: (arr == 0) | (arr > 2)
Out[x]: array([ True, False, False,   True, False, False])

In [x]: ~(arr > 1)        # the same as (arr <= 1)
Out[x]: array([ True,   True, False, False, False,   True])
```

Since the boolean quantities True and False are equal to the integer values 1 and
0, respectively, these boolean arrays can be passed to np.sum to count the number
of True values:

```
In [x]: np.sum(arr > 1)     # Number of elements greater than 1
Out[x]: 3
```

9.6 Reading Data Files into a NumPy Array

The main function for reading text files into a NumPy array is genfromtxt. The
call signature ("method prototype") for np.genfromtxt is quite extensive, but the
most important arguments and their defaults are described here:

```
np.genfromtxt(fname, dtype=float, comments='#', delimiter=None,
              skip_header=0, skip_footer=0, converters=None,
              missing_values=None, filling_values=None, usecols=None,
              unpack=False, invalid_raise=True, max_rows=None)
```

The arguments are as follows:

- fname: The only required argument, fname, which can be a filename or an open
  file object (see Chapter 8).
- dtype: The data type of the array defaults to float but can be set explicitly by
  the dtype argument.

- `comments`: Comments in a file are usually started by some character such as #
  (as with Python) or %. To tell NumPy to ignore the contents of any line following
  this character, use the `comments` argument – by default it is set to #.
- `delimiter`: The string used to separate columns of data in the file; by default it
  is `None`, meaning that any amount of whitespace (spaces, tabs) delimits the data.
  To read a comma-separated (csv) file, set `delimiter=','`.
- `skip_header`: An integer giving the number of lines at the start of the file to
  skip over before reading the data (e.g., to pass over header lines). Its default is 0
  (no header).
- `skip_footer`: An integer giving the number of lines at the end of the file to skip
  over (e.g., to ignore information given at the end of a table of data). Its default is
  0 (no footer).
- `missing_values`: The set of strings that indicate missing data in the input file
  (often `'*'` or `'-'`).
- `filling_values`: The set of values to be used as a default in the case of missing
  data. If not specified, `nan` is used.
- `converters`: An optional dictionary mapping the column index to a function
  converting string values in that column to data (see below).
- `usecols`: A sequence of column indexes determining which columns of the file
  to return as data; by default it is `None`, meaning all columns will be parsed and
  returned.
- `unpack`: By default, the data table is returned in a single two-dimensional array
  of rows and columns reflecting the structure of the file read in. Setting `unpack=`
  `True` will transpose this array so that individual columns can be picked off and
  assigned as one-dimensional arrays to different variables (see below).
- `invalid_raise`: By default an exception is raised if a line is inconsistent in the
  number of columns expected based on the delimiter specified. If the `invalid`
  `_raise` argument is set to `False` a warning is given and these lines are skipped
  instead.
- `max_rows`: The maximum number of rows to read in (useful for exploring large
  files). This argument cannot be used at the same time as `skip_footer`.

If the file you want to read consists simply of whitespace delimited values, with
no invalid data or header rows, you don't need most of these arguments. For example, the following text file, `H2-virial-coeffs-B.txt`, gives the second virial
coefficient (in $cm^3.mol^{-1}$) of hydrogen at different temperatures (in K):

```
25 -106.2
30  -80.7
40  -50.3
50  -33.4
```

```
100    -2.5
110.4  0
120    2.0
```

Calling `genfromtxt` with just the filename gives a two-dimensional array with both columns of data:

```
In [x]: np.genfromtxt('H2-virial-coeffs-B.txt')
Out[x]: array([[  25. ,  -106.2],
               [  30. ,   -80.7],
               [  40. ,   -50.3],
               [  50. ,   -33.4],
               [ 100. ,    -2.5],
               [ 110.4,     0. ],
               [ 120. ,     2. ]])
```

In this case, it might be more useful to unpack the columns into two one-dimensional arrays:

```
In [x]: T, B = np.genfromtxt('H2-virial-coeffs-B.txt', unpack=True)

In [x]: T
Out[x]: array([ 25. ,   30. ,   40. ,   50. ,  100. ,  110.4,  120. ])

In [x]: B
Out[x]: array([-106.2,   -80.7,   -50.3,   -33.4,    -2.5,     0. ,     2. ])
```

Data files are rarely this clean, however. For example, consider the following comma-separated file which gives the third virial coefficient for most (but not all) of the temperatures and includes header, footer and comment lines:

```
Virial Coefficients of H2
T /K, B /cm3.mol-1, C /10^-2 cm6.mol-1
 25, -106.2, 14.0
 30,  -80.7, 16.0
 40,  -50.3, 12.1
 50,  -33.4,  9.6
100,   -2.5,  6.1
// 110.4 K is the Boyle Temperature
110.4,   0,     *
120,    2.0,   5.7
------
Data from US National Bureau of Standards Publication
```

Here, there are two header lines, two footer lines and a comment line (preceded by the character sequence "//". The data itself is in comma-delimited columns and missing data is denoted with an asterisk ("*"). Examining the column header, it can be seen that the values in the last column are not in units of cm^6 mol^{-1} but have been scaled by a factor of 10^{-2}. If we want account for this factor, we can run the entries in this column through a converter function that will apply the scaling;[1] since we will probably only use this function in this one case, it is a good candidate for a `lambda` (anonymous) function (see Section 6.5):

```
In [x]: T, B, C = np.genfromtxt('H2-virial-coeffs.txt',
                    comments='//',
                    delimiter=',',
                    skip_header=2,
                    skip_footer=2,
                    missing_values='*',
                    converters={2: (lambda x: float(x) / 100)},
                    unpack=True)

In [x]: T
Out[x]: array([ 25. ,   30. ,   40. ,   50. ,  100. ,  110.4, 120. ])

In [x]: B
Out[x]: array([-106.2,  -80.7,  -50.3,  -33.4,   -2.5,    0. ,    2. ])

In [x]: C
Out[x]: array([0.14 , 0.16 , 0.121, 0.096, 0.061,    nan, 0.057])
```

9.7 Examples

E9.1

Question

A certain sample of the polymer polyisobutene, $(CH_2C(CH_3)_2)_n$, has a distribution of molar masses, M_i, given in the following table, where x_i is the mole fraction of polymers with total molar masses in the stated range.

M_i /kg mol^{-1}	x_i
6–10	0.07
10–14	0.16
14–18	0.20
18–22	0.28
22–24	0.18
24–28	0.08
28–32	0.03

---

[1] The numbers in this table may be interpreted as dimensionless values obtained by manipulating the physical quantity specified in the header, a process known as *quantity calculus* (see Section 14.1.2). For example, the first entry in the third column may be parsed as $C/10^{-2}cm^6 \, mol^{-1} = 14.0$, that is $C = 14.0 \times 10^{-2} cm^6 \, mol^{-1}$.

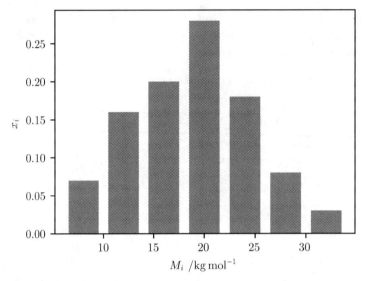

M_i /kg mol^{-1}

For this sample, estimate the number average molar mass,

$$\bar{M}_n = \frac{\sum_i N_i M_i}{\sum_i N_i},$$

and the mass average molar mass,

$$\bar{M}_w = \frac{\sum_i N_i M_i^2}{\sum_i N_i M_i},$$

where N_i is the number of polymers with molar mass M_i.

Also calculate the dispersivity (a measure of the how broad the distribution of polymer sizes is),

$$Đ = \frac{\bar{M}_w}{\bar{M}_n},$$

and the degree of polymerization,

$$DP_n = \frac{\bar{M}_n}{m}.$$

where the monomer molar mass of polyisobutene is $m = 56$ g mol^{-1}.

Solution

Take the average molar mass in each range and create NumPy arrays M and x.

```
import numpy as np

M = np.array([8, 12, 16, 20, 24, 28, 32])
x = np.array([0.07, 0.16, 0.2, 0.28, 0.18, 0.08, 0.03])
```

We have mole fractions, $x_i = N_i/N_{tot}$, where $N_{tot} = \sum_i N_i$, rather than the absolute number of polymers within each mass range, so the relevant formulas for the average molar masses are

$$\bar{M}_n = \sum_i x_i M_i, \quad \text{and } \bar{M}_w = \frac{\sum_i x_i M_i^2}{\sum_i x_i M_i}.$$

```
Mn = np.sum(x * M)
Mw = np.sum(x * M**2) / Mn
```

```
print(f'Mn = {Mn:.1f} kg.mol-1')
print(f'Mw = {Mw:.1f} kg.mol-1')
```

```
Mn = 18.8 kg.mol-1
Mw = 20.6 kg.mol-1
```

As an aside, note that these quantities can also be calculated as weighted averages with numpy.average:

```
np.average(M, weights=x), np.average(M, weights=M*x)
```

```
(18.800000000000004, 20.62978723404255)
```

The dispersivity and degree of polymerization can also be calculated:

```
# Dispersivity, D = Mw / Mn
D = Mw / Mn
print(f'Dispersivity, D = {D:.1f}')

# Monomer molar mass, g.mol-1.
m = 56
# Calculate the degree of polmerization, DPn = Mn / m.
DPn = Mn * 1000 / m
print(f'Degree of polymerization, DPn = {DPn:.0f}')
```

```
Dispersivity, D = 1.1
Degree of polymerization, DPn = 336
```

The distribution of polymer sizes is fairly narrow, and there are just over 330 monomer units per polymer on average.

E9.2

Statistics on Lead in Drinking Water

The comma-separated data file, Flint-Pb.csv, available at https://scipython .com/chem/ggc/, contains data collected by a citizen science campaign to measure

the concentration of lead (in parts per billion, ppb) in the drinking water of Flint,
Michigan in the United States. We will determine the maximum, mean, median and
standard deviation of these data. The legal action level for lead concentrations is 15
ppb; we will also count the number of samples that exceed this level.

The first few lines of the file are as follows:

```
# Data from FlintWaterStudy.org (2015) "Lead Results from Tap Water Sampling
# in Flint, MI during the Flint Water Crisis"
# Lead concentration in parts per billion ("Bottle 1, first draw").
SampleID,Pb / ppb
1,0.344
2,8.133
4,1.111
5,8.007
6,1.951
7,7.2
8,40.63
9,1.1
12,10.6
13,6.2
14,-
15,4.358
16,24.37
17,6.609
18,4.062
19,2.484
...
```

There are four header lines, and missing data (e.g., for sample ID 14) is represented
by a hyphen. These data can be read by numpy.genfromtxt:

```
import numpy as np

sampleID, Pb_ppb = np.genfromtxt('Flint-Pb.csv', skip_header=4,
                                 unpack=True, delimiter=',')
```

We can look at the first 20 values, to see that the missing data appear as NaN
values:

```
Pb_ppb[:20]
```

```
array([ 0.344,  8.133,  1.111,  8.007,  1.951,  7.2  , 40.63 ,  1.1  ,
       10.6  ,  6.2  ,   nan,  4.358, 24.37 ,  6.609,  4.062,  2.484,
        0.438,  1.29 ,  0.548,  3.131])
```

There are two ways to obtain the necessary statistics. Most straightforward is to use
the NumPy methods that ignore NaN values:

```
maximum = np.nanmax(Pb_ppb)
mean = np.nanmean(Pb_ppb)
median = np.nanmedian(Pb_ppb)
```

```
std = np.nanstd(Pb_ppb)

print(f'[Pb] Maximum = {maximum:.1f} ppb')
print(f'[Pb] Mean = {mean:.1f} ppb')
print(f'[Pb] Median = {median:.1f} ppb')
print(f'[Pb] Standard deviation = {std:.1f} ppb')
```

```
[Pb] Maximum = 158.0 ppb
[Pb] Mean = 10.6 ppb
[Pb] Median = 3.5 ppb
[Pb] Standard deviation = 21.5 ppb
```

Alternatively, we can remove the NaN values from the `Pb_ppb` array and use the regular NumPy statistics methods:

```
Pb_ppb = Pb_ppb[~np.isnan(Pb_ppb)]        "not (is NaN)" => valid data
```

```
np.max(Pb_ppb), np.mean(Pb_ppb), np.median(Pb_ppb), np.std(Pb_ppb)
```

```
(158.0, 10.6459926199262, 3.521, 21.520960778768078)
```

Our code is clearest if we name the lead concentration "action level" of 15 ppb as a suitable variable:

```
ACTION_LEVEL_Pb_ppb = 15
n_high_Pb = np.sum(Pb_ppb > ACTION_LEVEL_Pb_ppb)
print(f'Number of samples exceeding the "action level": {n_high_Pb}'
      f' out of {len(Pb_ppb)}')
```

```
Number of samples exceeding the "action level": 45 out of 271
```

9.8 Exercises

P9.1 A $3 \times 4 \times 4$ array is created as follows:

```
import numpy as np
a = np.linspace(1,48,48).reshape(3,4,4)
a
```

```
array([[[ 1.,  2.,  3.,  4.],
        [ 5.,  6.,  7.,  8.],
        [ 9., 10., 11., 12.],
        [13., 14., 15., 16.]],

       [[17., 18., 19., 20.],
        [21., 22., 23., 24.],
```

```
     [25., 26., 27., 28.],
     [29., 30., 31., 32.]],

    [[33., 34., 35., 36.],
     [37., 38., 39., 40.],
     [41., 42., 43., 44.],
     [45., 46., 47., 48.]]])
```

Index or slice this array to obtain the following:

(a)

```
   20.0
```

(b)

```
   [  9.   10.   11.   12.]
```

(c) The 4 × 4 array:

```
      [[ 33.   34.   35.   36.]
       [ 37.   38.   39.   40.]
       [ 41.   42.   43.   44.]
       [ 45.   46.   47.   48.]]
```

(d) The 3 × 2 array:

```
    [[  5.,    6.],
     [ 21.,   22.],
     [ 37.,   38.]]
```

(e) The 4 × 2 array:

```
    [[ 36.   35.]
     [ 40.   39.]
     [ 44.   43.]
     [ 48.   47.]]
```

(f) The 3 × 4 array:

```
    [[ 13.    9.    5.    1.]
     [ 29.   25.   21.   17.]
     [ 45.   41.   37.   33.]]
```

P9.2 The file `KCl-yields.txt`, available at `https://scipython.com/chem/xga/`, contains the results of an experiment in which 30 students measured the yield of KCl resulting from the decomposition of 50 g of potassium chlorate, $KClO_3$, when it is heated in the presence of a manganese dioxide catalyst:

$$2\,KClO_3 \xrightarrow{\Delta,\,MnO_2} 2\,KCl + 3\,O_2$$

Read in these data with `numpy.genfromtxt` and identify missing data and mass values either equal to zero or equal to more than three standard deviations from the median value. Omitting these entries, determine the maximum and minimum yields reported. Also calculate the mean and standard deviation of the sample's valid entries.

Create a new NumPy array of the students' yields as a percentage of the theoretical maximum yield and output it as a sorted list. Take the molar masses of $KClO_3$ and KCl to be 122.6 and 74.6 g mol^{-1}, respectively.

10

Graph Plotting with Matplotlib

10.1 Line Plots and Scatter Plots

Matplotlib is the most popular plotting library for Python. It can be used to produce both quick visualizations of data and, with care, high-quality (i.e., publication-ready) graphs in a variety of formats. This chapter will focus on Matplotlib's `pyplot` functionality, which is used to create and annotate figures using a simple procedural interface. `pyplot` needs to be imported, is usually aliased to the abbreviated identifier, `plt`:

```
import matplotlib.pyplot as plt
```

A simple use to plot and label some data points is to call `plt.plot` with two lists or arrays of numbers as follows:

```
import numpy as np
import matplotlib.pyplot as plt

x = np.array([-2, -1.5, -1, -0.5, 0, 0.5, 1, 1.5, 2])
y = x**3
plt.plot(x, y)
```

In a Python script or interactive Python session, nothing will appear on the screen until you call `plt.show()`, so before doing that various annotations can be added, such as labels for the axes and a title (see Figure 10.1):

```
plt.xlabel('x')
plt.ylabel('y')
plt.title('y = x^3')
plt.show()
```

In Jupyter Notebook, the default behavior is for the plot to be shown as soon as the cell containing the `plt.plot()` statement is executed, so the above annotations should be in the same cell as this statement. To add to a single plot from multiple notebook cells, execute the `%matplotlib notebook` command first.

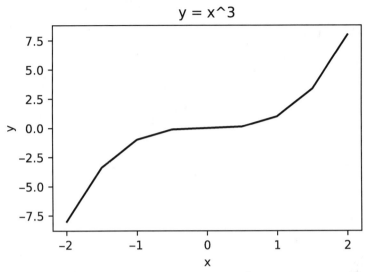

Figure 10.1   Matplotlib line plot of $y = x^3$ using nine points.

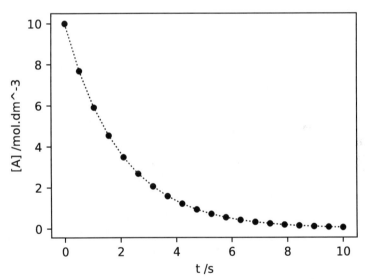

Figure 10.2   Matplotlib line plot of $[A] = [A]_0 e^{-kt}$.

Plots can be further customized with the use of markers and by specifying the line color, thickness and style with further arguments, as detailed in Tables 10.1–10.3. For example (Figure 10.2):

```
A0, k = 10, 0.5
t = np.linspace(0, 10, 20)
A = A0 * np.exp(-k * t)
plt.plot(t, A, c='m', lw=1, ls=':', marker='o')
```

```
plt.xlabel('t /s')
plt.ylabel('[A] /mol.dm^-3')
plt.show()
```

Table 10.1   Matplotlib color code
letters: the single-letter codes are
rather bright and saturated; the default
color cycle uses the more appealing
"Tableau 10" palette

Basic color codes	Tableau colors
b = blue	tab:blue
g = green	tab:orange
r = red	tab:green
c = cyan	tab:red
m = magenta	tab:purple
y = yellow	tab:brown
k = black	tab:pink
w = white	tab:gray
	tab:olive
	tab:cyan

Table 10.2   Some Matplotlib marker styles
(single-character string codes)

Code	Marker	Description
.	·	Point
o	○	Circle
+	+	Plus
x	×	Cross
D	◇	Diamond
v	▽	Downward triangle
^	△	Upward triangle
s	□	Square
*	★	Star

Calling `plt.plot` repeatedly with different data adds new lines to the plot. With
multiple plotted lines, it is helpful to distinguish them by passing a suitable string
to the `label` argument; in this case, `plt.legend()` must be called before showing
the plot (Figure 10.3):

```
A0 = 10
k = [0.2, 0.5, 1]
t = np.linspace(0, 10, 30)
rate_constants = [0.2, 0.5, 1]
for k in rate_constants:
```

```
    A = A0 * np.exp(-k * t)
    plt.plot(t, A, marker='o', label=f'k ={k:.1f} s-1'.format(k))
plt.xlabel('t /s')
plt.ylabel('[A] /mol.dm^-3')
plt.legend()
plt.show()
```

Table 10.3   Matplotlib line and marker properties

Argument	Abbreviation	Description
color	c	Line color
alpha		Line opacity: 0 (completely transparent) – 1 (totally opaque)
linestyle	ls	Line style: 'solid', 'dotted', 'dashed' 'dashdot'; alternatively '-', ':', '-' '-.'
linewidth	lw	Line width in points, the default is 1.5 pt.
markersize	ms	Marker size, in points
markevery		Set to a positive integer, N, to print a marker every N points; the default, None, prints a marker for every point
markerfacecolor	mfc	Fill color of the marker
markeredgecolor	mec	Edge color of the marker
markeredgewidth	mew	Edge width of the marker, in points

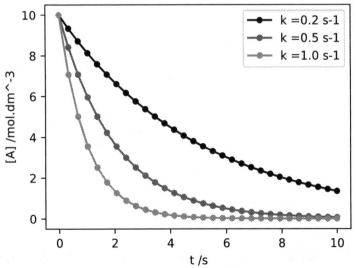

Figure 10.3   Matplotlib line plots of $[A] = [A]_0 e^{-kt}$ for different rate constants, k.

Note that it is not necessary to specify a new color for each call to `plt.plot`: Matplotlib will cycle through a predefined sequence of colors that distinguish the plotted lines.

By default, Matplotlib chooses the limits of the x and y axes to show all of the plotted data. To set these limits manually, use `plt.xlim` and `plt.ylim`.

For example, consider plotting the Lennard-Jones interatomic potential for argon,

$$V(r) = 4\epsilon \left[\left(\frac{\sigma}{r} \right)^{12} - \left(\frac{\sigma}{r} \right)^{6} \right],$$

where r is the interatomic distance, $\epsilon = 3.4 \, \text{kJ} \, \text{mol}^{-1}$ is the depth of the potential well and $\sigma = 3.4 \, \text{Å}$ is the separation distance at which the potential is zero: $V(\sigma) = 0$.

Simply plotting $V(r)$ on a grid of r values will not yield a very satisfactory graph (Figure 10.4):

```
import numpy as np
import matplotlib.pyplot as plt

rmax = 10
r = np.linspace(0.1, rmax, 1000)
# Lennard-Jones parameters for Ar: well depth (kJ/mol) and distance at which
# the interaction potential between the two atoms is zero (Angstroms).
E, sigma = 1, 3.4

def LJ(r, E, sigma):
    fac = (sigma / r)**6
    return 4 * E * (fac**2 - fac)

V = LJ(r, E, sigma)
plt.plot(r, V)
plt.xlabel('r / A')
plt.ylabel('V(r) / kJ.mol-1')
plt.title('Lennard-Jones Potential for Argon')
plt.show()
```

Here, the dominance of the repulsive term in the potential (r^{-12}) leads to extremely large values of $V(r)$ at small r, so that the attractive well region $(V(r) < 0)$ is not visible. The solution is to set the plot limits appropriately (Figure 10.5):

```
# Start the distance axis at 2 Angstroms.
plt.xlim(2, rmax)
# Limit the y-range to between the bottom of the well and a half the magnitude
# of the well depth.
plt.ylim(-E, 0.5*E)
```

There are some other common plot types provided by Matplotlib, including the scatter plot:

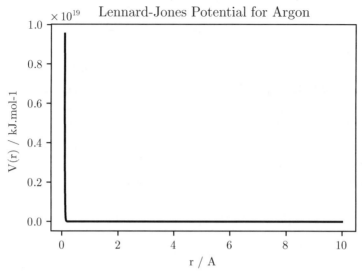

Figure 10.4   Naive plot of the Lennard-Jones potential for argon, without setting the plot limits.

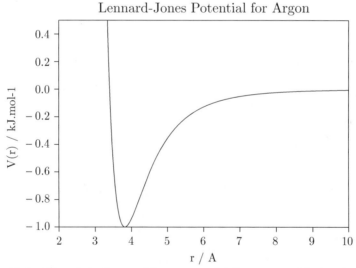

Figure 10.5   Plot of the Lennard-Jones potential for argon, with the plot limits set appropriately.

```
plt.scatter(x, y, s=None, c=None, marker=None}
```

makes a scatter plot of the (x, y) data with specified markers of size s (in points^2) and color(s) c (defaults are used if no values are provided for the these arguments). If s or c are sequences, the sizes and colors of the markers can be set per-data point.

To annotate a plot with text, call `plt.text`:

```
plt.text(x, y, s)
```

places the string s at the location (x, y) (in data-coordinates). Further arguments, ha (or `horizontalalignment`) and va (or `verticalalignment`) control how the text is anchored at this location: valid values are ha = 'left', 'center' or 'right' and va = 'top', 'center', 'baseline' or 'bottom'. For example, to ensure that the centre of the text label lies at (x, y), set ha='center', va='center'.

See Example E10.3, below, for a demonstration of the use of these two functions.

10.2 Examples

E10.1

The Temperature-Dependence of Potassium Halide Solubilities

The following data concern the solubilities of three potassium halides (in g of salt per 100 g of water) as a function of temperature:

T /°C	KCl	KBr	KI
0	28.2	54.0	127.8
20	34.2	65.9	144.5
40	40.3	76.1	161.0
60	45.6	85.9	176.2
80	51.0	95.3	191.5
100	56.2	104.9	208.0

We can plot these data on a single, labeled and titled chart, distinguishing the data points with different line styles and markers.

```python
import numpy as np
import matplotlib.pyplot as plt
```

```python
# Temperatures (degC) and solubilities (g / 100 g of water).
T = np.arange(0, 101, 20)
S_KCl = np.array((28.2, 34.2, 40.3, 45.6, 51, 56.2))
S_KBr = np.array((54, 65.9, 76.1, 85.9, 95.3, 104.9))
S_KI = np.array((127.8, 144.5, 161, 176.2, 191.5, 208))
```

```python
def plot_solubilities():
    """Plot the halide solubilities as a function of temperature."""
    plt.plot(T, S_KCl, c='k', ls='-', lw=2, marker='s', ms=8, label='KCl')
    plt.plot(T, S_KBr, c='k', ls='--', lw=2, marker='^', ms=8, label='KBr')
```

```
    plt.plot(T, S_KI, c='k', ls=':', lw=2, marker='o', ms=8, label='KI')
    plt.xlabel('T (degC)')
    plt.ylabel('S (g / 100 g H2O)')
    plt.legend()
    # Add a plot title: to ensure it fits, include a line break (\n).
    plt.title('Temperature dependence of the solubilities\n'
              'of some potassium salts')
plot_solubilities()
```

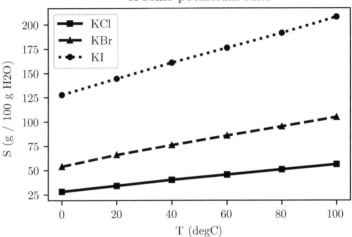

E10.2

Visualizing the Maxwell–Boltzmann Distribution

The Maxwell–Boltzmann distribution is a probability distribution for the speeds, v, of particles in an ideal gas at thermodynamic equilibrium at a temperature, T:

$$f(v) = \left(\frac{m}{2\pi k_B T} \right)^{3/2} 4\pi v^2 \exp\left(-\frac{mv^2}{2k_B T} \right),$$

where m is the particle mass.

A Python function to calculate this distribution is straightforward to write and use:

```
import numpy as np
import matplotlib.pyplot as plt
# Boltzmann constant, J.K-1, atomic mass unit (kg)
kB, u = 1.381e-23, 1.661e-27
```

```
def fMB(v, T, m):
    """

    Return value of the Maxwell-Boltzmann distribution for a molecule of
    mass m (in kg) at temperature T (in K), at the speed v (in m.s-1).

    """

    fac = m / 2 / kB / T
    return (fac / np.pi)**1.5 * 4 * np.pi * v**2 * np.exp(-fac * v**2)
```

```
# A grid of speeds between 0 and vmax.
vmax = 2000
v = np.linspace(0, vmax, 1000)
# The molecular mass (in kg) of the N2 molecule.
mN2 = 2 * 14 * u

# Calculate the Maxwell-Boltzmann distribution for a gas of N2 at two different
# temperatures (in K).
T1, T2 = 300, 600
f_N2_T1 = fMB(v, T1, mN2)
f_N2_T2 = fMB(v, T2, mN2)
```

The Matplotlib function `fill_between` can be used to fill the region under a line: Its basic arguments are x, the *x*-coordinates of the region to fill and y1 and y2 which are the bounds of the *y*-coordinates. If not given, y2=0, meaning that the region below the curve y1(x) to the *x*-axis will be filled. The color used to fill the region can be set with the argument `facecolor` (or `fc`); see Figure 10.6.

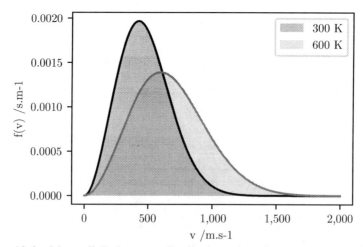

Figure 10.6   Maxwell–Boltzmann distribution plots for N_2 at two different temperatures.

```
def plot_MB():
    plt.plot(v, f_N2_T1)
    plt.plot(v, f_N2_T2, c='tab:red')
    plt.xlabel('v /m.s-1')
    plt.fill_between(v, f_N2_T1, alpha=0.2, label=f'{T1} K')
    plt.ylabel('f(v) /s.m-1')
    plt.fill_between(v, f_N2_T2, alpha=0.2, fc='tab:red', label=f'{T2} K')
    plt.title('Maxwell-Boltzmann speed distribution for N2 at two temperatures')
    plt.legend()
plot_MB()
```

(The argument `alpha=0.2` sets the fill opacity to 20%.)

E10.3

Question

The file `elements_TmTb.csv`, which can be downloaded from `https://scipython.com/chem/ghc/`, contains the values of the melting points and boiling points of the elements. Make a scatter plot of these data, coloring the markers differently for metals and non-metals. Also annotate the plot to indicate the element carbon and the element with the greatest difference between its melting point and boiling point.

For the purposes of this exercise, the nonmetallic elements may be taken to be those given in the following list.

```
nonmetals = ['H', 'He', 'C', 'N', 'O', 'F', 'Ne', 'S', 'P',
             'Ar', 'Se', 'Cl', 'Kr', 'Br', 'Xe', 'I', 'Rn']
```

Solution

```
import numpy as np
import matplotlib.pyplot as plt
```

The provided file contains comma-separated columns of element symbol, melting point and boiling point:[1]

```
!head elements_TmTb.csv
```

```
Element, Tm /K,  Tb /K
H,          14.0,   20.3
He,          1.8,    4.2
Li,        453.7, 1603.0
Be,       1560.0, 2742.0
B,        2349.0, 4200.0
```

---

[1] The command !head prints the first 10 lines of a file, at least on Unix-like operating systems.

```
C,       3800.0,  4300.0
N,         63.2,    77.4
O,         54.4,    90.2
F,         53.5,    85.0
```

The problem is that the first column consists of strings rather than numbers, and we need the element symbols. It is possible to use NumPy's genfromtxt method to read these columns into separate arrays, but we need the specify dtype=None (to force NumPy to infer the datatype and not just assume the first column is a float) and also to specify the encoding of the strings (ASCII or UTF-8 will do):

```
elements, Tm, Tb = np.genfromtxt('elements_TmTb.csv', delimiter=',',
                          skip_header=1, unpack=True, dtype=None,
                          encoding='utf8')
```

```
elements[:5], Tm[:5], Tb[:5]
```

```
(array(['H', 'He', 'Li', 'Be', 'B'], dtype='<U2'),
 array([1.400e+01, 1.800e+00, 4.537e+02, 1.560e+03, 2.349e+03]),
 array([  20.3,    4.2, 1603. , 2742. , 4200. ]))
```

We could either call plt.scatter twice (once for the metals and once for the non-metals) or construct a list of colors to use. If we take the latter approach, it turns out to be simplest to convert elements into a regular Python list first:

```
elements = list(elements)
n = len(elements)
# Metals will be indicated by blue circles.
c = ['b'] * n
for symbol in nonmetals:
    idx = elements.index(symbol)
    # Non-metals will be indicated by red circles.
    c[idx] = 'r'
```

Which element has the greatest difference between its melting and boiling point?

```
# NB there are NaN entries for elements without known values of Tm or Tb.
idx = np.nanargmax(Tb - Tm)
elements[idx]
```

```
'Np'
```

Neptunium, apparently.

```
plt.scatter(Tm, Tb, c=c)
plt.xlabel('Tm /K')
plt.ylabel('Tb /K')
```

```
def annotate_with_element_symbol(symbol):
    i = elements.index(symbol)
    plt.text(Tm[i], Tb[i], symbol)

annotate_with_element_symbol('C')
annotate_with_element_symbol('Np')
```

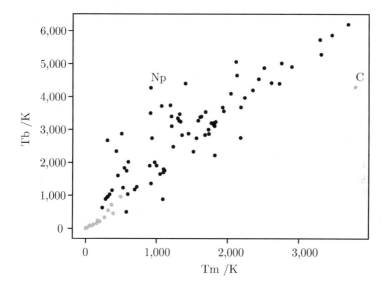

10.3 Exercise

P10.1 Write a function to plot the Maxwell–Boltzmann distribution of molecular speeds for a gas of particles of a given mass at a given temperature, indicating the modal speed ($v_\star$), mean ($\langle v \rangle$) and root mean square (rms, $\langle v^2 \rangle^{1/2}$) speeds with vertical lines.

Call this function for the atomic gases argon ($m = 40\,u$) and xenon ($m = 131\,u$) at 300 K.

Hints: The modal speed is the maximum of the probability distribution and can be found from df/dv. The mean and rms speeds can be obtained, respectively, from the integrals

$$\langle v \rangle = \int_0^\infty vf(v)\,dv \quad \text{and} \quad \langle v^2 \rangle = \int_0^\infty v^2 f(v)\,dv.$$

11

The Steady-State Approximation

The steady-state approximation is a commonly used approach for simplifying complex systems of chemical reactions where the concentrations of intermediate species remain fairly constant throughout much of the process (i.e., after an initial induction period during which they rise from zero).

Consider the simplest case of a multi-step reaction, $A \to B \to C$, in which a reactant, A, is converted, by a unimolecular process, to an intermediate, B, which then proceeds to form the product, C:

$$\frac{d[A]}{dt} = -k_1[A],$$

$$\frac{d[B]}{dt} = k_1[A] - k_2[B],$$

$$\frac{d[C]}{dt} = k_2[B].$$

In this case, the kinetics can be solved analytically. For the case $k_1 \neq k_2$, one obtains:

$$[A] = [A]_0 e^{-k_1 t}$$

$$[B] = \frac{k_1[A]_0}{k_2 - k_1} \left(e^{-k_1 t} - e^{-k_2 t} \right)$$

$$[C] = [A]_0 - [A] - [B]$$

(the last equation follows from mass balance).

To take a concrete example, consider $k_1 = 1 \, \text{s}^{-1}$ and $k_2 = 0.1 \, \text{s}^{-1}$:

```
import numpy as np
import matplotlib.pyplot as plt

t = np.linspace(0, 40, 200)
```

```
def plot_kinetics(k1, k2, A0=1):
    """Plot [A], [B] and [C] from the analytical rate expression."""
    A = A0 * np.exp(-k1*t)
    B = k1 * A0 / (k2 - k1) * (np.exp(-k1*t) - np.exp(-k2*t))
    C = A0 - A - B
    plt.plot(t, A, label='[A]')
    plt.plot(t, B, label='[B]', c='tab:red')
    plt.plot(t, C, label='[C]', c='tab:green')
def annotate_plot():
    plt.legend()
    plt.xlabel('t /s')
    plt.ylabel('conc. / mol.dm^-3')

k1, k2 = 1, 0.1
plot_kinetics(k1, k2)
annotate_plot()
plt.show()
```

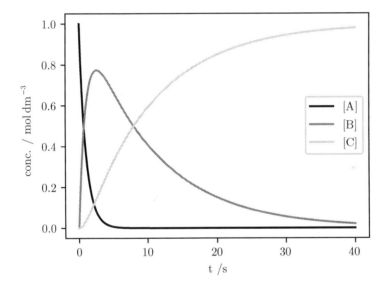

Clearly, the steady-state approximation cannot be expected to hold in this case, since the concentration of the intermediate, B, changes considerably with time. However, if $k_2 \gg k_1$ such that the rate constant for the reaction of the intermediate is much larger than that for its formation from A, then [B] is predicted to be small and slowly varying with time:

```
k1, k2 = 0.1, 1
plot_kinetics(k1, k2)
annotate_plot()
plt.show()
```

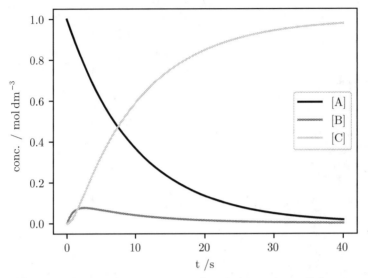

In this case, the steady-state approximation consists of the assumption that at a given instant in time:

$$\frac{d[B]}{dt} = k_1[A] - k_2[B] = 0$$

and hence:

$$\frac{d[C]}{dt} = k_2[B] = k_1[A]$$

That is,

$$[B]_{ss} = \frac{k_1}{k_2}[A]$$

and

$$\int_0^{[C]} d[C] = \int_0^t k_1[A]\,dt = [A]_0 \int_0^t e^{-k_1 t}\,dt$$
$$\Rightarrow [C]_{ss} = [A]_0 \left(1 - e^{-k_1 t}\right)$$

Note that the assumption of steady-state does not mean that [B] remains constant: Its value tracks that of [A] in proportion k_1/k_2 and since [A] is decreasing, after the induction period, so does [B].

```
def plot_kinetics_ss(k1, k2, A0=1):
    """Plot [A], [B] and [C] from the steady-state rate expression."""
    A = A0 * np.exp(-k1*t)
    Bss = k1 / k2 * A
    Css = A0 - A - Bss
    plt.plot(t, Bss, label='[B]ss', ls=':', c='tab:red')
    plt.plot(t, Css, label='[C]ss', ls=':', c='tab:green')
plot_kinetics(k1, k2)
plot_kinetics_ss(k1, k2)
annotate_plot()
plt.show()
```

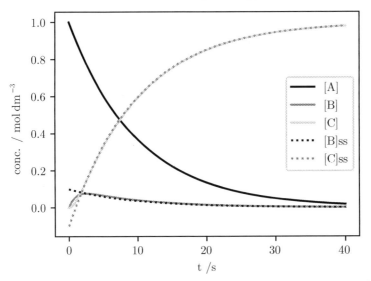

The induction period lasts about 3 seconds, after which the steady-state approximation (dotted lines) closely tracks the analytical solution. The functions `plt.xlim` and `plt.ylim` can be used to zoom in on this induction period:

```
plot_kinetics(k1, k2)
plot_kinetics_ss(k1, k2)
annotate_plot()
plt.xlim(0, 5)
plt.ylim(0, 0.2)
plt.show()
```

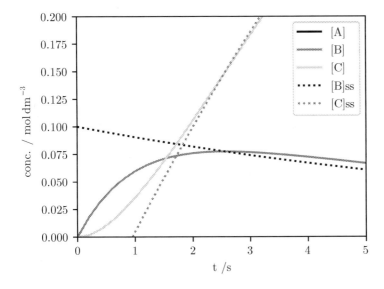

12

Liquid–Vapor Equilibrium

The partial vapor pressures of the components of an ideal solution of two liquids (a binary solution) are related to the liquid composition through:

$$p_1 = x_1 p_1^* \text{ and } p_2 = x_2 p_2^*,$$

where p_i^* is the vapor pressure of the pure liquid labeled i and the mole fractions of the liquid components are x_i. For a binary mixture, $x_1 + x_2 = 1$. The total vapor pressure of the mixture is thus

$$p = p_1 + p_2 = x_1 p_1^* + x_2 p_2^* = x_1 p_1^* + (1 - x_1) p_2^*,$$

and the partial vapor pressures are therefore

$$y_1 = \frac{p_1}{p} = \frac{x_1 p_1^*}{x_1 p_1^* + (1 - x_1) p_2^*}$$

$$\text{and } y_2 = \frac{p_2}{p} = \frac{x_2 p_2^*}{x_1 p_1^* + (1 - x_1) p_2^*}.$$

For ethanol, $p_1^* = 44.5$ mmHg and for water, $p_2^* = 23.8$ mmHg (both at 298 K). The above equations can be used to predict the composition of the vapor phase of a binary mixture of ethanol and water at different compositions, assuming ideal solution behavior.

```
import numpy as np
import matplotlib.pyplot as plt
```

```
# vapor pressures of pure ethanol (1) and water (2) at 298 K in mmHg
p1_star, p2_star = 44.5, 23.8
```

```
x1 = np.linspace(0, 1, 100)
y1 = x1 * p1_star / (x1*p1_star + (1-x1)*p2_star)
plt.plot(x1, y1, c='tab:green', label='Raoult')
# Draw a line at y1 = x1 for reference
```

```
plt.plot(x1, x1, c='k')
plt.xlabel('x1')
plt.ylabel('y1')
plt.xlim(0, 1)
plt.ylim(0, 1)
plt.legend()
```

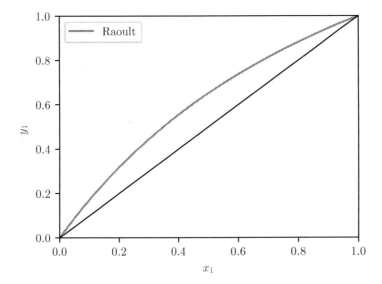

The *experimental* values of the vapor composition for different liquid mixtures of ethanol and water are provided in the file, `ethanol-water-vle-mass.txt`, available at `https://scipython.com/chem/caj/`. These data, however, consist of the percentage amounts of ethanol *by mass*, w_1, as a proportion of the total solution (or vapor) mass. The conversion to mole fraction is

$$x_1 = \frac{w_1/M_1}{w_1/M_1 + (100 - w_1)/M_2},$$

where $M_1 = 46.07$ g mol^{-1} and $M_2 = 18.02$ g mol^{-1} are the molar masses of ethanol and water, respectively.

```
wx, wy = np.genfromtxt('ethanol-water-vle-mass.txt', unpack=True, skip_header=4)
```

```
def get_mol_frac(w1, M1, M2):
    """Get the mole fraction of a component of a binary mixture.

    The provided w1 is the mass fraction as a percentage of solution mass;
    M1 and M2 are the molar masses of the component compounds.

    """

    return w1/M1 / (w1/M1 + (100-w1)/M2)
```

```
M1, M2 = 46.07, 18.02
x, y = get_mol_frac(wx, M1, M2), get_mol_frac(wy, M1, M2)
plt.plot(x, y, marker='.', lw=1, label='Experiment')
plt.fill_between(x, y, x, alpha=0.2)
# A reference line at y1 = x1.
plt.plot([0, 1],[0, 1], c='k', lw=1)
plt.plot(x1, y1, lw=1, label='Raoult')
plt.xlabel('x1')
plt.ylabel('y1')
plt.xlim(0, 1)
plt.ylim(0, 1)
plt.legend()
```

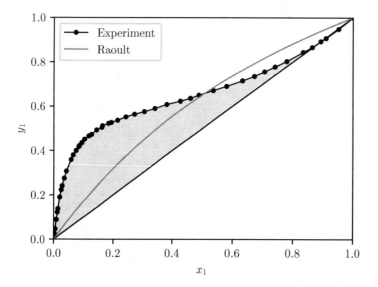

There is a considerable deviation from Raoult's law and, in fact, at about $x_1 = 0.89$ ($w_1 = 95.5\%$) the mixture forms an *azeotrope*: At this point, the compositions of the liquid and vapor phases are equal and fractional distillation by normal means cannot be used to further concentrate the ethanol, since condensing the vapor does not lead to a change in composition.

Zooming in on the region around the azeotropic composition shows this:

```
plt.plot(x, y, '-', marker='.', lw=1, label='Experiment')
plt.fill_between(x, y, x, alpha=0.2)
plt.plot([0, 1], [0, 1], lw=1, label='y1 = x1')
plt.plot(x1, y1, lw=1, label='Raoult')
plt.xlim(0.85, 1)
plt.ylim(0.85, 1)
plt.xlabel('x1')
plt.xlabel('y1')
plt.legend()
```

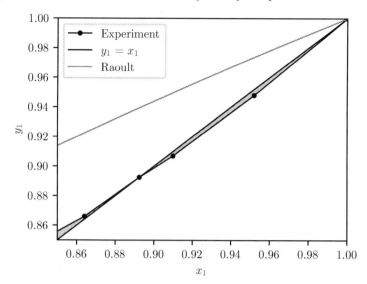

13

Jupyter Notebook

Jupyter Notebook provides an interactive environment for Python programming within a web browser. It enables Python code to be combined with documentation (including in rendered LaTeX – see Chapter 14), images and even rich media such as embedded videos. Jupyter Notebooks are increasingly being used by scientists to report their research by communicating the computations carried out on data as well as simply the results of those computations. The format makes collaboration between researchers easier and facilitates the reproduction of research results by allowing others to validate the analysis on the same data.

13.1 Jupyter Notebook Basics

13.1.1 Starting the Jupyter Notebook Server

If you have Jupyter installed, the server that runs its browser-based interface can be started from the command line with

```
jupyter notebook
```

or by opening the relevant application in graphical-based installations.

This will open a web browser window at the URL of the local Jupyter Notebook application (by default, at `http://localhost:8888`).

The Jupyter Notebook index page (Figure 13.1) contains a list of the notebooks currently available in the directory from which the notebook server was started. This is also the default directory to which notebooks will be saved (with the extension `.ipynb`), so it is a good idea to execute the above command somewhere convenient in your directory hierarchy for the project you are working on.

The index page contains three tabs: *Files* lists all the files, including Jupyter Notebooks and subdirectories within the current working directory; *Running* lists those notebooks that are currently active within your session (even if they are not

Figure 13.1   The Jupyter Notebook index page.

Figure 13.2   Jupyter with a new notebook document.

open in a browser window); *Clusters* provides an interface to Jupyter's parallel computing engine, which will not be described in this book.

From the index page, one can start a new notebook (by clicking on "New > Python 3 (ipykernel)") or open an existing notebook (by clicking on its name). To import an existing notebook into the index page, either click "Upload" at the top right of the page or drag the notebook file into the index listing from elsewhere on your operating system.

To stop the notebook server, press CTRL-C in the terminal window it was started from (and confirm at the prompt).

13.1.2 Editing a Jupyter Notebook

To start a new notebook, click the "New" button and select a notebook kernel (there should at least be one called "Python 3"). This opens a new browser tab containing the interface where you will write your code and connects it to a *kernel*, the computational engine responsible for executing the code and communicating with the browser interface.

The new notebook document (Figure 13.2) consists of a *title bar*, a *menu bar* and a *tool bar*, under which is a numbered prompt where you will type the code and markup (e.g., explanatory text and documentation) as a series of *cells*.

In the title bar, the name of the first notebook you open will probably be "Untitled"; click on it to rename it to something more helpful. The menu bar contains options for saving, copying and rearranging the Jupyter Notebook document. The tool bar consists of a series of icons that act as shortcuts for common operations that can also be achieved through the menu bar.

There are three types of input cells where you can write the content for your notebook:

- Code cells: The default type of cell, this type of cell consists of executable code. As far as this book is concerned, the code you write here will be Python, but Jupyter does provide a mechanism for executing code written in other languages.
- Markdown cells: This type of cell allows for a rich form of documentation for your code (see below). When executed, the input to a markdown cell is converted into HTML, which can include mathematical equations, font effects, lists, tables, embedded images and videos.
- Raw cells: Input into this type of cell is not changed by the notebook – its content and formatting is preserved exactly.

13.1.3 Running Cells

Each cell can consist of more than one line of input, and the cell is not interpreted until you "run" (i.e., execute) it. This is achieved by selecting the appropriate option from the menu bar (under the "Cell" drop-down submenu), by clicking the "Run cell" "play" button on the toolbar, or through the following keyboard shortcuts:

- `Shift-Enter`: Execute the cell, showing any output, and then *move the cursor* onto the cell below. If there is no cell below, a new, empty one will be created.
- `CTRL-Enter`: Execute the cell in place, but *keep the cursor* in the current cell.
- `Alt-Enter`: Execute the cell, showing any output, and then *insert and move the cursor to a new cell* immediately beneath it.

The menu bar, under the "Cell" drop-down submenu, provides many ways of running a notebook's cells: Usually, you will want to run the current cell individually or run it and all those below it. If things go wrong and your code "hangs" (the notebook becomes unresponsive) or the order of execution of the cells has become hopelessly muddled, then interrupt or restart the kernel from the "Kernel" drop-down submenu as appropriate.

Code Cells

You can enter anything into a code cell that you can from any other interactive Python session or when writing a Python program in an editor. Code in a given cell has access to objects defined in other cells (providing they have been run). For example,

```
In [ ]:    n = 100
```

Pressing Shift-Enter or clicking Run Cell executes this statement (defining n but producing no output) and opens a new cell underneath the old one:

```
In [1]:    n = 100
```

```
In [ ]:
```

Entering the following statements at this new prompt:

```
In [ ]:    triangular_number = n * (n+1) // 2
           print(f'1 + 2 + ... + {n} = {triangular_number}')
```

and executing as before produces output and opens a third empty input cell. The whole notebook document then looks like:

```
In [1]:    n = 100
```

```
In [2]:    triangular_number = n * (n+1) // 2
           print(f'1 + 2 + ... + {n} = {triangular_number}')
```

```
Out[2]:    1 + 2 + ... + 100 = 5050
```

```
In [ ]:
```

You can edit the value of n in input cell 1 and rerun the entire document to update the output. It is worth noting that it is also possible to set a new value for n *after* the calculation in cell 2:

```
In [3]:    n = 500
```

Running cell 3 and then cell 2 then leaves the output to cell 2 as:

```
Out[2]:    1 + 2 + ... + 500 = 125250
```

even though the cell above still defines n to be 100. That is, unless you run the entire document from the beginning, the output does not necessarily reflect the output of a script corresponding to the code cells taken in order.

13.2 Markdown Cells in Jupyter Notebook

In addition to containing Python code, Jupyter cells can also contain text and images to describe and document the Notebook. This content is formatted using a fairly simple syntax called Markdown which can produce content similar to that of a web page. Indeed, Markdown is a superset of HyperText Markup Language

(HTML), the standard markup language for documents displayed in a web browser. However, rather than having to insert HTML tags into the cell directly, Markdown has a simple set of conventions which it interprets to render the resulting formatted text. The following sections describe this syntax.

13.2.1 Paragraphs and Headings

A text paragraph is simply denoted by consecutive lines of text. Individual paragraphs are separated by one or more blank lines.

A document can be structured into a hierarchy of sections with headings of up to six levels. A heading is denoted by the corresponding number of hash symbols followed by a space and the heading text. For example,

In [x]:
```
# Section 1

## Section 1.1

This is the first paragraph of the section called
"Section 1.1" it consists of a few lines of text
and ends with a blank line.

This is the second paragraph of the section.
It is defined by four lines.
These will be merged into a single block of text because
they are not separated by blank lines.

## Section 1.2

And so on...
```

is rendered as follows:

Section 1

Section 1.1

This is the first paragraph of the section called "Section 1.1" it consists of a few lines of text and ends with a blank line.

This is the second paragraph of the section. It is defined by four lines. These will be merged into a single block of text because they are not separated by blank lines.

Section 1.2

And so on...

13.2.2 *Font Styles*

Italic, bold and underlined font styles are produced by surrounding text with one asterisk, *, two asterisks, **, and an underscore character, _, respectively. To display an actual asterisk or underscore, it may be necessary to escape it by preceding it with a backslash, \, as in the following example:

```
In [x]:    This is a paragraph of text with a variety of font styles,
           including *italic*, **bold** and _underlined_. If you want
           to surround some text with literal asterisks, you have to
           \*escape them\*.
```

This is a paragraph of text with a variety of font styles, including *italic*, **bold** and underlined. If you want to surround some text with literal asterisks, you have to *escape them*.

13.2.3 *Itemized and Enumerated Lists*

Bulleted (unordered) lists (which may be nested) are indicated with any of the characters *, - or + standing for the bullet. List items can consist of multiple paragraphs if they are separated by blank lines and indented to align with other text in the same item:

```
In [x]:    An unordered list:

           * Solids
               - Iron
               - Titanium

           * Liquids
               - Bromine

                   _Note_: Bromine is a reddish-brown liquid, with a
                   melting point of -7.2 °C and a boiling point of
                   58.8 °C. Its vapor is noticeable even at room
                   temperature.

               - Mercury

                   _Note_: Mercury is the only only metallic element
                   that is liquid at standard temperature and pressure.

           * Gases
               - Hydrogen
               - Oxygen
               - Chlorine
```

An unordered list:

- Solids
  - Iron
  - Titanium
- Liquids
  - Bromine

    Note: Bromine is a reddish-brown liquid, with a melting point of −7.2 °C and a boiling point of 58.8 °C. Its vapor is noticeable even at room temperature.
  - Mercury

    Note: Mercury is the only only metallic element that is liquid at standard temperature and pressure.
- Gases
  - Hydrogen
  - Oxygen
  - Chlorine

Enumerated (numbered) lists (which can also be nested) are produced in the same way, but by labeling items with a number followed by a period:

In [x]:
```
##### Instructions

1. Draw a thin line in pencil 1.5 cm from the bottom of the
   TLC plate.
2. Apply sample spots in equal distances across the line.
3. Pour the solvent into the TLC chamber to a depth of  1 cm.
4. Place the plate inside the champer with the pencil line
   just above the solvent surface.
5. Allow sufficient time for the development of the TLC spots.
```

Instructions

1. Draw at thin line in pencil 1.5 cm from the bottom of the TLC plate.
2. Apply sample spots in equal distances across the line.
3. Pour the solvent into the TLC chamber to a depth of 1 cm.
4. Place the plate inside the champer with the pencil line just above the solvent surface.
5. Allow sufficient time for the development of the TLC spots.

The list items are numbered starting at the number given for the first item and incremented by one for each subsequent item; any numbers at all can be given in the markup for these subsequent items, but they are not used.

13.2.4 Links

Links can be introduced into a Markdown cell in three ways:

Inline Links

An inline link provides a URL in round brackets after the text to be turned into a link surrounded in square brackets. Rendered within a Notebook, the link text is underlined and clickable to navigate to the URL.

In [x]:
```
This is a link to the [Project Jupyter
    Website](https://jupyter.org/).
```

> This is a link to the Project Jupyter Website.

Reference Links

Reference links are similar, but give a name in square brackets after the text to be turned into a link. This name should be associated with a URL elsewhere in the document with the syntax [*name*]: *url*.

In [x]:
```
Some resources for free chemical data include
    [Wikipedia][wikipedia], [ChemSpider][chemspider] and
    the [Dortmund Data Bank][ddb].

[wikipedia]: [https://wikipedia.org]
[chemspider]: [https://chemspider.com]
[ddb]: [http://www.ddbst.com/free-data.html]
```

> Some resources for free chemical data include Wikipedia, ChemSpider and the Dortmund Data Bank.

Automatic Links

If the clickable text is the same as the link URL itself, just surround the URL in angle brackets to make an automatic link:

In [x]:
```
My website is <https://scipython.com>.
```

My website is https://scipython.com.

Links to Local Files

To link to a resource on your local file system, give the file path as the URL, relative to the Notebook directory, prefixed with `files/`. Links open in a new browser tab or a dialog is opened prompting a download when clicked.

In [x]:
```
Here is [my local file, data.txt](files/data.txt)
```

Here is my local file, data.txt.

Literal Text and Code Examples

Markdown text that is indented by four spaces is rendered in a Notebook in a monospaced font, representing literal text:

In [x]:
```
This is an example of literal text:

    id   x   y
     1   0   2
     2   1   4
     3   2   8
     4   4  16
```

Produces:

This is an example of literal text:

```
id   x   y
 1   0   2
 2   1   4
 3   2   8
 4   4  16
```

The same result can be achieved without indenting by "fencing" the literal text between a pair of three backticks, ``` ``` ... ``` ```; this is appropriate for code examples, where the code language can be specified after the opening backticks; in this case, Jupyter will provide syntax highlighting:

In [x]:

```
This is a code example:

```python
# A code example
import numpy as np
y = np.sin(np.pi / 3)

def display_scaled_y(a):
    print(a * y)
```
```

This is a code example:

```
# A code example
import numpy as np
y = np.sin(np.pi / 3)

def display_scaled_y(a):
    print(a * y)
```

13.2.5 Tables

Tables are represented in Markdown with columns separated by the pipe character, | and rows on separate lines. A header row can be indicated by separating it from the table body with a row of at least three dashes in each cell, as in the following example:

```
| Alloy | Melting point | Composition |
| ----- | ------------- | ----------- |
| Field's metal | 62 °C | Bi (32.5\%) / Sn (16.5\%) / In (51\%) |
| Roses's metal | 98 °C | Bi (50\%) / Pb (25\%) / Sn (25\%) |
| Wood's metal | 70 °C | Bi (50\%) / Pb (26.7\%) / Sn (13.3\%) / Cd (10\%) |
| Cerrosafe | 74 °C | Bi (42.5\%) / Pb (37.7\%) / Sn (11.3\%) / Cd (8.5\%) |
```

Alloy	Melting point	Composition
Field's metal	62 °C	Bi (32.5%) / Sn (16.5%) / In (51%)
Roses's metal	98 °C	Bi (50%) / Pb (25%) / Sn (25%)
Wood's metal	70 °C	Bi (50%) / Pb (26.7%) / Sn (13.3%) / Cd (10%)
Cerrosafe	74 °C	Bi (42.5%) / Pb (37.7%) / Sn (11.3%) / Cd (8.5%)

The columns do not have to align in the raw Markdown, but there do have to be a consistent number of them in each row. Colons can be used in the row separating the table header from the body to align the column text to the left ($|:-|$), center($|:-:|$) or right ($|-:|$):

```
| Nuclide | mass /Da | Nuclear spin | Natural abundance |
|:-------:|---------:|:------------:||:---------|
|$^{46}$Ti | 45.953  | 0 | 0.0825 |
|$^{47}$Ti | 46.952  | 5/2 | 0.0744 |
|$^{48}$Ti | 47.948  | 0 | 0.7372 |
|$^{49}$Ti | 48.948  | 7/2 | 0.0541 |
|$^{50}$Ti | 49.945  | 0 | 0.0518 |
```

Nuclide	mass /Da	Nuclear spin	Natural abundance
^{46}Ti	45.953	0	0.0825
^{47}Ti	46.952	5/2	0.0744
^{48}Ti	47.948	0	0.7372
^{49}Ti	48.948	7/2	0.0541
^{50}Ti	49.945	0	0.0518

13.2.6 Images and Video

Links to image and other media files work in the same way as ordinary links (and can be inline or reference links), but are preceded by an exclamation mark, !. The text in square brackets between the exclamation mark and the link URL acts as *alt text*: an alternative description of the resource for screen reader software, allowing visually impaired users to interact with the resource.

In [x]:
```
![Structure of COVID-19 virus spike receptor-binding domain
    complexed with a neutralizing
    antibody](files/spike-receptor.png)
Figure 13.3: The structure of the COVID-19 virus spike
    receptor-binding domain complexed with a neutralizing
    antibody (Worldwide Protein Data Bank entry 7BZ5, authors
    Y. Wu, J. Qi and F. Gao).
```

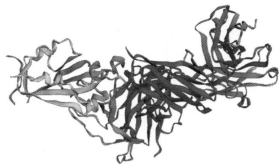

Figure 13.3   The structure of the COVID-19 virus spike receptor-binding domain complexed with a neutralizing antibody (Worldwide Protein Data Bank entry 7BZ5, authors Y. Wu, J. Qi and F. Gao).

Video links must use the HTML5 <video> tag, but note that not all browsers support all video formats. For example,

In [x]:
```
<video controls style="width: 500px; margin: 0 auto; display:
    block;" src="files/reaction-animated.mp4" />
```

The data constituting images, video and other locally linked content are not *embedded* in the notebook document itself: these files must be provided with the notebook when it is distributed.

14

LaTeX

14.1 Mathematics with LaTeX

LaTeX (sometimes styled LaTeX) is a free software system for the preparation of scientific and technical documents, with powerful features for typesetting mathematics in a clear and consistent way.

Mathematical equations can be written in LaTeX format in a Jupyter Markdown cell and are rendered using the JavaScript library, MathJax. Equations that are to appear inline with the text are delimited by single dollar signs; "displayed" equations that appear on their own, are indicated by doubled dollar signs:

In [x]:
```
An inline equation appears within a sentence of text, as in
    the definition of the function $f(\theta) =
    \sin(\theta)$; displayed equations get their own line(s)
    between lines of text:
$$\sum_{i=1}^n i = \frac{n(n+1)}{2}.$$
```

An inline equation appears within a sentence of text, as in the definition of the function $f(\theta) = \sin(\theta)$; displayed equations get their own line(s) between lines of text:

$$\sum_{i=1}^n i = \frac{n(n+1)}{2}.$$

In addition to representing mathematical formulas in Notebook cells, this book relies on LaTeX to label Matplotlib plots and charts appropriately.

This section will provide a broad overview of the use of LaTeX to markup mathematics in a Jupyter Notebook; the following Section (14.2) will detail the use of the mhchem package to render chemical equations and reactions.

Table 14.1   Basic LaTeX relations and operators

\le, \leq	≤	\ge, \geq	≤
\neq	≠	\ll	≪
\gg	≫	\approx	≈
\propto	∝	\equiv	≡
\sim	∼	\simeq	≃
\pm	±	\mp	∓
\times	×	\cdot	·
\oplus	⊕	\otimes	⊗
\ominus	⊖	\circ	∘
\hbar	ℏ	\ell	ℓ
\infty	∞	\dagger	†
\partial	∂	\nabla	∇
\Re	ℜ	\Im	ℑ
\parallel	∥	\perp	⊥
\forall	∀	\exists	∃
\rightarrow	→	\Rightarrow	⇒
\rightleftharpoons	⇌	\leftrightharpoons	⇋

Table 14.2   Lower case Greek letters

α	\alpha	β	\beta	γ	\gamma	δ	\delta
ϵ	\epsilon	ε	\varepsilon	ζ	\zeta	η	\eta
θ	\theta	ϑ	\vartheta	ι	\iota	κ	\kappa
λ	\lambda	μ	\mu	ν	\nu	ξ	\xi
o	o	π	\pi	ϖ	\varpi	ρ	\rho
ϱ	\varrho	σ	\sigma	ς	\varsigma	τ	\tau
υ	\upsilon	φ	\phi	φ	\varphi	χ	\chi
ψ	\psi	ω	\omega				

14.1.1 Basic Syntax

The basic syntax that LaTeX follows to represent mathematical expressions can be summarized in a few principles.

Commands and Symbols

The mathematical symbols below, which can be found on a standard Western keyboard, can be entered directly into formulas.

+ - = < > / : [ ] ( ) ! ' |

Greek letters, symbols and most common mathematical functions are represented as named commands, preceded by a backslash, for example, \alpha, \cos and \sum. Commands for the more important symbols are given in Table 14.1, whilst Tables 14.2 and 14.3 list the Greek letter commands, and Table 14.4 lists some common mathematical functions.

Table 14.3   Upper case Greek letters. Letters not listed in this table are the same as their Latin counterparts, for example, the upper case Greek letter α is A

Γ	\Gamma	Δ	\Delta	Θ	\Theta	Λ	\Lambda
Ξ	\Xi	Π	\Pi	Σ	\Sigma	ϒ	\Upsilon
Φ	\Phi	Ψ	\Psi	Ω	\Omega		

Table 14.4   Some common mathematical functions

arccos	\arccos	arcsin	\arcsin	arctan	\arctan	cos	\cos
cosh	\cosh	cot	\cot	coth	\coth	csc	\csc
det	\det	dim	\dim	exp	\exp	inf	\inf
lim	\lim	ln	\ln	log	\log	max	\max
min	\min	sec	\sec	sin	\sin	sinh	\sinh
tan	\tan	tanh	\tanh				

If you don't find the symbol you are looking for listed in this chapter, there are many useful online resources that can help. For offline reference, the official and free "Comprehensive LaTeX Symbol List" (in PDF format) is helpful to have downloaded (from `http://tug.ctan.org/info/symbols/comprehensive/symbols-a4.pdf`). The website `https://detexify.kirelabs.org/classify.html` allows the user to draw the approximate shape of the symbol they want and suggests likely matches. Be aware that some of the suggestions require extra LaTeX packages and may not be available by default within a Jupyter Notebook.

Whitespace in LaTeX Source

Whitespace is handled by the LaTeX engine itself and has no effect within the source except where required to separate commands; for example, $a+b$ and $a + b$ are typeset identically as $a + b$, but the space between `alpha` and + in $\alpha + \beta$ is required.

Exponents and Indices

The caret symbol, ˆ, indicates that the following symbol should be raised: that is, set in superscript (e.g., as an exponent). The underscore symbol, _, indicates that the following symbol should be lowered: that is, set in subscript (e.g., as an index to a variable).

Where exponents and indices are combined, they are set intelligently: a_i^2 displays as a_i^2.

Braces

Parts of a formula that form a necessary part of a mathematical expression or are to be typeset together are grouped within braces. For example, `y^2n` produces y^2n whereas `y^{2n}` produces y^{2n}.

Font Styles

Non-numeric text entered directly is displayed, by default, in an italic font as is universal practice in mathematical typesetting. It is therefore a good idea to use the appropriate LaTeX commands to achieve the correct rendering of formulas. For example, `$sin(x)$` renders as $sin(x)$ – here, it is best to use `$\sin(x)$` to get the result $\sin(x)$.

If you need a sequence of characters to be output in an upright (Roman) font, use the `\mathrm` command: `$\mathrm{lcm(4,6)}$` produces $\mathrm{lcm}(4, 6)$. For longer sequences of upright text, one can use `\text` in the same way: the difference is that, whereas whitespace is ignored within the `\mathrm` command, it is preserved with `\text`. Compare:

```
$$
V(x) = 0 \mathrm{for all} x
$$
```

$$
V(x) = 0 forall x
$$

with:

```
$$
V(x) = 0 \text{ for all } x
$$
```

$$
V(x) = 0 \text{ for all } x
$$

Fractions

The `\frac` command produces fractions and takes two arguments: the numerator and denominator expressions. Inline fractions are rendered in a slightly smaller font to better match the surrounding text: `$\frac{x}{2}$` produces $\frac{x}{2}$, whereas

```
$$
\frac{x}{2}
$$
```

produces:

$$
\frac{x}{2}
$$

Fractions can be nested to any reasonable depth:

```
$$
H = \frac{2}{\frac{1}{x_1} + \frac{1}{x_2}}
$$
```

renders as:

$$H = \frac{2}{\frac{1}{x_1} + \frac{1}{x_2}}$$

Roots

A square root in "radical" notation is produced by the `\sqrt` command: for example, `$$\frac{1}{\sqrt{x+1}$$` produces:

$$\frac{1}{\sqrt{x+1}}$$

For a root with different index, use the notation `\sqrt[n]x`; for example, for a cube root: `$\sqrt[3]{8}$` gives $\sqrt[3]{8}$. This use of square brackets for optional arguments is quite common in LaTeX.

Sums, Products and Integrals

The commands for the summation, product and integral symbols are `\sum`, `\prod` and `\int`, respectively. The upper and lower limits are optional and indicated using superscripts and subscripts. For example, `$$\int_0^2 x^2 dx = \frac{8}{3}$$` yields:

$$\int_0^2 x^2 dx = \frac{8}{3}$$

Brackets

Regular parentheses (square or round brackets) can be inserted directly into the LaTeX source, but they do not resize automatically to match the expressions they enclose: `$$(\frac{1}{x})$$` gives

$$(\frac{1}{x})$$

To force the brackets to be rendered at a matching size, use the `\left( ... \right)` syntax: `$$\left( \frac{1}{x} \right)$$`:

$$\left(\frac{1}{x} \right)$$

Table 14.5   LaTeX environments for various matrix expressions

No brackets	`\begin{matrix}`
	`-1 & 0 & 2\\`
	`z & y & z\\`
	`\end{matrix}`
		$\begin{matrix} -1 & 0 & 2 \\ x & y & z \end{matrix}$

Round brackets	`\begin{pmatrix}`
	`-1 & 0 & 2\\`
	`z & y & z\\`
	`\end{pmatrix}`
		$\begin{pmatrix} -1 & 0 & 2 \\ x & y & z \end{pmatrix}$

Square brackets	`\begin{bmatrix}`
	`-1 & 0 & 2\\`
	`z & y & z\\`
	`\end{bmatrix}`
		$\begin{bmatrix} -1 & 0 & 2 \\ x & y & z \end{bmatrix}$

Pipes, e.g. determinant	`\begin{vmatrix}`
	`-1 & 0 & 2\\`
	`z & y & z\\`
	`\end{vmatrix}`
		$\begin{vmatrix} -1 & 0 & 2 \\ x & y & z \end{vmatrix}$

Other brackets and modulus symbols can be used in the same way, for example, `\left[ ... \right]` and `\left| ... \right|`, but braces (curly brackets) must be escaped: `\left\{ ... \right\}`.

Vectors

Vector quantities may be written with an arrow over the symbol with the `\vec` command, in bold using `\boldsymbol` or underlined with `\underline`. For example, `$$\vec{r}, \boldsymbol{r}, \underline{r}$$`:

$$\vec{r}, \boldsymbol{r}, \underline{r}$$

Matrices

The easiest way to markup matrices is to use one of the matrix environments `\matrix`, `\pmatrix`, `\bmatrix` or `\vmatrix` according to the bracket type required, as illustrated in Table 14.5. The rows of the matrix are separated by a double backslash, `\\`, and columns are separated by ampersands, `&`.

Fine-Tuning

The LaTeX processor generally does a good job of placing elements where they need to be to set the formatted mathematics in a clear and appealing way. However,

Table 14.6   LaTeX spacing
commands: A quad is the width of
the upper case letter M

\!	$-\frac{3}{18}$ of a quad
\,	$\frac{3}{18}$ of a quad
\:	$\frac{4}{18}$ of a quad
\;	$\frac{5}{18}$ of a quad
\quad	1 quad
\qquad	2 quads

it is occasionally necessary to help it out by tweaking the spacing or customizing the size of fonts.

Horizontal spacing can be controlled with the commands given in Table 14.6 in terms of fractions of a *quad*, the width of a capital letter M in the current font.

The font size commands `\displaystyle` and `\textstyle` can be used to force the font size used to that of displayed or inline text formulas, respectively. These commands do not take any arguments: They apply until another sizing command is encountered. One common use is where a displayed fraction is too large. As an example of these fine-tuning techniques, compare these displayed formulas:

```
$$
\frac{1}{2}\int\int_D x^2 + y^2 \mathrm{d}x\mathrm{d}y
$$
```

$$\frac{1}{2}\int\int_D x^2 + y^2 \mathrm{d}x\mathrm{d}y$$

with

```
$$
\textstyle \frac{1}{2} \displaystyle \int\!\!\!\!\int_D x^2 + y^2
    \, \mathrm{d}x\,\mathrm{d}y
$$
```

$$\tfrac{1}{2}\iint_D x^2 + y^2 \, \mathrm{d}x\,\mathrm{d}y$$

Here, the integrals have been brought closer together with negative spacing and the differentials spread out a little with an extra thin space. The fraction $\frac{1}{2}$ has been brought down to inline font size to with `\textstyle`, which makes it look slightly better.

Table 14.7   Annotations above a symbol

Command	Example	Usage
\hat{A}	$\hat{A}$	Operators, unit vectors
\bar{x}	$\bar{x}$	Sample mean, negation, complex conjugation
\dot{x}	$\dot{x}$	First derivative (Newton's notation, often for derivatives with respect to time)
\ddot{x}	$\ddot{x}$	Second derivative
\tilde{\nu}	$\tilde{\nu}$	Wavenumber (spectroscopic notation)

Annotating Symbols

In chemistry notation it is often necessary to annotate symbols in various ways. The commands in Table 14.7 provide the most commonly used functionality.

To add a prime to a quantity, use a single quote, as in `A'`. To add a double prime, use two single quotes, `A''`, rather than a double quote, `A"`; that is,

```
A' \;\mathrm{and}\; A'' \;\text{but not}\; A".
```

$$A' \text{ and } A'' \text{ but not } A".$$

14.1.2 Units

This book follows the recommendations laid out in the IUPAC standards document, "Quantities, Units and Symbols in Physical Chemistry" (the "Green Book").[1] A *physical quantity*, Q, is considered to be a product of a *numerical value*, Q, and a set of *units*, $[Q]$:

$$Q = Q\,[Q]. \tag{14.1}$$

With very few exceptions, the symbol denoting the physical quantity is set in italic text (which is, in any case, the default within the LaTeX math environment), whereas the units should be in a Roman (upright) font. For example,

$$p = 101.325 \, \text{kPa}$$

However, subscripts and superscripts that are not themselves physical quantities or numbers (e.g., indexes) but rather labels are printed in Roman text. That is,

---

[1] E. R. Cohen et al., IUPAC Green Book, Third Edition, IUPAC & RSC Publishing Cambridge (2008). Downloadable from https://iupac.org/wp-content/uploads/2019/05/IUPAC-GB3-2012-2ndPrinting-PDFsearchable.pdf.

$$p_{atm} = 101.325 \, \text{kPa}$$

but

$$C_V = 15.26 \, \text{J} \, \text{K}^{-1} \, \text{mol}^{-1}$$

and

$$p_i = 0.521 \, \text{bar}.$$

The name or symbol used for a physical quantity does not imply any particular use of unit: The units must always be specified explicitly.

The units themselves are, for the most part, expressed using exponent notation in this book. That is, kJ mol^{-1} instead of kJ/mol. This convention removes any potential ambiguity possible with multiple reciprocal units (e.g., we use J K^{-1} mol^{-1} instead of the ambiguous J/K/mol or the ugly J/(K.mol).) The units are separated from the numerical value by a single space (which must be specified explicitly as \; in LaTeX and components of the unit are separated by the \, thin space:

```
$$
E = 10 \; \mathrm{kJ\,mol^{-1}}
$$
```

produces:

$$E = 10 \, \text{kJ} \, \text{mol}^{-1}$$

When providing the values of physical quantities, particularly in tables of numerical values, it is helpful to use the method known as *quantity calculus*: This considers Equation 14.1 to be a true algebraic equation that may be rearranged for convenience. In this convention, the following are equivalent:

$$p_{atm} = 101.325 \, \text{kPa}$$
$$p_{atm}/\text{kPa} = 101.325$$

In tables and in graph axes, sometimes a power of ten is factored out and included in the label as well. As an illustration, the values in Table 14.8 are dimensionless numbers to be interpreted as physical quantities using the table headings. For example, the first value in the second column of this table, 1.667 is equal to 10^3 K/T: that is, the quantity $1/T = 1.667 \times 10^{-3}$ K^{-1}. This approach also has the merit of handling logarithms nicely: $\ln(k/\text{cm}^3 \, \text{mol}^{-1} \, \text{s}^{-1}) = -2.005$ makes it clear that the physical quantity, k, divided by its stated units gives a numerical value whose natural logarithm is -2.005. That is, $k = e^{-2.005} \, \text{cm}^3 \, \text{mol}^{-1} \, \text{s}^{-1} = 0.1347 \, \text{cm}^3 \, \text{mol}^{-1} \, \text{s}^{-1}$.

Table 14.8   Example Arrhenius plot data

T/K	$10^3\,K/T$	$k\,/cm^3\,mol^{-1}\,s^{-1}$	$\ln(k/cm^3\,mol^{-1}\,s^{-1})$
600	1.667	0.1347	-2.005
650	1.538	0.1572	-1.850
750	1.333	0.2012	-1.604

14.2 Chemical Equations

The LaTeX commands and environments introduced above can be used to mark up chemical formulas and equations, but it can be cumbersome to ensure element symbols are in Roman font and to wrap subscripts and superscripts in braces. For example, the chromate ion, CrO_4^{2-} (aq) requires the following source:

```
$$
\mathrm{CrO_4^{2-}(aq)}
$$
```

The package mhchem can be used to make this easier. Within a Jupyter Notebook, however, it must first be included in the document by running a Markdown cell with the contents:

```
$$\require{mhchem}$$
```

This tells the software that renders the Notebook's Markdown content that the mhchem package must be used to interpret instructions within the \ce{...} command. With this in place, formulas can be specified (for the most part) without extra braces and without expliciting invoking \mathrm:

```
$$
\ce{CrO4^2-(aq)}
$$
```

Note that \ce{...} only functions within the maths environment, that is, delimited by $ (inline content) or $$ (block content).

Entire chemical reactions can be set within a single \ce{...} command, using -> and <=> to denote the reaction arrow and equilibrium symbol, respectively:

```
$$
\ce{W^74+ + e- -> W^73+}
$$
```

produces:

$$W^{74+} + e^- \longrightarrow W^{73+}$$

Table 14.9   Examples of marking up chemical reactions with mhchem

Mark up source	Result
`\ce{^209_83Bi}`	$^{209}_{83}\text{Bi}$
`\ce{CuSO4.5H2O}`	$CuSO_4 \cdot 5\,H_2O$
`\ce{Ph-OH}`	$Ph{-}OH$
`\ce{H2C=O}`	$H_2C{=}O$
`\ce{HC#CH}`	$HC{\equiv}CH$
`\ce{CO(g) + H2O(g) <=> CO2(g) + H2(g)}`	$CO(g) + H_2O(g) \rightleftharpoons CO_2(g) + H_2(g)$
`\ce{2O3(g) <=>> 3O2(g)}`	$2\,O_3(g) \underset{}{\overset{}{\rightleftharpoons}} 3\,O_2(g)$
`\ce{N2(g) + O2(g) <<=> 2NO(g)}`	$N_2(g) + O_2(g) \rightleftharpoons 2\,NO(g)$
`\ce{AgNO3 + KCl -> AgCl v + KNO3}`	$AgNO_3 + KCl \longrightarrow AgCl \downarrow + KNO_3$
`\ce{2Al + 2NaOH + 2H2O -> 2NaAlO2 + 3H2 ^}`	$2\,Al + 2\,NaOH + 2\,H_2O \longrightarrow 2\,NaAlO_2$ $+\,3\,H_2 \uparrow$
`\ce{Alcohol ->[\ce{KMnO4}] Acid}`	$Alcohol \xrightarrow{KMnO_4} Acid$
`\ce{2KClO3(s) ->[][\Delta] 3O2(g) + 2KCl(s)}`	$2\,KClO_3(s) \xrightarrow[\Delta]{} 3\,O_2(g) + 2\,KCl(s)$
`\ce{$n$C2H4 -> (CH2CH2)_{$n$}}`	$n\,C_2H_4 \longrightarrow (CH_2CH_2)_n$
`\ce{A <=>[\alpha][\beta] B}`	$A \underset{\beta}{\overset{\alpha}{\rightleftharpoons}} B$

```
$$
\ce{ Cr2O7^2- + 14H+ + 6e- -> 2Cr^3+ + 7H2O }
$$
```

yields

$$Cr_2O_7{}^{2-} + 14\,H^+ + 6\,e^- \longrightarrow 2\,Cr^{3+} + 7\,H_2O$$

Further examples of the use of the mhchem package are given in Table 14.9.

14.3 Example

E14.1

Question

The standard electrode potential at 298 K, under alkaline conditions, for the reduction of ClO_4^- (aq) to ClO_3^- (aq) is 0.37 V. Under acidic conditions, it is 1.20 V. Use this information to deduce the value for the autoionization constant of water, K_w, and its decimal cologarithm, pK_w.

Solution

(The Markdown used to produce the chemical and mathematical equations is given below; when run, these cells produce the indicated output.) The corresponding Jupyter Notebook can be downloaded from https://scipython.com/chem/gla/. The half-cell reactions are as follows. Under alkaline conditions ($E^\ominus = 0.37$ V):

```
$\require{mhchem}$
```

```
$$
\ce{ClO4^{-}(aq) + H2O(l) + 2e- -> ClO3^-(aq) + 2OH^{-}(aq)}
$$
```

$$\text{ClO}_4^-(\text{aq}) + \text{H}_2\text{O(l)} + 2\,\text{e}^- \longrightarrow \text{ClO}_3^-(\text{aq}) + 2\,\text{OH}^-(\text{aq})$$

Under acidic conditions ($E^\ominus = 1.20\,\text{V}$):

```
$$
\ce{ClO4^{-}(aq) + 2H^+(aq) + 2e- -> ClO3^{-}(aq) + H2O(l)}
$$
```

$$\text{ClO}_4^-(\text{aq}) + 2\,\text{H}^+(\text{aq}) + 2\,\text{e}^- \longrightarrow \text{ClO}_3^-(\text{aq}) + \text{H}_2\text{O(l)}$$

A cell constructed with the first half-cell on the right-hand side and the second on the left-hand side will have a conventional cell reaction:

```
$$
\ce{2H2O(l) -> 2OH^{-}(aq) + 2H^{+}(aq)}
$$
```

$$2\,\text{H}_2\text{O(l)} \longrightarrow 2\,\text{OH}^-(\text{aq}) + 2\,\text{H}^+(\text{aq})$$

The standard EMF of this cell is:

```
Estd = 0.37 - 1.20
Estd
```

```
-0.83
```

That is, $E^\ominus = -0.83\,\text{V}$. The corresponding standard molar Gibbs Free Energy change is:

```
import numpy as np

# Faraday constant (C.mol-1) and gas constant (J.K-1.mol-1)
F, R = 96485, 8.314
# Temperature, K
T = 298
DrG_std = -F * Estd
DrG_std
```

```
80082.55
```

That is, $\Delta_r G_m^{\ominus} = 80.08\,\text{kJ}\,\text{mol}^{-1}$. The autoionization constant (ionic product) of water is the equilibrium constant of the above cell reaction, written as:

$$K_w = [\text{OH}^-][\text{H}^+]$$

since the concentrations are so small they can approximate the ion activities well. This is related to $\Delta_r G_m^{\ominus}$ through:

$$\Delta_r G_m^{\ominus} = -RT \ln K_w.$$

Therefore,

```
Kw = np.exp(-DrG_std / R / T)
Kw
```

```
9.168560848141593e-15
```

This value is usually expressed as the negative of the decimal logarithm (the decimal cologarithm), pK_w:

```
pKw = -np.log10(Kw)
pKw
```

```
14.03769882842608
```

That is, $pK_w = 14$ for pure water at 298 K.

15

Chemistry Databases and File Formats

Many different formats for storing and transmitting chemical information have been proposed since computers and the internet have been widely adopted in research and education. Some of these are specific to particular domains, for example, protein chemistry or molecular dynamics simulation, some encode molecular structure to a greater or lesser extent, whilst others are simply identifiers (references to a record in a database that may or may not be publicly available). Some formats are machine-readable only; others are (at least partially) human-readable.

This chapter will summarize a few of the more popular file and identifier formats and some of the online services and software packages that can be used to search for and convert between them.

15.1 Formats

15.1.1 SMILES

The Simplified Molecular-Input Line-Entry System (SMILES) represents chemical species as short, plain text (ASCII) strings.

- Atoms are identified by their standard element symbols, which are placed within square brackets, for example, [Pt] (the square brackets are optional for the normal usage of elements common in organic chemistry).
- In most cases, hydrogen atoms need not be explicitly stated: They are assumed to be added to fulfill the valence requirements of each atom in the molecule. For example, C represents methane, CH_4. Radicals and ions sometimes require the hydrogens to be provided with the atom, inside square brackets: for example, [CH3] for the methyl radical, $CH_3^{\cdot}$, and [NH4+] for the ammonium ion, NH_4^+.
- Bonds are indicated by the characters - (for a single bond), = (double bonds) and # (triple bonds). A single bond is assumed if omitted: For example, CC and C-C are both valid representations of ethane, CH_3CH_3.
- Components of a chemical species that are not covalently bonded together can be indicated with a period (full stop), .. For example, sodium chloride, [Na+].[Cl-].

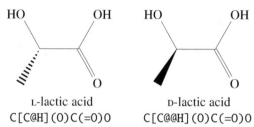

L-lactic acid   D-lactic acid
C[C@H](O)C(=O)O   C[C@@H](O)C(=O)O

Figure 15.1   The structures and SMILES representations of the enantiomers of lactic acid.

- Ring structures are realized by picking an atom at which to break the ring, labeling it with a number and then "reconnecting" it to another atom by using the same number. For example, cyclohexane is C1CCCCC1. The Kekulé structure of pyridine may be written as C1=CC=NC=C1 or C1=CN=CC=C1.
- Aromatic bonds are indicated either by using the : separator or by writing the constituent atoms in lowercase. For example, benzene can be written either as C1:C:C:C:C:C1 or c1ccccc1.
- Branches are indicated by using parentheses around the branching group (or non-hydrogen atom): For example, ethanoic acid may be written as CC(O)=O or CC(=O)O; *tert*-butyl alcohol, $(CH_3)_3COH$, as CC(C)(C)O.
- The stereoconfiguration around double bonds may be optionally specified using the characters \ and /. For example Br/C=C\Br and Br/C=C/Br are *cis-* and *trans*-1,2-dibromoethene, respectively.
- The stereochemistry of tetrahedral chiral centers can be specified using the notation @ and @@ to distinguish between the anticlockwise and clockwise ordering of the atoms or groups after the first bond to a chiral center. For example, C[C@H](O)C(=O)O is L-(+)-lactic acid: Looking along the $CH_3 - C$ bond, the substituents H, OH and COOH appear anticlockwise; conversely C[C@@H](O) C(=O)O is D-(−)-lactic acid (see Figure 15.1). Note that in this case, the hydrogen at the chiral carbon must be specified explicitly and that there are alternative, equivalent choices for the ordering of the substituents.

No information about the atomic coordinates is encoded in a SMILES string, though this can usually be looked up in one of many online databases (see Section 15.2).

Some more examples are given in Table 15.1.

15.1.2 XYZ

The XYZ file format (extension .xyz) is a simple and loosely defined standard for specifying the geometry of a molecule or collection of atoms (e.g., a liquid

Table 15.1   Some examples of SMILES identifiers for chemical species

N#N	N_2	molecular nitrogen
C(=O)=O	CO_2	carbon dioxide
CC(C)O	OH (isopropanol structure)	isopropanol
CC(=O)C	O (acetone structure)	acetone
[Ca+2].[F-].[F-]	CaF_2	calcium fluoride
[K+].[O-][Mn](=O)(=O)=O	$KMnO_4$	potassium permanganate
c1c2ccccc2ccc1	(naphthalene structure)	naphthalene
C1CC[C@H]([C@@H](C1)Cl)Cl	Cl / Cl (structure)	*trans*-1,2-dichlorocyclohexane
N[C@@H](C)C(=O)O	H_2N OH O (structure)	L-alanine

or solid in a molecular dynamics simulation of their motion). The basic format consists of:

1. A header line giving the number of atoms, n;
2. Another header line (which may be blank) containing a title, filename or comments;
3. n lines with at least four columns giving each atom's position, consisting of the element symbol and three-dimensional atomic coordinates, in Å. The origin of the coordinate system used is unspecified but is frequently the molecular center of mass.

No explicit information about the atom connectivity is given (though additional processing through various software packages may be able to infer it). The file may

Table 15.2   Overview of the Molfile format

Header block	Title line
	Name of file source, timestamp, etc.
	Comment line
Counts line	Number of atoms and bonds; other information
Atom block	x, y, z coordinates (in Å); element symbol; optional additional information
Bond block	Index to first atom, index to second atom; bond type (1, 2, 3, 4 for single, double, triple and aromatic bonds, respectively); stereochemistry information
Properties block	Additional molecule properties: one per line, as values following the character M and a string encoding the property
End of record	M END

contain additional columns depending on how it was created and for what purpose. For molecular dynamics simulations, it is not unusual for many XYZ records at different times to be concatenated in a single (sometimes very large) file from which then can be turned into an animation.

An XYZ file representing water might take the form:

```
3
Water, H2O
O          0.0000        0.0000        0.1178
H          0.0000        0.7555       -0.4712
H          0.0000       -0.7555       -0.4712
```

This water molecule lies in the yz plane, with the origin at the molecular center of mass.

15.1.3 Molfile

The widely used MDL Molfile format (extension .mol) was created by MDL Information Systems to describe the bonds, connectivity and coordinates of the atoms in a molecule. As with the SMILES format, hydrogen atoms can be assumed if not explicitly listed. It is a plain-text format, consisting of blocks of data encoding the molecular properties as summarized in Table 15.2.

For example, here is a Molfile representing the structure of water:

```
Water, H2O
Marvin   01211112152D

  3  2  0  0  0  0            999 V2000
   -0.4125    0.7145    0.0000 H   0  0  0  0  0  0  0  0  0  0  0  0
    0.0000    0.0000    0.0000 O   0  0  0  0  0  0  0  0  0  0  0  0
   -0.4125   -0.7145    0.0000 H   0  0  0  0  0  0  0  0  0  0  0  0
  2  1  1  0
  2  3  1  0
M   END
```

The header block contains three lines: a title, some rather cryptic metadata indicating that this structure was exported from some version of the Marvin chemical editor and a comment line which is blank. There follows the "Counts line": 11 fields, each of three characters in length, followed by one that is six characters long giving the version of the Molfile format being used (in this case, V2000); only the first, second and fifth fields are of much interest: They give the number of atoms (3), number of bonds (2) and whether the molecule is chiral (1) or not (0).

There are three entries in the atom block: each line gives the atomic coordinates (x, y, z) and the element symbol. In this example, the subsequent fields are all 0, but in some cases these are used to represent charges, isotopes, valence, etc.

The bond block contains two lines: one for each O − H bond. In the first line, 2 1 1 indicates that atom 2 (the oxygen atom) is connected to atom 1 (a hydrogen) through a single bond. The second line, 2 3 1 connects the oxygen to the second hydrogen atom (index 3), also through a single bond. The remaining field is zero here, but used in some molecules to indicate stereochemistry (1 for a "wedged" bond, 6 for a "dashed" one.

Finally, there is no block for additional properties, and M END indicates the end of the record.

As another example, here is a Molfile for L-(+)-lactic acid:

```
 6  5  0  0  1  0  0  0  0  0  0999 V2000
    1.2386    0.0000    0.0000 O   0  0  0  0  0  0  0  0  0  0  0  0
    0.0000   -1.9069    0.0000 O   0  0  0  0  0  0  0  0  0  0  0  0
    3.4989   -1.3961    0.0000 O   0  0  0  0  0  0  0  0  0  0  0  0
    1.1931   -1.3296    0.0000 C   0  0  0  0  0  0  0  0  0  0  0  0
    2.3338   -2.0259    0.0000 C   0  0  0  0  0  0  0  0  0  0  0  0
    2.3618   -3.3589    0.0000 C   0  0  0  0  0  0  0  0  0  0  0  0
  1  4  2  0
  2  4  1  0
  5  3  1  1
  4  5  1  0
  5  6  1  0
M  END
```

In this file, the header block is entirely blank, and the hydrogen atoms have been left implicit. The molecule is chiral (as indicated by the flag 1 in the Counts line), and the third bond listed (a single bond between atoms 5 and 3) is described as "wedged." D-(-)-Lactic acid would have 5 3 1 6 for this entry.

15.1.4 Chemical Markup Language

Chemical Markup Language (CML) is a self-describing format based on Extensible Markup Language (XML) which is both machine- and human-readable. It is portable, can be read and produced by a variety of software tools and, as an open standard, was designed to encourage interoperability and the reuse of data. However, XML-based formats have been criticized for being inflexible, redundant, verbose and even insecure, and they have been declining in popularity over the last few years in favor of formats perceived as being more "agile." CML is not as widely used as Molfile, and the community built around its support and development has been quiet for almost a decade. The CML document consists of a set of nested elements (or "tags"), with labels and additional data associated through key-value pairs called attributes. Individual elements (e.g., those representing individual atoms) may be given a unique id property which can be used to refer to it elsewhere in the document (for instance, in bonds). For example, The CML representation of D-(-)-lactic acid is as follows:

```
<?xml version="1.0"?>
<molecule xmlns="www.xml-cml.org/schema">
 <atomArray>
  <atom id="a1" elementType="O" hydrogenCount="0" x2="1.1503" y2="0.0000"/>
  <atom id="a2" elementType="O" hydrogenCount="1" x2="0.0000" y2="-1.9967"/>
  <atom id="a3" elementType="O" hydrogenCount="1" x2="3.4545" y2="-1.3334"/>
  <atom id="a4" elementType="C" hydrogenCount="0" x2="1.1503" y2="-1.3334"/>
  <atom id="a5" elementType="C" hydrogenCount="1" x2="2.3041" y2="-1.9932">
   <atomParity atomRefs4="a5 a3 a6 a4">1</atomParity>
  </atom>
  <atom id="a6" elementType="C" hydrogenCount="3" x2="2.3041" y2="-3.3267"/>
 </atomArray>
 <bondArray>
  <bond atomRefs2="a1 a4" order="2"/>
  <bond atomRefs2="a2 a4" order="1"/>
  <bond atomRefs2="a5 a3" order="1"/>
  <bond atomRefs2="a4 a5" order="1"/>
  <bond atomRefs2="a5 a6" order="1"/>
 </bondArray>
</molecule>
```

The <atomArray> and <bondArray> elements contain information about the atomic coordinates and bonding, respectively; chirality is supported by using the <atomParity> element.[1] In this example, only two-dimensional coordinates, x2 and y2, are given, for depicting the molecule in printed figures and online images; an alternative representation uses the attributes x3, y3 and z3 for giving the coordinates in three dimensions.

15.1.5 InChI and InChIKey Identifiers

The International Chemical Identifier (InChI) is an open, text-based standard for identifying and encoding information about molecules. It was originally developed by the International Union of Pure and Applied Chemistry (IUPAC) and the US

---

[1] See the full specification at www.xml-cml.org/spec/ for full details about the CML format.

National Institute of Standards and Technology (NIST). The atoms, their connectivity, stereochemistry, isotopic composition and charges are described in "layers": Parsable strings, delimited by the forward slash character, /, and identified by a prefix character. InChI strings can, in principle, be interpreted by a human, but for all but the most simple molecules, require parsing by one of the several software tools available. Standard InChI strings start with `InChI=1S/`; the more important layers and sublayers that can follow are:

- Main layer
  - Chemical formula (mandatory; no prefix): the stoichiometric composition of the molecule;
  - Connection (prefix `c`): the bonds between the atoms identified by their position in the chemical formula (apart from hydrogens);
  - Hydrogen atoms (prefix `h`): the number of hydrogen atoms attached to each of the other atoms;
- Charge layer (prefix `q` or `p` for certain hydrides): the species charge;
- Stereochemical layer (prefixes `b` and `t`): information on double-bond and tetrahedral stereochemistry (other, rarer types of stereochemistry are also supported);
- Isotopic layer (prefix `i`): the specific isotope of particular atoms in the structure, where relevant.

For example, the InChI string for ethanol (CH_3CH_2OH) is

$$InChI=1S/C2H6O/c1-2-3/h3H,2H2,1H3$$

The chemical formula is C2H6O, the non-hydrogen atoms being labeled 1 (C), 2 (C) and 3 (O). The connections between them are given by the layer /c1-2-3. Finally, the hydrogen layer, /h3H,2H2,1H3, connects three hydrogen atoms to atom 1 (the methyl carbon), two to atom 2 (the other carbon) and one to atom 3 (the oxygen).

   Isopropanol, $CH_3CH(OH)CH_3$ is

$$InChI=1S/C3H7OH/c1-3(2)4/h3-4H,1-2H3$$

Atom 3 (the central carbon) is bonded to atoms 1 (a methyl carbon), 2 (another methyl carbon and atom 4(the oxygen). Branching is described by placing atom identifiers in parentheses in the connection layer. Ranges of atom identifiers are allowed in the hydrogen layer: /h3-4H,1-2H3 attaches one hydrogen atom to atoms 3 and 4 (the oxygen and central carbon) and three hydrogen atoms to atoms 1 and 2 (the methyl carbons).

   The InChI standard is fairly comprehensive, with many features that require advanced knowledge to interpret, and it is not practical to generate InChI strings by hand for any but the most simple organic molecules. Luckily, there are many services and databases that can generate these strings on demand.

The InChIKey is a 27-character hashed version of the full InChI string which was designed as a unique identifier for chemical compounds without encoding any explicit information about their composition or structure. The full InChI string is, in many cases, too long to be easily used in online searches. Chemical structures represented as images (e.g., PNG or GIF files) cannot be used as useful search terms, and so the InChIKey was developed as a practical way to refer to chemical species in databases. There are three parts to the InChIKey, which takes the form XXXXXXXXXXXXXX-YYYYYYYYFV-P: The first 14 characters are a hash of the main layer and /q (charge) sublayer, and identify the stoichiometry and molecular structure; the second part consists of eight characters deriving from a hash of the remaining layers, followed by some characters identifying the InChI version. The final character indicates the protonation of the parent structure (important for some species). In this way, it is possible to identify from the InChIKey different isotopologues of the same molecule.

For example, consider the InChIKeys for the molecules, HCl and DCl:

Molecule	InChI	InChIKey
HCl	InChI=1S/ClH/h1H	VEXZGXHMUGYJMC-UHFFFAOYSA-N
DCl	InChI=1S/ClH/h1H/i/hD	VEXZGXHMUGYJMC-DYCDLGHISA-N

In this way, all isotopologues of a molecule can be identified from the first part of the InChIKey.

InChIKeys are stored in many databases online, and software tools exist that can both convert an InChI string to an InChIKey and look up an InChIKey to retrieve a molecular structure in the other formats discussed above.

15.2 Online Services

There are several free online services and databases that can be searched for information about chemical compounds and which return molecular structures in a variety of different formats. Some of these are summarized in this section.

15.2.1 Wikipedia

The free, online encyclopedia Wikipedia (wikipedia.org) contains information and data relating to common chemical substances, sourced and curated by volunteers. Most searches for common trivial names will resolve to a single page for a molecule, with a sidebar listing important identifiers and chemical properties. For example, a search for the terms CH3COOH, CH3CO2H and ethanoic acid all resolve to the page titled *Acetic acid*; a search on the term C2H4O2 yields a page listing links to pages on a range of molecules with that stoichiometric formula, including acetic acid.

Many Wikipedia pages on chemical substances include the SMILES, InChI and InChIKey identifiers, in addition to links to their entries in external databases. Typically, properties such as the molar mass, density, melting point, boiling point, pK_a, heat capacity, standard molar enthalpy of formation and so on are also available.

A more limited (and somewhat eclectic) range of substances have dedicated, and more comprehensive data pages devoted to them (see https://en.wikipedia .org/wiki/Category:Chemical_data_pages for a list).

15.2.2 ChemSpider

ChemSpider (`www.chemspider.com`) is an online database of molecules owned and maintained by the Royal Society of Chemistry. At the time of writing, it contains over 100 million structures, many of them obtained from other databases through crowdsourced contributions. Species can be searched for by name (including systematic name, synonyms, trade names and alternative, trivial names), formula, SMILES string, InChI string and InChIKey. The data returned include the mass, a structure image (in static two-dimensional form and interactive, rotatable three-dimensional representation), alternative names (including those in a variety of languages), chemical and physical properties and, in some cases, IR, UV/visible and proton NMR spectra.

Saving a structure produces an MDL Molfile (see Section 15.1.3), usually with supplemental data in structure-data file (SDF) format (a Molfile with some additional information appended to it).

ChemSpider has an Application Programming Interface (API), allowing searches to be performed from code without the need to manually query and download the database in a web browser. ChemSpiPy is one popular Python library used to access the ChemSpider API; this is the topic of Example E15.1 at the end of this chapter.

15.2.3 PubChem

PubChem (`https://pubchem.ncbi.nlm.nih.gov/`) is a large database of substances, including chemical compounds, maintained by the US National Institutes of Health (NIH). In addition to identifiers and properties, structural information can be downloaded in a variety of formats, including Molfile / SDF and JSON. Note that the XML format offered is based on PubChem's own XML Schema which is totally different from CML.

15.2.4 NIST

The US National Institute of Standards and Technology (NIST) online Chemistry WebBook, at `https://webbook.nist.gov/chemistry/` contains data on the thermodynamic, spectroscopic and other properties of a wide range of compounds. The search interface allows queries by chemical formula, name, InChI, InChIKey and even by Molfile file upload.

Data are generally presented from a range of sources, usually with included uncertainties, but there is no API and care should be taken to read the comments attached to each entry because in many cases, multiple values are presented for the same physical quantity (sometimes at different temperatures). Thermochemistry data is divided into *gas-phase* and *condensed-phase* data, and care should also be taken to select the correct data set for a given application.

15.3 Example

E15.1

ChemSpiPy and the ChemSpider API

ChemSpiPy is a Python library that allows access to the ChemSpider Application Programming Interface (API). Python scripts can use it to retrieve information about chemical species by querying the ChemSpider service over the internet and to automatically import data without having to visit the ChemSpider website itself.

Access to the ChemSpider API is, at the time of writing, free for a maximum of 1,000 queries a month. An account with the Royal Society of Chemistry's Developers' website and a valid API Key are required.

First, register for and activate an account at `https://developer.rsc.org/accounts/create`. Once logged in, pull down your email address and select "Apps": from this page you can create a new App (name it whatever you want) and click the button labeled "ADD KEY." The key to add, at the time of writing, is named "Compounds v1 (Trial)": This will grant you an API Key that takes the form of a string of 32 seemingly random letters and numbers. This acts as a token to identify you to the ChemSpider service.

This example uses a Python library called ChemSpiPy, which can be installed using a package manager such as the one included with Anaconda or from the command line using pip:

```
pip install chemspipy
```

To use it, first import the ChemSpiPy library:

```
from chemspipy import ChemSpider
```

Create a ChemSpider object and connect to the API, using the Key:

```
cs = ChemSpider('<INSERT YOUR 32-CHARACTER KEY HERE>')
```

Each substance in ChemSpider has its own unique ID and, knowing this, the corresponding Compound object can be retrieved:

```
c = cs.get_compound(171)   # 171 is acetic acid
```

```
c
```

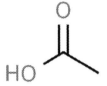

The c object contains a lot of useful data and identifiers for the compound:

```
common_name = c.common_name
formula = c.molecular_formula
M = c.molecular_weight
inchi = c.inchi
inchikey = c.inchikey
smiles = c.smiles

print('Common name:', common_name)
print('Molecular formula:', formula)
print(f'Molecular weight: {M} g.mol-1')
print('InChI:', c.inchi)
print('InChIKey:', c.inchikey)
print('SMILES:', c.smiles)
```

```
Common name: Acetic acid
Molecular formula: C_{2}H_{4}O_{2}
Molecular weight: 60.052 g.mol-1
InChI: InChI=1S/C2H4O2/c1-2(3)4/h1H3,(H,3,4)
InChIKey: QTBSBXVTEAMEQO-UHFFFAOYSA-N
SMILES: CC(=O)O
```

Both 2D and 3D Molfiles for the structure of this molecule are available:

```
mol_2d = c.mol_2d
print(mol_2d)
```

```
  ACD/Labs04281710402D

  4  3  0  0  0  0  0  0  0  0  1 V2000
    1.1482    0.0000    0.0000 O   0  0  0  0  0  0  0  0  0  0  0  0
    0.0000   -1.9973    0.0000 O   0  0  0  0  0  0  0  0  0  0  0  0
    1.1482   -1.3300    0.0000 C   0  0  0  0  0  0  0  0  0  0  0  0
    2.3010   -1.9973    0.0000 C   0  0  0  0  0  0  0  0  0  0  0  0
```

```
    1   3   2   0   0   0   0
    2   3   1   0   0   0   0
    3   4   1   0   0   0   0
M   END
```

```
mol_3d = c.mol_3d
print(mol_3d)
```

```
176
  Marvin   12300703363D

  8   7   0   0   0   0               999 V2000
      1.3733      0.7283     -0.0000 O   0   0   0   0   0   0   0   0   0   0   0   0
     -0.4454      1.7645     -0.0000 O   0   0   0   0   0   0   0   0   0   0   0   0
     -0.5806     -0.5463      0.0000 C   0   0   0   0   0   0   0   0   0   0   0   0
      0.1495      0.6974     -0.0000 C   0   0   0   0   0   0   0   0   0   0   0   0
     -1.2111     -0.6068      0.8883 H   0   0   0   0   0   0   0   0   0   0   0   0
     -1.2111     -0.6068     -0.8883 H   0   0   0   0   0   0   0   0   0   0   0   0
      0.0879     -1.4100      0.0000 H   0   0   0   0   0   0   0   0   0   0   0   0
      1.8376     -0.0203     -0.0000 H   0   0   0   0   0   0   0   0   0   0   0   0
  1   4   1   0   0   0   0
  1   8   1   0   0   0   0
  2   4   2   0   0   0   0
  3   4   1   0   0   0   0
  3   5   1   0   0   0   0
  3   6   1   0   0   0   0
  3   7   1   0   0   0   0
M   END
$$$$
```

Of course, it may well be that you do not know the ChemSpider ID. In this case, it is possible to search the database by compound name in the same way as on the website. The search is carried out asynchronously, meaning that it is performed in the background while the rest of your Python code continues to run. That is why the immediate result of a query will be reported as Results(Created): The search object has been created but has not yet been processed nor have any results returned.

```
res = cs.search('tryptophan')
res
```

```
Results(Created)
```

The search does not usually take long (in most cases a second or so), and its status can be checked directly:

```
res.status
```

```
'Complete'
```

The possible statuses are `Created`, `Failed`, `Unknown`, `Suspended`, and `Complete`. The search results are returned in a `search.Results` container object:

```
res
```

```
Results([Compound(1116), Compound(6066), Compound(8707)])
```

Three matching `Compounds` were found, with IDs 1116, 6066 and 8707. The `Results` objects can be iterated over and indexed:

```
for c in res:
    print(f'{c.csid}: {c.common_name}, {c.molecular_formula}')

res[1]
```

```
1116: DL-Tryptophan, C_{11}H_{12}N_{2}O_{2}
6066: L-Tryptophan, C_{11}H_{12}N_{2}O_{2}
8707: D-(+)-tryptophan, C_{11}H_{12}N_{2}O_{2}
```

The three compounds are the amino acid tryptophan considered as a racemic mixture (1116) and as the pure form of its two optical isomers (6066 and 8707).

It is also possible to search by chemical formula, SMILES string, InChI string and InChIKey. At the time of writing, the syntax for this is slightly different, and the process less direct than searching by compound name. The functions `filter_formula`, `filter_smiles`, `filter_inchi` and `filter_inchikey` return a *query ID*, a token representing the search that can subsequently be passed to the further functions `filter_status` and `filter_results` as illustrated below:

```
qid = cs.filter_formula('C6H14')
qid
```

```
'dba38d65-f690-4649-915a-16c3ee006968'
```

```
cs.filter_status(qid)
```

```
{'count': 6, 'message': '', 'status': 'Complete'}
```

`filter_status` has returned a dictionary indicating that the search is complete and six compounds match the formula searched for.

```
res = cs.filter_results(qid)
res
```

```
[6163, 6340, 7010, 7604, 7767, 4956268]
```

This is a list of ChemSpider IDs corresponding to the compounds matching the search. To retrieve the Compound objects themselves, call get_compounds as follows:

```
# cs.get_compounds takes a list of ChemSpider IDs and retrieves the
# Compound objects for each of them.
compounds = cs.get_compounds(res)
compounds
```

```
[Compound(6163),
 Compound(6340),
 Compound(7010),
 Compound(7604),
 Compound(7767),
 Compound(4956268)]
```

```
for c in compounds:
    print(f'{c.common_name:>32}: {c.smiles}')
```

```
              2,2-Dimethylbutane: CCC(C)(C)C
                    Diisopropyl: CC(C)C(C)C
                  3-Methylpentane: CCC(C)CC
                       isohexane: CCCC(C)C
                         Hexane: CCCCCC
      Ethylene - isobutane (1:1): CC(C)C.C=C
```

```
# We can also index into the compounds list:
c = compounds[2]    # 3-Methylpentane
c
```

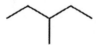

Resolving SMILES strings, InChI strings and InChIKeys is similar. Typically, these represent a single species only and so the Results sequence is of length one:

```
qid = cs.filter_smiles('c1ccoc1')
res = cs.filter_results(qid)
res
```

```
[7738]
```

```
c = cs.get_compound(res[0])
print(f'{c.common_name}, {c.molecular_formula}: {c.molecular_weight} g.mol-1')
c
```

```
qid = cs.filter_inchikey('CZMRCDWAGMRECN-UGDNZRGBSA-N')
res = cs.filter_results(qid)
res
```

```
c = cs.get_compound(res[0])
print(f'{c.common_name}:, {c.molecular_formula}, {c.molecular_weight} g.mol-1')
c
```

```
Sucrose:, C_{12}H_{22}O_{11}, 342.2965 g.mol-1
```

There may still be ambiguities or structural variations to resolve in the results, for example

```
qid = cs.filter_inchi('InChI=1S/6FH.U/h6*1H;/q;;;;;;+6/p-6')
res = cs.filter_results(qid)
res
```

```
[22966, 23253295]
```

```
compounds = cs.get_compounds(res)
for c in compounds:
    print(f'{c.common_name:>32}: {c.common_name}')
```

```
       Uranium hexafluoride: Uranium hexafluoride
    Uranium(6+) hexafluoride: Uranium(6+) hexafluoride
```

```
compounds[0]
```

compounds[1]

$$U^{6+} \; F^- \; F^- \; F^- \; F^- \; F^- \; F^-$$

The conventional representation of uranium hexafluoride as a covalently bonded molecule is the first of these.

15.4 Exercises

P15.1 Write the SMILES strings for the following molecules:

1. Toluene, $Ph - CH_3$
2. The trans fatty acid, vaccenic acid:

3. Glycine, $H_2NCH2COOH$
4. D-Limonene:

5. Hydrindane (bicyclo[4.3.0]nonane):

P15.2 Deduce the structure of the following molecules from their InChI strings:

1. InChI=1S/C6H6O/c7-6-4-2-1-3-5-6/h1-5,7H
2. InChI=1S/C5H12O/c1-3-5(2)4-6/h5-6H,3-4H2,1-2H3
3. InChI=1S/C6H10/c1-6-4-2-3-5-6/h4H,2-3,5H2,1H3
4. InChI=1S/C5H9NO2/c7-5(8)4-2-1-3-6-4/h4,6H,1-3H2,(H,7,8)

   Hint: the notation (H,7,8) means that a single hydrogen atom is shared by atoms 7 and 8.

16

More NumPy and Matplotlib

This chapter is a short survey of and reference for some useful functionality of NumPy and Matplotlib that has not been covered already. Both packages are large, and a full description would be impossible in a book this size, but the commonly used functions are demonstrated here.

16.1 NumPy

16.1.1 NumPy Array Manipulation

This section describes the more important array manipulation functions in NumPy, with some examples. Where a function returns a *view* on the original array, this means that the returned array shares its data with the original: changing one will change the other.[1] For example:

```
In [x]: arr1 = np.arange(0, 12)

In [x]: arr1
Out[x]: array([ 0,  1,  2,  3,  4,  5,  6,  7,  8,  9, 10, 11])

In [x]: arr2 = arr.reshape((2, 2, 3))

In [x]: arr2
Out[x]:
array([[[ 0,  1,  2],
        [ 3,  4,  5]],

       [[ 6,  7,  8],
        [ 9, 10, 11]]])

In [x]: arr2[1, 0, 2] = -99
```

[1] Note that because of the way that the data must be stored for some complex arrays, it may not be possible to return a view, in which case the data is copied and the two objects will be independent.

163

```
In [x]: arr2
Out[x]:
array([[[  0,    1,    2],
        [  3,    4,    5]],

        [[  6,    7,  -99],
         [  9,   10,   11]]])

In [x]: arr1          # arr1 shares its data with arr2 so it too has changed:
Out[x]: array([  0,    1,    2,    3,    4,    5,    6,    7,  -99,    9,   10,   11])
```

Some of the functions below are called from the NumPy library (usually imported as np), others are methods belonging to NumPy array objects themselves (formally, these are ndarray objects).

ndarray.shape

Return the shape of an array as the length of its dimensions in a tuple, for example:

```
In [x]: arr = np.array([[1, 2, 3], [4, 5, 6]])

In [x]: arr
Out[x]:
array([[1, 2, 3],
       [4, 5, 6]])

In [x]: arr.shape
Out[x]: (2, 3)              # two rows, three columns
```

ndarray.reshape(newshape)

Reshape an array to the shape given by the tuple, *newshape*. Usually, a view on the original array (sharing its data) is returned:

```
In [x]: arr = np.array([1, 2, 3, 4, 5, 6])  # one dimensional array, shape (6,)

In [x]: arr.reshape((3, 2))                  # reshape to three rows, two columns
Out[x]:
array([[1, 2],
       [3, 4],
       [5, 6]])
```

ndarray.ravel()

Return a flattened array, usually as a view on the original array, sharing its data:

```
In [x]: arr = np.array([[1, 2, 3], [4, 5, 6]])

In [x]: arr
Out[x]:
array([[1, 2, 3],
       [4, 5, 6]])
```

```
In [x]: arr.ravel()
Out[x]: array([1, 2, 3, 4, 5, 6])
```

ndarray.transpose(), *ndarray.T*

Return the transpose of an array (reverses its axes); usually returns a view on the original array, sharing its data:

```
In [x]: arr = np.array([[1, 2], [3, 4]])

In [x]: arr
Out[x]:
array([[1, 2],
       [3, 4]])

In [x]: arr.T        # also arr.transpose()
Out[x]:
array([[1, 3],
       [2, 4]])
```

numpy.atleast_1d(), *numpy.atleast_2d()*, *numpy.atleast_3d()*

This family of methods returns a view on a given array with at least the specified number of dimensions: Dimensions are added as needed. For example,

```
In [x]: arr = np.array([1, 2, 3, 4])

In [x]: arr.shape           # arr is 1D with four elements
Out[x]: (4,)

In [x]: arr2 = np.atleast_2d(arr)

In [x]: arr2
Out[x]: array([[1, 2, 3, 4]])

In [x]: arr2.shape          # an extra axis has been added to make arr2 2D
Out[x]: (1, 4)
```

numpy.concatenate()

Join a sequence of two or more arrays, provided as a `tuple`:

```
In [x]: arr1 = np.array([1, 2, 3])
In [x]: arr2 = np.array([4, 5, 6, 7])
In [x]: arr3 = np.array([8])
In [x]: np.concatenate((arr1, arr2, arr3))
Out[x]: array([1, 2, 3, 4, 5, 6, 7, 8])
```

numpy.stack()

Join a sequence of two or more arrays along a new axis (by default, the first dimension, `axis=0`):

```
In [x]: arr1 = np.array([1, 2, 3])

In [x]: arr2 = np.array([4, 5, 6])

In [x]: np.stack((arr1, arr2))
Out[x]:
array([[1, 2, 3],
       [4, 5, 6]])

In [x]: np.stack((arr1, arr2), axis=1)        # stack as columns
Out[x]:
array([[1, 4],
       [2, 5],
       [3, 6]])
```

numpy.hstack()

Stack arrays in sequence column-wise ("horizontally"). For one-dimensional arrays, this is the same as **numpy.concatenate**:

```
In [x]: arr1 = np.array([1, 2, 3])

In [x]: arr2 = np.array([4, 5, 6])

In [x]: np.hstack((arr1, arr2))
Out[x]: array([1, 2, 3, 4, 5, 6])

In [x]: arr1 = np.array([[1], [2], [3]])        # 3x1 "column" vector

In [x]: arr2 = np.array([[4], [5], [6]])

In [x]: np.hstack((arr1, arr2))
Out[x]:
array([[1, 4],
       [2, 5],
       [3, 6]])
```

numpy.vstack()

Stack arrays in sequence row-wise ("vertically").

```
In [x]: arr1
Out[x]: array([1, 2, 3])

In [x]: arr2
Out[x]: array([4, 5, 6])

In [x]: np.vstack((arr1, arr2))
```

```
Out[x]:
array([[1, 2, 3],                    # 2x3 array
       [4, 5, 6]])
```

numpy.tile(A, reps)

Build an array by repeating the array *A* by the number of times given by *reps*, which may be a tuple:

```
In [x]: arr = np.array([1, 2, 3])

In [x]: np.tile(arr, 4)
Out[x]: array([1, 2, 3, 1, 2, 3, 1, 2, 3, 1, 2, 3])

In [x]: np.tile(arr, (4, 3))
Out[x]:
array([[1, 2, 3, 1, 2, 3, 1, 2, 3],
       [1, 2, 3, 1, 2, 3, 1, 2, 3],
       [1, 2, 3, 1, 2, 3, 1, 2, 3],
       [1, 2, 3, 1, 2, 3, 1, 2, 3]])
```

numpy.meshgrid(x, y)

Create a rectangular grid out of one-dimensional arrays *x* and *y*, which is useful for creating coordinate matrices:

```
In [x]: x = np.linspace(0, 4, 5)

In [x]: y = np.linspace(0, 3, 4)

In [x]: X, Y = np.meshgrid(x, y)

In [x]: X
Out[x]:
array([[0., 1., 2., 3., 4.],
       [0., 1., 2., 3., 4.],
       [0., 1., 2., 3., 4.],
       [0., 1., 2., 3., 4.]])

In [x]: Y
Out[x]:
array([[0., 0., 0., 0., 0.],
       [1., 1., 1., 1., 1.],
       [2., 2., 2., 2., 2.],
       [3., 3., 3., 3., 3.]])
```

The arrays X and Y can *each* be indexed with indexes (i, j): the x array is repeated as rows down X and the y array as columns across Y. A function of two coordinates can therefore be evaluated on the grid as simply f(X, Y) – see Example E16.1.

16.1.2 NumPy Polynomials

NumPy provides a number of sub-packages for representing and manipulating different kinds of polynomial function: in addition to power series, there are classes for various other common polynomial basis functions: Chebyshev, Legendre, Laguerre series and so on.[2] This section will describe the ordinary power series, represented as an instance of the `numpy.polynomial.Polynomial` class. It is helpful to alias the name of this class to something short, like P:

```
In [x]: from numpy.polynomial import Polynomial as P
```

To create a polynomial, $P(x)$, either pass its coefficients (in order of *ascending* power of x) in a sequence or, if you know its roots, use `P.fromroots`:

```
In [x]: p = P((-6, -7, 0, 1))        # p(x) = x^3 - 7x - 6
In [x]: p = P.fromroots([-2, -1, 3])     # p(x) = (x+2)(x+1)(x-3)
```

In fact, these are actually the same polynomial:

```
In [x]: p
Out[x]: Polynomial([-6., -7.,  0.,  1.], domain=[-1,  1], window=[-1,  1])
```

(The `domain` and `window` attributes will be important for polynomial fitting.)
    To evaluate a `Polynomial`, call it:

```
In [x]: p(2)
Out[x]: -12.0                    # 2^3 - 7*2 - 6 = -12
```

Polynomial objects can be added (+), subtracted (−), multiplied (*), divided (//) and substituted into one another:

```
In [x]: p + P([0, 0, 0, 2])              # p(x) + x^3
Out[x]: Polynomial([-6., -7.,  0.,  3.])    # = 3x^3 - 7x - 5=6

In [x]: p * P([0, 1])                    # multiply p(x) by x
Out[x]: Polynomial([ 0., -6., -7.,  0.,  1.]) # x^4 - 7x^2 - 6x

In [x]: p // P([0, 2, 1])                # divide p(x) by x^2 + 2x = x - 2
Out[x]: Polynomial([-2.,  1.])

In [x]: q = P([1, 1])                    # q(x) = x + 1
In [x]: p(q)                             # p(x+1) = x^3 + 3x^2 - 4x - 12
Out[x]: Polynomial([-12.,  -4.,   3.,   1.])
```

The roots of a Polynomial, if they are not already known, are returned by `Polynomial.roots()`:

---

[2] The package documentation at `https://numpy.org/doc/stable/reference/routines.polynomials.package.html` provides full details.

```
In [x]: p.roots()                    # x^3 - 7x - 6 = (x+2)(x+1)(x-3)
Out[x]: array([-2., -1.,   3.])
```

Differentiation and integration of `Polynomials` is supported by the `deriv` and `integ` methods:

```
In [x]: p.deriv()                    # dp/dx = 3x^2 - 7
Out[x]: Polynomial([-7.,   0.,   3.])

In [x]: # To take the mth derivative, pass m as an argument to deriv:
In [x]: p.deriv(m=2)                  # d2p/dx2 = 6x
Out[x]: Polynomial([0., 6.])

In [x]: p.integ()                    # -6x - (7/2)x^2 + (1/4)x^4
Out[x]: Polynomial([ 0.  , -6.  , -3.5 ,  0.  ,  0.25])

In [x]: p.integ(k=4)                 # k defines the constant of integration
Out[x]: Polynomial([ 4.  , -6.  , -3.5 ,  0.  ,  0.25])
```

Fitting an nth order polynomial, $P_n(x) = a_0 + a_1 x + a_2 x^2 + \ldots + a_n x^n$, to a set of data points, (x_i, y_i) involves finding the coefficients a_j that minimize the average difference between $P_n(x_i)$ and y_i in some sense (see also Chapter 19). However, since the monomial basis functions, $1, x, x^2, x^3$ become very large in magnitude outside the "window" $-1 \le x \le 1$ for higher powers of x, it is usually a good idea to map the "domain" of the data (the range of the actual values of the independent variable, x_i) to this window.

For example, suppose you wish to fit a quartic polynomial to the Lennard-Jones potential of Xe_2:

$$V(r) = 4\epsilon \left[\left(\frac{\sigma}{r} \right)^{12} - \left(\frac{\sigma}{r} \right)^6 \right] = \frac{A}{r^{12}} - \frac{B}{r^6},$$

where $\epsilon = 1.77\,\text{kJ mol}^{-1}$ and $\sigma = 4.1\,\text{Å}$. The interesting region of this function is around $4.2 \le r \le 5.5\,\text{Å}$:

```
In [x]: E, sigma = 1.77, 4.1
In [x]: A, B = 4 * E * sigma**12, 4 * E * sigma**6

In [x]: def V(r):
   ...:      return A / r**12 - B / r**6
   ...:

In [x]: r = np.linspace(4.2, 5.5, 1000)
In [x]: pfit = P.fit(r, V(r), 4)
In [x]: pfit
Out[x]: Polynomial([-1.6393034 ,  0.6043512 ,  0.31373957, -0.67033907,
                     0.40460482], domain=[4.2, 5.5], window=[-1.,   1.])
```

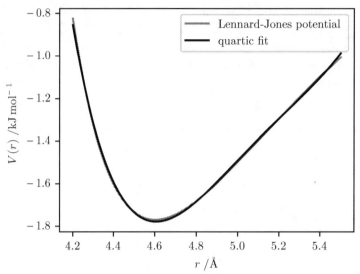

Figure 16.1　The Lennard-Jones potential and a fit to it using a quartic polynomial (see text).

NumPy has automatically set the domain of the fitted polynomial to the range of the internuclear distance variable, r, and the coefficients are of similar magnitude. But note although we can call the `pfit` function to generate values of the fitted polynomial, the coefficients themselves apply *after* the linear transformation from the domain to the window has been applied. That is, the best-fit polynomial is not $p_{\text{fit}} = a_0 + a_1 x + a_2 x^2 + a_3 x^3 + a_4 x^4$ where $a_0 = -1.639304$ etc., but $p_{\text{fit}} = a_0 + a_1 x' + a_2 x'^2 + a_3 x'^3 + a_4 x'^4$ where x' is related to x by the linear map from the domain to the window ($x' = \mu x + \chi$). To return the coefficients of a polynomial that can be used directly, use the `convert()` function:

```
In [x]: pfit.convert()
Out[x]:
Polynomial([ 1543.92526335, -1212.86206256,   356.15664798,   -46.41324243,
          2.26661429], domain=[-1.,   1.], window=[-1.,   1.])
```

The Lennard-Jones potential and the quartic fit are compared in Figure 16.1.

```
In [x]: plt.plot(r, V(r), label='Lennard-Jones potential')
In [x]: plt.plot(r, pfit(r), label='quartic fit')
In [x]: plt.xlabel(r'$r\;/\mathrm{\AA}$')
In [x]: plt.ylabel(r'$V(r)\;/\mathrm{kJ\,mol^{-1}}$')
In [x]: plt.legend()
In [x]: plt.show()
```

There is a legacy function, `numpy.polyfit` which performs the fit directly, without this mapping from domain to window, but it is, in general, less numerically stable and is no longer recommended:

```
In [x]: np.polyfit(r, V(r), 4)
Out[x]:
array([    2.26661429,   -46.41324243,    356.15664798, -1212.86206256,
         1543.92526334])
```

(Note that it also outputs the polynomial coefficients in the opposite order to the `Polynomial` class, i.e. in *descending* powers of *x*.)

16.1.3 NumPy Comparisons

Section 9.5 introduced some basic NumPy comparison operators, for example:

```
In [x]: arr1 = np.array([1, 2, 3, 4, 5])

In [x]: arr2 = np.array([5, 4, 3, 2, 1])

In [x]: arr1 == arr2
Out[x]: array([False, False,  True, False, False])
```

However, because of the finite precision of floating point arithmetic, this may give unexpected answers when equality is expected:

```
In [x]: 0.1**2 == 0.01
Out[x]: False
```

The cause of this is the fact that 0.1 is not exactly representable, and the approximate number used gives a square that is not equal to the computer's representation of 0.01:

```
In [x]: 0.1**2
Out[x]: 0.010000000000000002
```

NumPy provides a method, `isclose` to determine if two values are equal to within some tolerance. The documentation[3] provides more detail, but the default usage of this method works well in most cases:

```
In [x]: np.isclose(0.1**2, 0.01)
Out[x]: True
```

Note that in most interfaces, NumPy truncates the output so the lack of precision in its numbers is not always obvious. But this can be customized as follows:

```
In [x]: # Set the number of digits of precision in the output.
In [x]: np.set_printoptions(precision=17)
```

---

[3] https://numpy.org/doc/stable/reference/generated/numpy.isclose.html.

```
In [x]: arr1 = np.array([(x / 10) ** 2 for x in range(6)])
In [x]: arr1
Out[x]:
array([0.                , 0.01                , 0.04000000000000001,
       0.09                , 0.16000000000000003, 0.25               ])

In [x]: arr1 == [0.0, 0.01, 0.04, 0.09, 0.16, 0.25]
Out[x]: array([ True, False, False,  True, False,  True])

In [x]: np.isclose(arr1, [0.0, 0.01, 0.04, 0.09, 0.16, 0.25])
Out[x]: array([ True,  True,  True,  True,  True,  True])
```

The related method, `allclose`, returns a single boolean value according to whether *all* of the elements in two arrays are equal or not:

```
In [x]: np.allclose(arr1, [0.0, 0.01, 0.04, 0.09, 0.16, 0.25])
Out[x]: True
```

16.2 Physical Constants (SciPy)

The package, `(scipy.constants)`[4] provides the numerical values of many of the constants used in physics and chemistry. The more common ones, in SI units, can be imported directly (see Table 16.1). For example:

```
In [x]: from scipy.constants import e, h

In [x]: e                    # elementary charge, C
Out[x]: 1.602176634e-19

In [x]: h                    # Planck constant, J.s
Out[x]: 6.62607015e-34
```

The values of these physical constants, at the time of writing, are the CODATA 2018 Internationally Recommended values.[5]

A more complete database of physical constants, along with uncertainties and explicit strings defining their units is accessible by importing the `scipy.constants` package. This provides a dictionary, `scipy.constants.physical_constants` of tuples in the format: `(value, unit, uncertainty)`, keyed by identifying strings. For example:

```
In [x]: import scipy.constants as pc

In [x]: pc.physical_constants["proton mass"]
Out[x]: (1.67262192369e-27, 'kg', 5.1e-37)        # value, units, uncertainty
```

---

[4] See the documentation at https://docs.scipy.org/doc/scipy/reference/constants.html. for full details of this package.

[5] https://physics.nist.gov/cuu/Constants/index.html.

```
In [x]: pc.physical_constants["Planck constant"]
Out[x]: (6.62607015e-34, 'J Hz^-1', 0.0)
```

Table 16.1   Physical constants that can be directly imported from
scipy.constants in SI units

c	299792458.0	speed of light in vacuum, m s^{-1}
epsilon_0	8.8541878128e-12	vacuum permittivity, ϵ_0, F m^{-1}
h	6.62607015e-34	Planck constant , h, J s
hbar	1.0545718176461565e-34	$\hbar = h/(2\pi)$, J s
e	1.602176634e-19	elementary charge, C
R	8.314462618	molar gas constant, J K^{-1} mol^{-1}
N_A	6.02214076e+23	Avogadro constant, mol^{-1}
k	1.380649e-23	Boltzmann constant, J K^{-1}
Rydberg	10973731.56816	Rydberg constant, m^{-1}
m_e	9.1093837015e-31	electron mass, kg
m_p	1.67262192369e-27	proton mass, kg
m_n	1.67492749804e-27	neutron mass, kg

(Note that the value of h was fixed to this value by definition in 2018, so its uncertainty is zero.)

There is a simple search functionality for the constant names:

```
In [x]: pc.find("permittivity")
Out[x]: ['atomic unit of permittivity', 'vacuum electric permittivity']
```

There are two matching keys; the second of these is the quantity usually given the symbol ϵ_0:

```
In [x]: pc.physical_constants["vacuum electric permittivity"]
Out[x]: (8.8541878128e-12, 'F m^-1', 1.3e-21)
```

The scipy.constants package also defines the values of SI prefixes (kilo-, mega- and so on) and conversion factors to SI units for commonly used alternative units:

```
In [x]: pc.nano          # SI nano-prefix (10^-9)
Out[x]: 1e-09

In [x]: pc.kilo          # SI kilo-prefix (10^3)
Out[x]: 1000.0

In [x]: pc.atm           # 1 atm in Pa; see also pc.bar, pc.torr, pc.mmHg
Out[x]: 101325.0

In [x]: pc.zero_Celsius  # 0 degC in K
Out[x]: 273.15

In [x]: pc.eV            # 1 eV in J
Out[x]: 1.602176634e-19
```

16.3 More Matplotlib

16.3.1 Error Bars

Scientific data is often associated with uncertainties, and wherever possible these should be indicated on plots and graphs. A dedicated Matplotlib method, errorbar allows this. Its call signature contains many optional arguments,[6] but the most useful of them are:

```
errorbar(x, y, yerr, xerr, fmt, ecolor, elinewidth, capsize, capthick)
```

Here, x and y are the arrays of N data points to be plotted. yerr and xerr, if given, determine the size of the error bars. If set to single, scalar value all the error bars along that axis will have the same magnitude. If set to a one-dimensional vector of length N, then symmetric errors for each point can be set separately; if set to an array of shape 2, N, then the lower and upper limits for each point can be different (asymmetric error bars). The other arguments customize the plot as follows:

- fmt: the format, for example, marker style of the plotted points, as for plt.plot;
- ecolor: the color of the error bar lines; if not set, this defaults to the same as the plotted line;
- elinewidth: the width of the error bar lines;
- capsize: the length of the error bar caps in points (defaults to 0: no errorbar caps);
- capthick: the thickness of the error bar caps.

For example (see also Example E20.1):

```
import numpy as np
import matplotlib.pyplot as plt

# The data is y = exp(-x/2)
x = np.array([1, 2, 3, 4, 5])
y = np.exp(-x / 2)

# A 10% error in the x-variable: the absolute error increases with increasing x.
xerr = x * 0.1
# A constant absolute error of 0.05 in the y-variable
yerr = 0.05

plt.errorbar(x, y, yerr, xerr, fmt='o', capsize=4)
```

This code produces Figure 16.2.

---

[6] See the documentation at https://matplotlib.org/stable/api/_as_gen/matplotlib.pyplot
.errorbar.html for details.

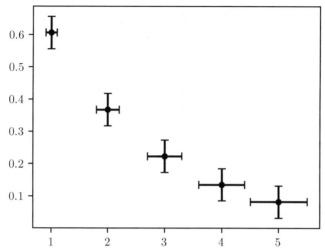

Figure 16.2   An example error bar plot with uncertainties indicated on both *x* and *y* axes.

16.3.2 Contour Plots and Surface Plots

Two popular ways of visualizing a function of two variables, $f(x, y)$, are as a contour plot and as a "three-dimensional" surface plot. This section will briefly introduce the Matplotlib functionality that creates these types of plots, with reference to some atomic orbital wavefunctions.

We will focus on the 2s and 2p orbitals. In polar coordinates:

$$\psi_{2s} = N(2 - r)e^{-r/2},$$
$$\psi_{2p} = N're^{-r/2}\cos\theta.$$

These wavefunctions are independent of the azimuthal angle, ϕ: Both orbitals have cylindrical symmetry about the z-axis, so any plane containing this axis will look the same. Practically, we evaluate the wavefunctions for polar angles in the range $-\pi \leq \theta \leq \pi$ to cover the whole of this plane.

To create the coordinate arrays for the function, the `np.meshgrid` function is useful (see Section 16.1.1).

```
In [x]: import numpy as np
In [x]: import matplotlib.pyplot as plt

In [x]: r = np.linspace(0, 12, 100)
In [x]: theta = np.linspace(-np.pi, np.pi, 100)
In [x]: R, Theta = np.meshgrid(r, theta)
```

The unnormalized wavefunctions themselves are easy to code:

```
def psi_2s(r, theta):
    """Return the value of the 2s orbital wavefunction.

    This orbital is spherically-symmetric so does not depend on
    the angular coordinates theta or phi. The returned wavefunction
    is not normalized.

    """

    return (2 - r) * np.exp(-r/2)

def psi_2pz(r, theta):
    """Return the value of the 2pz orbital wavefunction.

    This orbital is cylindrically-symmetric so does not depend on
    the angular coordinate phi. The returned wavefunction
    is not normalized.

    """

    return r * np.exp(-r/2) * np.cos(theta)
```

To make a contour plot of the radial distribution function, $r^2|\psi|^2$, we can use plt.contour, which takes two-dimensional arrays, X and Y, and plots contours of the height values provided in a third two-dimensional array, Z. Therefore, we need to transform from our polar coordinates back into Cartesian coordinates first:

```
In [x]: X, Y = R * np.cos(Theta), R * np.sin(Theta)
In [x]: Z_2s = R**2 * psi_2s(R, Theta)**2
```

Before plotting the contours, there is one more thing to be aware of: Matplotlib will not necessarily set the aspect ratio (the relative scales of the *x* and *y* axes) to 1, meaning that circles do not necessarily appear circular on the plot. We can enforce this by calling plt.axis('square') and turning off the Axes frame and labeling altogether with plt.axis('off'):

```
In [x]: plt.contour(X, Y, Z_2s, colors='k')
In [x]: plt.axis('square')
In [x]: plt.axis('off')
```

The result is Figure 16.3. By default, a *colormap* is applied to the contours; to override this, explicitly set the colors argument to a Matplotlib color specifier. Sim-

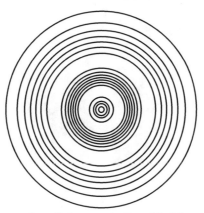

Figure 16.3   A contour plot of the atomic 2*s* orbital in a plane containing the nucleus.

ilarly, the contour lines can be customized with arguments such as `linestyles`, `linewidths` and so on.[7]

A related function, `plt.contourf`, creates contour plots filled according to a specified colormap. The default colormap is "viridis," a blue-to-yellow scale that accommodates most forms of color blindness, is perceptually uniform (without the distracting bands of high luminosity colors at intermediate values that occur with, for example, the rainbow or historically popular "jet" colormaps) and prints well in greyscale. Other colormaps are available – see the online documentation[8] and countless, earnest blog posts for details.

`plt.contour` and `plt.contourf` can be used together to plot contour lines over a background color scale (Figure 16.4):

```
In [x]: Z_2pz = R**2 * psi_2pz(R, Theta)**2
In [x]: plt.contour(X, Y, Z_2pz, colors='k')
In [x]: plt.contourf(X, Y, Z_2pz)
In [x]: plt.axis('square')
In [x]: plt.axis('off')
```

Three-dimensional surface plots can be made with `plot_surface`, but there a few syntax differences to regular two-dimensional plots. First, `Axes3D` must be imported as follows:

```
In [x]: from mpl_toolkits.mplot3d import Axes3D
```

---

[7] See the documentation at `https://matplotlib.org/stable/api/_as_gen/matplotlib.pyplot`
`.contour.html` for more details, including on how to set the values at which the contours appear with a list passed to the argument `levels`.

[8] `https://matplotlib.org/stable/api/_as_gen/matplotlib.pyplot.contourf.html`.

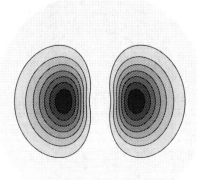

Figure 16.4   A filled contour plot of the $2p_z$ orbital in a plane containing the z axis.

Second, an `Axes` object must be created as a subplot on a Matplotlib `Figure` with the `projection` argument set to '3d' (the procedural `pyplot` interface we have been using so far will not work):

```
In [x]: fig = plt.figure()
In [x]: ax = fig.add_subplot(projection='3d')
```

With this in place, a surface plot can be added directly to the Axes. For example, the following two-dimensional Gaussian function is plotted in Figure 16.5:

$$f(x, y) = \exp\left[-\frac{(x - x_0)^2}{a^2} - \frac{(y - y_0)^2}{b^2} \right]$$

```
In [x]: x = y = np.linspace(0, 1, 40)
In [x]: X, Y = np.meshgrid(x, y)
In [x]: x0 = y0 = 0.5
In [x]: a, b = 0.2, 0.4
In [x]: f = np.exp(-(X-x0)**2 / a**2 - (Y-y0)**2 / b**2)

In [x]: ax.plot_surface(X, Y, f)
```

16.3.3 Multiple Axes

A Matplotlib `Figure` may have multiple axes; in the simplest case, they can be arranged in a grid of rows and columns as follows:

```
In [x]: fig = plt.figure()
In [x]: axes = fig.add_subplot(nrows=3, ncols=2)
In [x]: # Alternatively: fig, axes = plt.subplots(nrows=3, ncols=2)
```

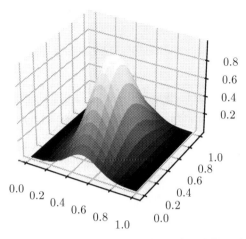

Figure 16.5   A surface plot of a two-dimensional Gaussian function.

In this example, axes is an array, with dimensions 3×2, of Axes objects, which can each be independently plotted on.

For example, to illustrate the effect of increasing numbers of terms in the Fourier Series expansions of the square-wave and sawtooth functions with frequency v:

$$f_{sq}(t) = \frac{4}{\pi} \sum_{k=1}^{\infty} \frac{1}{2k-1} \sin\left[2\pi(2k-1)vt\right]$$

$$f_{saw}(t) = \frac{1}{2} - \frac{1}{\pi} \sum_{k=1}^{\infty} \frac{(-1)^k}{k} \sin\left[2\pi kvt\right].$$

```
# Number of time sample points.
n = 2048
# Period, s.
T = 1
# Frequency, s-1.
nu = 1 / T
# Length sampling, s.
duration = 2
t = np.arange(0, duration, duration/n)

# Square-wave, _|-|_|-|
f_sq = np.ones(n)
f_sq[(t >= T/2) & (t < T)] = -1
f_sq[t >= T] = f[t < T]

# Saw-tooth, /|/|/|
f_saw = (t+0.5) % 1

def fourier_expansion_square(nterms):
    fourier_square = np.zeros(n)
    for k in range(1, nterms+1):
        fac = 2 * k - 1
```

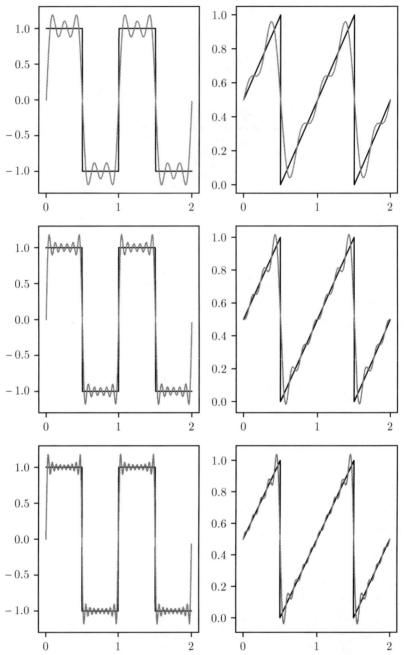

Figure 16.6   Fourier Series expansions for the square wave (left) and sawtooth
(right) periodic functions with 3, 6 and 9 terms (top, middle and bottom rows).

```
            fourier_square += np.sin(2 * np.pi * fac * nu * t)/ fac
        return 4 / np.pi * fourier_square

def fourier_expansion_sawtooth(nterms):
    fourier_sawtooth = np.zeros(n)
    for k in range(1, nterms+1):
        fourier_sawtooth += (-1)**k * np.sin(2 * np.pi * k * nu * t) / k
    return 0.5 - 1/np.pi * fourier_sawtooth
```

The left-hand column will hold the square wave plots; the right hand the sawtooth plots (as shown in Figure 16.6):

```
fig, axes = plt.subplots(nrows=3, ncols=2)
for irow in range(3):
    axes[irow, 0].plot(t, f_sq)
    nterms = (irow+1) * 3        # 3, 6 and 9 terms in the expansion
    axes[irow, 0].plot(t, fourier_expansion_square(nterms))

    axes[irow, 1].plot(t, f_saw)
    axes[irow, 1].plot(t, fourier_expansion_sawtooth(nterms))
plt.tight_layout()
```

16.4 Example

E16.1

The wavefunction for a particle moving in a two-dimensional potential well defined by $V(x, y) = 0$ for $0 \leq x \leq L_x$ and $0 \leq y \leq L_y$ and infinity elsewhere is:

$$\psi_{n_x,n_y}(x, y) = \frac{2}{\sqrt{L_x L_y}} \sin\left(\frac{n_x \pi x}{L_x}\right) \sin\left(\frac{n_y \pi y}{L_y}\right),$$

where the state is defined by the quantum numbers $n_x = 1, 2, \ldots$ and $n_y = 1, 2, \ldots$.

These functions can be visualized either as a contour plot or a surface plot, as shown below (Figures 16.7 and 16.8).

```
import numpy as np
import matplotlib.pyplot as plt
from mpl_toolkits.mplot3d import Axes3D
```

```
# Choose to use a square well and define a meshgrid of coordinate points.
Lx = Ly = 1
npts = 101
x = np.linspace(0, Lx, npts)
y = np.linspace(0, Ly, npts)
X, Y = np.meshgrid(x, y)
```

```
def psi_nxny(x, y, nx, ny):
    """The two-dimensional particle-in-a-box wavefunction for state nx, ny.

    x, y can be a single point or compatible arrays of such points.

    """

    N = 2 / np.sqrt(Lx * Ly)
    return N * np.sin(nx * np.pi * x / Lx) * np.sin(ny * np.pi * y / Ly)
```

```
def contour_plot(ax, nx, ny):
    """
    Make a contour plot of the wavefunction defined by (nx, ny) on
    the provided Axes object, ax.

    """

    ax.contourf(X, Y, psi_nxny(X, Y, nx, ny))
    # Ensure that the x- and y-axes appear with the same aspect
    # ratio: i.e. squares appear square.
    ax.axis('square')

def surface_plot(ax, nx, ny):
    """
    Make a surface plot of the wavefunction defined by (nx, ny) on
    the provided Axes object, ax. Axes3D must have been imported
    from mpl_toolkits.mplot3d for this to work!

    """

    # Plot the wavefunction in "3D", colouring the the surface
    # by value according to the "hot" colormap.
    ax.plot_surface(X, Y, psi_nxny(X, Y, nx, ny), cmap='hot')
```

```
# A 2x2 grid of Axes.
fig, axes = plt.subplots(nrows=2, ncols=2)
contour_plot(axes[0, 0], 1, 1)
contour_plot(axes[1, 0], 2, 1)
contour_plot(axes[0, 1], 1, 2)
contour_plot(axes[1, 1], 2, 2)
# To prevent the axis labels overlapping, it is sometimes
# necessary to call tight_layout() to add some padding around
# subplot Axes:
plt.tight_layout()
```

```
# Create a 3D figure, and plot the wavefunction as a surface.
fig = plt.figure()
ax = fig.add_subplot(projection='3d')
nx, ny = 2, 4
surface_plot(ax, nx, ny)
plt.show()
```

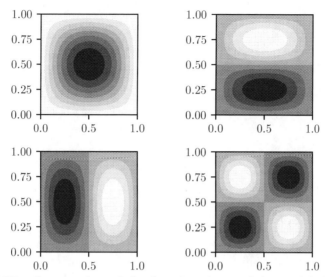

Figure 16.7   Contour plots of the first four states of the particle in a two-dimensional square box.

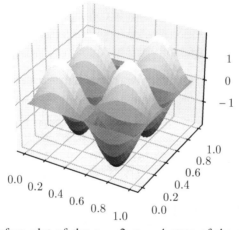

Figure 16.8   Surface plot of the $n_x = 2, n_y = 4$ state of the particle in a two-dimensional square box.

17

Thermodynamic Cycles

17.1 Internal Energy and the First Law

The internal energy, U, of a thermodynamic system is the total energy, potential and kinetic, contained within it. It does not include the kinetic energy of the system as a whole relative to some external frame or the potential energy due to external force fields: The internal energy of a system sitting on the lab bench at sea level is the same if the system is placed on an airliner cruising at 1,000 km/hr, 10 km above sea level as long as its internal *state* is the same. For systems in equilibrium, the internal state can be thought of as entirely defined by the temperature, T, volume, V, pressure, p, and composition, N_i, of the system. It might be helpful to review some properties of the internal energy:

- Internal energy is a *state function*: It depends only on the state of the system and not on the history of states it may have passed through to get there.
- Internal energy is an *extensive* property of the system: It depends on the system's size (quantity of matter), as do other extensive properties such as mass and volume; the related *intensive* properties are *specific internal energy, $u = U/m$* (in $J\,kg^{-1}$) and *molar internal energy, $U_m = U/n$* (in $J\,mol^{-1}$).
- For an *isolated system* (one that does not exchange either matter or energy with its surroundings), the internal energy is *conserved*: At the microscopic level, its constituent particles may be zipping around with individual kinetic and potential energies that change over time as they interact and collide with each other, but the sum of these energies is a constant, U.
- The internal energy is only defined to within an additive constant: We could agree to add the same, arbitrary constant to all measures of U, and it wouldn't change any of our predictions about the system's properties.
- Thermodynamics is concerned with the change in U (and other functions) between states: $\Delta U = U_2 - U_1$; since U is a state function, the path between these states does not affect the value of ΔU: We can (and will) choose the path that is easiest to calculate ΔU for, knowing that this quantity is the same for all paths.

An ideal gas has no interactions between its constituent particles and so no potential energy: Its internal energy is simply the sum of the particles' kinetic energies.

$$U = \frac{1}{2}m \sum_i v_i^2.$$

At equilibrium, the particles are not all moving at the same speed but have speeds given by the Maxwell–Boltzmann distribution:

$$f(v) = \left(\frac{m}{2\pi k_B T}\right)^{3/2} \exp\left(-\frac{mv^2}{2k_B T}\right)$$

For a mole of ideal gas, there are so many particles that the above sum can be replaced with an integral over this distribution with the result that

$$U_m = N_A \int_0^\infty v^2 f(v)\, dv = \tfrac{3}{2}RT.$$

That is, the temperature of an ideal gas is a direct measure of its internal energy.

The First Law of Thermodynamics expresses the ways in which a system's internal energy can change through interaction with its surroundings:

$$\Delta U = q + w. \tag{17.1}$$

A couple of definitions and clarifications are needed at this point. Equation 17.1 applies to a *closed* system: one that can exchange energy but not matter with its surroundings. If the system being considered is undergoing a chemical reaction that gives off gaseous products, then that gas must be considered part of the system, even if in this case, the boundary between the system and the surroundings may be difficult to define.

Equation 17.1 states that there are two ways that the internal energy of a closed system can change:

1. It can be heated: If the system is placed in thermal contact with its surroundings, net energy will flow between the two if they have different temperatures. The total amount of energy transferred in this way during the process of interest is called the *heat*, q. If $q = 0$ (e.g., the system is insulated from its surroundings) the process is called *adiabatic*.
2. It can do work: There are various forms of work that a system could do (expansion work, electrical work, etc.), but they all involve a displacement or deformation of the system against a force. w is defined as the work done *on* the system: A positive w would result from the surroundings compressing the system, generally causing an increase in its internal energy (think of pushing in a bicycle pump); if the system expands against the force exerted by the pressure of the atmosphere (e.g., when a liquid evaporates), it has to do work in pushing back the atmosphere and $w < 0$.

Equation 17.1 is often expressed in differential form:

$$dU = dq + dw$$

since during a process, the state of the system will typically be changing continuously.

We will consider two specific processes:

1. Constant-volume ("isochoric") heating of the system: The system is placed in thermal contact with its surroundings, which have a higher temperature, but in such a way that it cannot expand. In this case, it can do no work and so $dU = dq$. It is an experimental fact that for a fixed amount of energy supplied through heating in this way, the temperature rise experienced is different for different materials: The coefficient that relates this temperature rise to the increase in internal energy is the (constant volume) *heat capacity*:

$$C_V = \left(\frac{\partial U}{\partial T}\right)_V \tag{17.2}$$

That is, a system with a large heat capacity (e.g., a block of iron) can receive a relatively large amount of heat without its temperature increasing very much; a system with a small heat capacity (e.g., a gas) will reach a higher temperature for the same heat input.

2. Constant-temperature ("isothermal") expansion. First, consider an ideal gas expanding against an externally applied pressure, p_{ex}. If this external pressure is constant, the instantaneous work done *on* the gas is $dw = -p_{ex}dV$, and the total work done

$$w = -\int_{V_1}^{V_2} p_{ex} \, dV = -p_{ex}(V_2 - V_1). \tag{17.3}$$

Note that if $V_2 < V_1$ (the gas is compressed), then $w > 0$ (work is done *on* the gas). If the gas expands into a vacuum ($p_{ex} = 0$), then it does no work.

The change of state we have just considered is sometimes called *irreversible* expansion: Imagine confining the gas to a cylinder with a piston weighted down by a brick; as long as the brick is in place, the pressure of the gas is countered by the combined pressure of the atmosphere and the gravitational force due to the weight of the brick. If the brick is removed, then there is a spontaneous and immediate tendency for the gas to push back the piston as it expands to regain mechanical equilibrium, this time at a lower pressure.

Now, consider what would happen if, instead of the brick, the piston were held down by the same weight of sand. If the grains of sand are removed slowly, one by one, the expansion process would be between the same initial and final volumes (it would just take a very long time). At each stage of the process, the gas would expand by a tiny amount, due to the miniscule difference in internal and external pressure. As the weight of the grains is thought to approach zero, each step can be thought of as an *infinitesimal* change and the overall expansion approaches the idealization of a so-called *reversible* process. Since the external pressure matches (i.e., differs only infinitesimally from) the gas pressure at each stage of the expansion, $p_{ex} = p$ and the work done is

$$w = -\int_{V_1}^{V_2} p\, dV = -\int_{V_1}^{V_2} \frac{nRT}{V}\, dV.$$

This integral can be evaluated in the case of constant temperature (an *isothermal* expansion) to give

$$w = -nRT \ln \frac{V_2}{V_1}. \tag{17.4}$$

The reversible expansion clearly does the most work that can be obtained from the system for these values of V_1 and V_2: if the external pressure were any higher, it wouldn't expand at all.

Since the temperature is constant the internal energy does not change in the isothermal expansion of an ideal gas: $\Delta U = 0$. This is possible because as the gas does its work it is gaining the same amount of energy as heat from its surroundings: from the First Law, $q = w$.

17.2 Example

E17.1

The Stirling Engine

The *Stirling engine* is a potentially efficient heat engine that may be represented by the *idealized Stirling cycle*. There are four steps to the cycle, starting with a compressed, ideal gas. With reference to Figure 17.1, which is generated by the code below, the steps are:

A → B: *Reversible, isothermal expansion.* The gas expands, doing work $w_1 < 0$; in order for the temperature to stay constant, energy has to flow into the system from the surroundings ($q_1 > 0$).

B → C: *Isochoric cooling.* The system is placed in thermal contact with a cooler reservoir; energy flows out of it ($q_2 < 0$) as its temperature falls, at constant volume, to that of the reservoir. The system does no work in this step, $w_2 = 0$.

C → D: *Reversible, isothermal compression.* The gas is compressed back to its original volume ($w_3 > 0$), at constant temperature; energy must, therefore, flow out of the system ($q_3 < 0$).

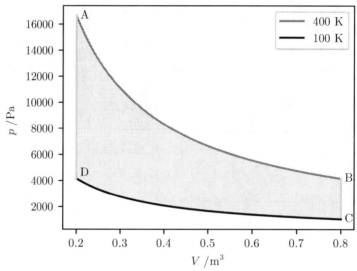

Figure 17.1   An idealized Stirling Cycle. The shaded area represents the work done by the engine.

$D \to A$: *Isochoric heating.* The system is placed in thermal contact with a hotter reservoir; energy flows in ($q_4 > 0$) as it is heated, to the higher temperature of the reservoir. Since this is a constant-volume step, no work is done ($w_4 = 0$).

The Stirling cycle may be parameterized by the temperatures of the hot and cold reservoirs, T_{hot} and T_{cold}, respectively, and by the compressed and expanded volumes, V_1 and V_2. For the purposes of illustration, we will work with 1 mol of ideal gas as the working fluid.

```python
import numpy as np
from scipy.constants import R
import matplotlib.pyplot as plt

def get_p(V, T):
    """Return the pressure of 1 mol of an ideal gas at (V, T)."""
    return R * T / V
```

```python
def stirling_cycle(V1, Tcold, V2, Thot):
    """Plot a p-V diagram for a Stirling cycle defined by V1, Tcold, V2, Thot."""

    # Define variables for the state of the gas at each stage of the cycle.
    VA, TA, pA = V1, Thot, get_p(V1, Thot)
    VB, TB, pB = V2, Thot, get_p(V2, Thot)
    VC, TC, pC = V2, Tcold, get_p(V2, Tcold)
    VD, TD, pD = V1, Tcold, get_p(V1, Tcold)

    # A -> B: isothermal expansion at Thot.
    Vhot = np.linspace(VA, VB, 1000)
    phot = get_p(Vhot, Thot)
```

```
      plt.plot(Vhot, phot, c='r', label='{} K'.format(Thot))

      # B -> C: isochoric cooling from Thot to Tcold
      plt.plot([VB, VC], [pB, pC], c='k')

      # C -> D: isothermal compression at Tcold
      Vcold = np.linspace(VC, VD, 1000)
      pcold = get_p(Vcold, Tcold)
      plt.plot(Vcold, pcold, c='b', label='{} K'.format(Tcold))

      # D -> A: isochoric heating from Tcold to Thot
      plt.plot([VD, VA], [pD, pA], c='k')

      # Fill the region between the isotherm curves.
      plt.fill_between(Vcold[::-1], pcold[::-1], phot, color='gray')

      # Label the four states A, B, C and D.
      plt.text(VA, pA, 'A')
      plt.text(VB, pB, 'B')
      plt.text(VC, pC, 'C')
      plt.text(VD, pD, 'D')

      plt.xlabel('$V\;/\mathrm{m^3}$')
      plt.ylabel('$p\;/\mathrm{Pa}$')
      plt.legend()

V1, Tcold = 0.2, 100
V2, Thot = 0.8, 400
stirling_cycle(V1, Tcold, V2, Thot)
```

17.3 Exercise

P17.1   An important alternative way the gas can expand is through an adiabatic process: if the cylinder is insulated so it is thermally isolated from its surroundings, $q = 0$ and as it does expansion work its internal energy must decrease, $\Delta U = w < 0$ and its temperature drops. The reverse process, adiabatic compression, is familiar from inflating a bicycle tyre: As the air in the pump is compressed (rapidly enough that there is little time for heat flow to the surroundings) its temperature rises.

In an adiabatic expansion, $dU = dw$ and so $C_V dT = -p dV$ from the definition of the heat capacity, C_V (Equation 17.2). If the system is an ideal gas,

$$\frac{C_V}{nR}\frac{dT}{T} = -\frac{dV}{V}, \tag{17.5}$$

which can be integrated for a reversible expansion between the state (V_1, T_1) and (V_2, T_2) to give

$$\frac{C_V}{nR} \ln \frac{T_2}{T_1} = -\ln \frac{V_2}{V_1}. \tag{17.6}$$

This expression is usually presented in the equivalent form $T_1 V_1^c = T_2 V_2^c$ where $c = \frac{C_V}{nR}$. The corresponding relation between pressure and volume is found from the fact that, for an ideal gas

$$\frac{T_2}{T_1} = \frac{p_2 V_2}{p_1 V_1},$$ (17.7)

which leads to

$$p_1 V_1^\gamma = p_2 V_2^\gamma \quad \text{where } \gamma = \frac{c+1}{c} = \frac{5}{3}.$$ (17.8)

for a (monatomic) ideal gas. That is, pV^γ is constant along an adiabat.

Another important thermodynamic cycle, the Carnot cycle, is based on a sequence of reversible isothermal and adiabatic processes between four states, A, B, C and D:

A $\rightarrow$ B: *Reversible, isothermal expansion* from a state (p_A, V_A) to (p_B, V_B) at a constant temperature, T_{hot}. During this step, heat must flow into the system from an external reservoir.

B $\rightarrow$ C: *Reversible, adiabatic expansion* from state (p_B, V_B) to (p_C, V_C). During this step, the system is somehow insulated from its surroundings so that $q = 0$; its temperature therefore falls to T_{cold}.

C $\rightarrow$ D: *Reversible, isothermal compression* from state (p_C, V_C) to (p_D, V_D) at a constant temperature T_{cold}. This time the system is placed in thermal contact with a reservoir which receives heat.

D $\rightarrow$ A: *Reversible, adiabatic compression* from state (p_D, V_D) back to (p_A, V_A). Again, the system is insulated so no heat can flow and its temperature returns to T_{hot}

Adapt the code from the previous Example to illustrate the Carnot Cycle on a labeled plot for states A, B, C and D determined through the following parameters:

- $V_1 = 0.2 \text{ m}^3$;
- $T_{hot} = 800 \text{ K}$;
- Isothermal compression ratio, $r_i = V_B/V_A = 2$;
- Adiabatic compression ratio, $r_a = V_C/V_B = 2$.

Demonstrate that the efficiency, defined as $\eta = w/q_{hot}$ where w is the work done and q_{hot} is the heat put into the system in step A $\rightarrow$ B, is given by

$$\eta = 1 - \frac{T_{cold}}{T_{hot}}.$$

18

Vectors, Matrices and Linear Algebra

18.1 NumPy Arrays as Vectors

18.1.1 Vectors and Vector Operations

A one-dimensional array can be used to store the components of a vector, with each component of the vector being a number in the array. Vector addition, subtraction and multiplication by a scalar use the usual operators, +, - and *:

```
In [x]: a = np.array([1, 0, -0.5])

In [x]: b = np.array([2.5, 1, 2])

In [x]: a + b
Out[x]: array([3.5, 1. , 1.5])

In [x]: a - b
Out[x]: array([-1.5, -1. , -2.5])

In [x]: a * 4         # or 4 * a
Out[x]: array([ 4., 0., -2.])
```

The vector dot (scalar) product, however, is implemented using the .dot() method or with the @ infix operator:

```
In [x]: a @ b
Out[x]: 1.5

In [x]: a.dot(b)         # or np.dot(a, b)
Out[x]: 1.5
```

Multiplying two arrays together with the * operator results in the elementwise product:

```
In [x]: a * b
Out[x]: array([ 2.5, 0. , -1. ])
```

The vector (cross) product requires the `np.cross` function:

```
In [x]: np.cross(a, b)
Out[x]: array([ 0.5 , -3.25,  1.  ])
```

18.1.2 Column Vectors

In mathematical terms, a column vector is the transpose of a row vector, with its entries arranged in the rows of a single column instead of the columns of a row. The transpose of a NumPy array, however, is formed by reversing the order of its axes. Since the vectors above have been represented by one-dimensional arrays, this transpose has no effect:

```
In [x]: v = np.array([1, 2, 3])
In [x]: v.T          # or v.transpose() or np.transpose(v)
Out[x]: array([1, 2, 3])
```

That is, `v` is an array of shape `(3,)` and `v.T` has the same shape.

A column vector in NumPy is represented by a $n \times 1$ two-dimensional array.

```
In [x]: c = np.array([[1], [2], [3]])
In [x]: c
Out[x]:
array([[1],
       [2],
       [3]])
```

Here, `c` has been defined to have the shape `(3,1)`: three rows in one column.

To create a column, vector, therefore, requires the addition of another axis, something that can be achieved with the use of `np.newaxis`:

```
In [x]: v
Out[x]: array([1, 2, 3])

In [x]: c = v[:, np.newaxis]     # insert an axis in the "column" position.
In [x]: c
Out[x]:
array([[1],
       [2],
       [3]])
```

In fact, `np.newaxis` is just the `None` special value, and so `v[:, None]` would work here as well.

The transpose of this column vector (array of shape `(3,1)`) is one way to represent a row vector:

```
In [x]: r = c.T
In [x]: r
Out[x]: array([[1, 2, 3]])
```

That is, an array of shape $(1,3)$.

For algebra involving only vectors, row vectors represented as one-dimensional arrays work fine. However, for manipulations involving both vectors and matrices, both column and row vectors may be used, as shown in the following section.

18.2 NumPy Arrays as Matrices

18.2.1 Defining Matrices

An obvious use for a two-dimensional NumPy array is to represent a matrix, and there are several functions available to manipulate arrays in this way. First, as we have seen, a matrix can be defined simply from a nested sequence of lists or tuples:

```
In [x]: M = np.array([[0, 2, 3],
                      [4, 1, 9],
                      [5, 8, 6]
                     ])
```

A one-dimensional array can be reshaped into a two-dimensional matrix:

```
In [x]: M = np.array([1, 2, 3, 10, 20, 30]).reshape((2, 3))
In [x]: M
Out[x]:
array([[ 1,  2,  3],
       [10, 20, 30]])
```

Note that NumPy stores arrays in *row-major* order: When reshaping, consecutive elements of the row are taken from adjacent locations in the original, one-dimensional array. Fortran users might be more used to their arrays being stored in *column-major* order, which can be enforced with the argument `order='F'`:

```
In [x]: M = np.array([1, 2, 3, 10, 20, 30]).reshape((2, 3), order='F')
In [x]: M
Out[x]:
array([[ 1,  3, 20],
       [ 2, 10, 30]])
```

This is not the same thing as the array *transpose* which swaps the order of the axes themselves:

```
In [x]: A = np.array([1, 2, 3, 4, 5, 6]).reshape((2,3))

In [x]: A
Out[x]:
array([[1, 2, 3],
       [4, 5, 6]])

In [x]: A.T
Out[x]:
```

```
array([[1,  4],
       [2,  5],
       [3,  6]])
```

The *identity matrix* of a given size is returned by the `np.eye` function:

```
In [x]: np.eye(3)
Out[x]:
array([[1.,  0.,  0.],
       [0.,  1.,  0.],
       [0.,  0.,  1.]])
```

18.2.2 Matrix Multiplication

Matrix multiplication is achieved with the `@` infix operator or with `np.dot`:

```
In [x]: A = np.arange(4).reshape((2,2)) + 1

In [x]: A
Out[x]:
array([[1,  2],
       [3,  4]])

In [x]: B = np.array([[0,  1], [1,  0]])
Out[x]:
array([[0,  1],
       [1,  0]])

In [x]: A @ B        # or A.dot(B)
Out[x]:
array([[2,  1],
       [4,  3]])

In [x]: B @ A        # or B.dot(A)
Out[x]:
array([[3,  4],
       [1,  2]])
```

Again, the `*` operator multiplies the matrices in an elementwise fashion:

```
In [x]: A * B
Out[x]:
array([[0,  2],
       [3,  0]])
```

18.2.3 Matrix and Vector Multiplication

A matrix may be multiplied on the left or the right by either a column or a row vector, providing its dimensions are compatible. In practical terms, this means that an array of shape (n, m), representing an $n \times m$ matrix, may be multiplied (on the right) by either a one-dimensional array of shape $(m,)$ or a two-dimensional array of shape $(m,\ k)$:

$$\begin{pmatrix} 1 & 2 & 3 \\ 4 & 5 & 6 \end{pmatrix} \begin{pmatrix} 0.5 \\ 0.2 \\ 0.1 \end{pmatrix} = \begin{pmatrix} 1.2 \\ 3.6 \end{pmatrix}$$

```
In [x]: A = np.array([1, 2, 3, 4, 5, 6]).reshape((2,3))

In [x]: A
Out[x]:
array([[1, 2, 3],
       [4, 5, 6]])

In [x]: b = np.array([0.5, 0.2, 0.1])

In [x]: b
Out[x]: array([0.5, 0.2, 0.1])

In [x]: A @ b
Out[x]: array([1.2, 3.6])
```

Mathematically, this shouldn't work if b is considered to be a row vector, but NumPy's flexible rules allow it. The result is another one-dimensional array, [1.2, 3.6], that could be interpreted as a vector.

Multiplication of the matrix on the right by a column vector generates the expected column vector explicitly:

```
In [x]: c = b[:, np.newaxis]
In [x]: c
Out[x]:
array([[0.5],
       [0.2],
       [0.1]])

In [x]: A @ c
Out[x]:
array([[1.2],
       [3.6]])
```

Multiplication of a matrix on the left by a vector works in a similar way:

$$(10 \quad 20) \begin{pmatrix} 1 & 2 & 3 \\ 4 & 5 & 6 \end{pmatrix} = (90 \quad 120 \quad 150)$$

```
In [x]: d = np.array([10, 20])

In [x]: d @ A             # OK:
Out[x]: array([ 90, 120, 150])
```

18.2.4 Broadcasting

Broadcasting describes the rules that NumPy follows to carry out operations when two arrays have different shapes. An operation is only possible on arrays with *compatible* dimensions, where (starting with the last dimension and working backward) two dimensions are compatible if they are *equal* or *one of them is 1*. For example, it is possible to add two 3×2 arrays because both dimensions are the same (the elements are matched up by their positions in the array):

```
In [x]: a = np.array(((0, 4), (1, 4), (-2, 7)))

In [x]: a
Out[x]:
array([[ 0,  4],
       [ 1,  4],
       [-2,  7]])

In [x]: b = np.array(((-1, -1), (0, 2), (-2, 3)))
In [x]: b
Out[x]:
array([[-1, -1],
       [ 0,  2],
       [-2,  3]])

In [x]: a + b        # no problem: both arrays are 3x2
Out[x]:
array([[-1,  3],
       [ 1,  6],
       [-4, 10]])
```

It is also possible to add a 3×2 array to a one-dimensional array of length 2: In this case, the operation is carried out over the two columns *for each of the 3 rows*: it is broadcast along the remaining dimension:

```
In [x]: c = np.array((5, 4))
In [x]: a + c        # no problem: (3,2) is compatible with (2,)
Out[x]:
array([[ 5,  8],
       [ 6,  8],
       [ 3, 11]])
```

However, a one-dimensional array of length 3 is not compatible with a 3×2 array since their last dimensions are different, and so broadcasting is not possible:

```
In [x]: d = np.array((0, 6, -2))
In [x]: a + d        # won't work: (3,2) is not compatible with (3,)
------------------------------------------------------------------------
ValueError                            Traceback (most recent call last)
Input In [x], in <module>
----> 1 a + d

ValueError: operands could not be broadcast together with shapes (3,2) (3,)
```

To broadcast this operation over the columns, we need to make the last dimension of d compatible by introducing another axis with dimension 1:

```
In [x]: d[:, np.newaxis].shape
Out[x]: (3, 1)

In [x]: a + d[:, np.newaxis]     # OK: (3,2) is compatible with (3,1)
Out[x]:
array([[ 0,  4],
       [ 7, 10],
       [-4,  5]])
```

In this way, the three elements of the array d have been added to each column of the 3×2 array, a.

18.3 Linear Algebra

Many useful linear algebra functions are available in NumPy's linalg submodule. This is already imported when NumPy itself is imported; for example, if NumPy is imported with import numpy as np, the submodule's functions have the prefix np.linalg.

For example, the magnitude of a vector represented by a NumPy array is returned by the np.linalg.norm function:

```
In [x]: a = np.array((3, 4))     # e.g. the vector 3i + 4j
In [x]: np.linalg.norm(a)        # sqrt(3^2 + 4^2)
Out[x]: 5.0
```

18.3.1 Matrix Powers, Determinant and Inverse

Matrix Powers

The Python exponentiation operator, ** acts *elementwise* on a NumPy array:

```
In [x]: A = np.array([[1, 2], [3, 4]])

In [x]: A
Out[x]:
array([[1, 2],
       [3, 4]])

In [x]: A**3
Out[x]:
array([[ 1,  8],
       [27, 64]])
```

To raise a (square) matrix to some integer power, n (i.e, repeat the operation of matrix multiplication n times), use np.linalg.matrix_power:

```
In [x]: np.linalg.matrix_power(A, 3)      # i.e. A @ A @ A
Out[x]:
array([[ 37,   54],
       [ 81,  118]])
```

If *n* is negative, the matrix inverse of the array is calculated and raised to the power
$|n|$. This ensures that, for example, $\mathbf{A}^{-n}\mathbf{A}^n = \mathbf{I}$, the identity matrix:

```
In [x]: np.linalg.matrix_power(A, -3) @ np.linalg.matrix_power(A, 3)
Out[x]:
array([[ 1.00000000e+00,  -6.92779167e-14],
       [ 4.52970994e-14,   1.00000000e+00]])
```

(this is the identity matrix, to within numerical round-off).

If the provided matrix is not square (or not invertible in the case of negative *n*),
a `LinAlgError` exception is raised.

Determinants and the Matrix Inverse

The determinant of a square matrix is returned by the function `np.linalg.det`:

```
In [x]: A = np.array([[ 2,   0,  -1],
                      [-2,   0,   2],
                      [ 1,  -1,   0]])

In [x]: np.linalg.det(A)
Out[x]: 2.0000000000000004
```

The inverse, $\mathbf{A}^{-1}$, of a square matrix, $\mathbf{A}$, is the matrix for which $\mathbf{A}^{-1}\mathbf{A} = \mathbf{A}\mathbf{A}^{-1} = \mathbf{I}$,
the identity matrix. If it exists, the matrix inverse is returned by the function
`np.linalg.inv`:

```
In [x]: Ainv = np.linalg.inv(A)

In [x]: Ainv
Out[x]:
array([[ 1. ,   0.5,   0. ],
       [ 1. ,   0.5,  -1. ],
       [ 1. ,   1. ,   0. ]])

In [x]: Ainv  @ A              # also A @ Ainv
Out[x]:
array([[1., 0., 0.],
       [0., 1., 0.],
       [0., 0., 1.]])
```

A singular matrix, which has no inverse, will give rise to a `LinAlgError` exception:

```
In [x]: B = np.array([[1, 2,   3],
   ...:               [2, 4,   6],
```

```
   ...:                        [0, 7, -1]
   ...:                    ])

In [x]: np.linalg.inv(B)
------------------------------------------------------------------------
LinAlgError                            Traceback (most recent call last)
...
LinAlgError: Singular matrix

In [x]: np.linalg.det(B)
Out[x]: 0.0
```

Here, the matrix B has two linearly dependent rows (the second is twice the first), so its determinant is zero and no inverse exists.

In general, calculating the matrix inverse has the potential to be numerically unstable and slow. When combined with the inherent finite precision of floating point numbers, it is almost always better to look for another way to solve a problem that seems to require the matrix inverse. The common case of solving a set of simultaneous equations is covered in Section 18.3.3, below. The least squares solution ("best fit") to an overdetermined problem is covered in Chapter 19.

18.3.2 Eigenvalues and Eigenvectors

The eigenvalues and eigenvectors of a matrix have many applications in chemistry, including the analysis of complex reactions, determining the normal vibrational modes of molecules and, of course, quantum mechanics.

In NumPy, the eigenvalues and (right) eigenvectors of a square matrix are returned by the function np.linalg.eig. The eigenvalues are not returned in any particular order, but the eigenvalue v[i] corresponds to the eigenvector P[:, i]: That is, the eigenvectors are arranged in the *columns* of a two-dimensional array, corresponding to the order of the eigenvalues. If, for some reason, the eigenvalue calculation does not converge, a LinAlgError is raised.

For example,

```
In [x]: A = np.array([[1, 2],
                       [3, 0]])

In [x]: v, P = np.linalg.eig(A)

In [x]: v                     # eigenvalues first
Out[x]: array([ 3., -2.])

In [x]: P                     # read the eigenvectors from the columns of P
Out[x]:
array([[ 0.70710678, -0.5547002 ],
       [ 0.70710678,  0.83205029]])
```

In this case both eigenvalues are real. The first, $\lambda_1 = 3$ is associated with the normalized eigenvector $(\frac{1}{\sqrt{2}}, \frac{1}{\sqrt{2}})$, and the second is $\lambda_2 = -2$ with an eigenvector that turns out to be $(-\frac{2}{\sqrt{13}}, \frac{3}{\sqrt{13}})$.

The matrix $\mathbf{P}$, consisting of columns of the eigenvectors of $\mathbf{A}$ diagonalizes $\mathbf{A}$ through the similarity transformation $\mathbf{D} = \mathbf{P}^{-1}\mathbf{A}\mathbf{P}$, where $\mathbf{D}$ is the matrix with the eigenvalues of $\mathbf{A}$ along its diagonal:

```
In [x]: Pinv = np.linalg.inv(P)

In [x]: Pinv @ A @ P          # this matrix should be diagonal
Out[x]:
array([[ 3.00000000e+00,  -6.66133815e-16],
       [ 0.00000000e+00,  -2.00000000e+00]])
```

It is also true that $\mathbf{A} = \mathbf{P}\mathbf{D}\mathbf{P}^{-1}$: The matrix $\mathbf{A}$ can be reconstructed from its eigenvalues and eigenvectors:

```
In [x]: D = np.eye(2) * v      # generate D directly from the eigenvalues

In [x]: P @ D @ Pinv           # this matrix should be equal to A
Out[x]:
array([[1.00000000e+00, 2.00000000e+00],
       [3.00000000e+00, 2.15735746e-16]])
```

The function `np.linalg.eigvals` returns only the eigenvalues and not the eigenvectors, and may be faster and use less storage under some circumstances.

If your matrix happens to be *Hermitian* (equal to its own conjugate transpose, $\mathbf{A} = \mathbf{A}^{T*}$, so that its entries satisfy $a_{ij} = a_{ji}^*$ which is always true for a real symmetric matrix), then the functions `np.linalg.eigh` and `np.linalg.eigvalsh` can be used and then the calculation is faster since only the upper or lower triangular part of the matrix is needed. Which part is used is determined by the argument UPLO, which takes the values `'L'` (the default) or `'U'`. Hermitian matrices have real eigenvalues.

For example, the following matrix is Hermitian:

$$\begin{pmatrix} -1 & i & 2-i \\ -i & 0 & -1+i \\ 2+j & -1-j & 4 \end{pmatrix}$$

The eigenvalues can be found from the full matrix with `np.linalg.eigvals`:

```
In [x]: A = np.array([[-1    ,    1j, 2-1j],
   ...:               [  -1j, 0    , -1+1j],
   ...:               [ 2+1j, -1-1j,  4]
   ...:              ])

In [x]: np.linalg.eigvals(A)
Out[x]:
```

```
array([ 5.17102979-4.35518074e-16j,  -2.48261292+8.05588614e-18j,
        0.31158313+2.05417583e-16j])
```

(The very small complex parts are due to numerical round-off errors.)

In this case, it is faster and more accurate to use the function, `np.linalg.eigvalsh`, which only requires the upper or lower triangular part of the matrix:

```
In [x]: A = np.array([[-1,    1j,  2-1j],
   ...:               [ 0,  0   , -1+1j],
   ...:               [ 0,  0   ,  4]
   ...:              ])

In [x]: v = np.linalg.eigvalsh(A, UPLO='U')
In [x]: v
Out[x]: array([-2.48261292,  0.31158313,  5.17102979])
```

The *trace* of a matrix (i.e., the sum of its diagonal elements) can be calculated with `np.trace`) and is equal to the sum of its eigenvalues:

```
In [x]: np.sum(v)
Out[x]: 2.9999999999999996

In [x]: np.trace(A)
Out[x]: (3+0j)
```

18.3.3 Solving a Set of Linear Equations

A set of linear equations is a collection of equations that can be written in the form

$$
\begin{aligned}
a_{11}x_1 + a_{12}x_2 + \cdots a_{1n}x_n &= b_1 \\
a_{21}x_1 + a_{22}x_2 + \cdots a_{2n}x_n &= b_2 \\
&\vdots \\
a_{m1}x_1 + a_{m2}x_2 + \cdots a_{mn}x_n &= b_m,
\end{aligned}
$$

where x_j are the n unknown variables, a_{ij} are their coefficients in each of the m equations, and b_i are constants. This system of equations can be written compactly in matrix form as $\mathbf{Ax} = \mathbf{b}$:

$$
\begin{pmatrix}
a_{11} & a_{12} & \cdots & a_{1n} \\
a_{21} & a_{22} & \cdots & a_{2n} \\
\vdots & & \ddots & \vdots \\
a_{m1} & a_{m2} & \cdots & a_{mn}
\end{pmatrix}
\begin{pmatrix}
x_1 \\ x_2 \\ \vdots \\ x_n
\end{pmatrix}
=
\begin{pmatrix}
b_1 \\ b_2 \\ \vdots \\ b_m
\end{pmatrix}.
$$

If $n = m$ the matrix $\mathbf{A}$ is square and there are the same number of equations as unknowns. If the system of equations has a solution, the unknown variables x_j can be found using `np.linalg.solve` (this is more reliable than multiplying the matrix equation by $\mathbf{A}^{-1}$).

For example,

$$4x - y + 2z = 33$$
$$2y + 3z = 4$$
$$-2x + 3y - z = -19$$

can be written in matrix form as

$$\begin{pmatrix} 4 & -1 & 2 \\ 0 & 2 & 3 \\ -2 & 3 & -1 \end{pmatrix} \begin{pmatrix} x \\ y \\ z \end{pmatrix} = \begin{pmatrix} 33 \\ 4 \\ -19 \end{pmatrix}.$$

The solution can be found as follows:

```
In [x]: A = np.array([[ 4, -1,  2],
   ...:               [ 0,  2,  3],
   ...:               [-2,  3, -1]
   ...:              ])

In [x]: b = np.array([33, 4, -19])

In [x]: np.linalg.solve(A, b)
Out[x]: array([ 7., -1.,  2.])
```

That is, $x = 7$, $y = -1$ and $z = 2$.

If the array `A` is not square or is singular (implying that two or more rows are linearly dependent so there is not enough information in the system to determine a unique solution), then a `LinAlgError` is raised. In the common case for which there are more equations than unknowns ($m > n$), the system of equations is called *overdetermined* and (almost always) has no solution (values of x_j that satisfy all the equations); in this case, one can find an approximate solution using the method of least squares (see Chapter 19).

18.4 Examples

E18.1

Balancing a Combustion Reaction

As an example of balancing a chemical reaction, take the combustion of octane, C_8H_{18}. We wish to find the value of the coefficients a, b, c, and d in the following equation:

$$aC_8H_{18} + bO_2 \longrightarrow cCO_2 + dH_2O.$$

The stoichiometric constraints can be written as a sequence of three equations, one for each of the atoms C, H and O:

$$8a - c = 0$$
$$18a - 2d = 0$$
$$2b - 2c - d = 0$$

In matrix form, these equations take the form $\mathbf{Mc} = \mathbf{0}$:

$$\begin{pmatrix} 8 & 0 & -1 & 0 \\ 18 & 0 & 0 & -2 \\ 0 & 2 & -2 & -1 \end{pmatrix} \begin{pmatrix} a \\ b \\ c \\ d \end{pmatrix} = 0,$$

where $\mathbf{M}$ is the chemical composition matrix.

This system is underdetermined (there are four coefficients to find, but only three equations). However, since we can clearly multiply every coefficient by the same constant and keep the equation balanced, we can choose to fix one of them to a value we choose, say $a = 1$. Then the system to be solved can be written $\mathbf{Ac} = \mathbf{x}$:

$$\begin{pmatrix} 8 & 0 & -1 & 0 \\ 18 & 0 & 0 & -2 \\ 0 & 2 & -2 & -1 \\ 1 & 0 & 0 & 0 \end{pmatrix} \begin{pmatrix} a \\ b \\ c \\ d \end{pmatrix} = \begin{pmatrix} 0 \\ 0 \\ 0 \\ 1 \end{pmatrix},$$

where $\mathbf{M}$ has been augmented to form $\mathbf{A}$ by the addition of the final row corresponding to the constraint $a = 1$. This system of equations can be solved uniquely as follows.

```
import numpy as np
A = np.array([[ 8, 0, -1,  0],    # C
              [18, 0,  0, -2],    # H
              [ 0, 2, -2, -1],    # O
              [ 1, 0,  0,  0]     # constraint: a = 1
             ])
x = np.array([0, 0, 0, 1])
```

```
coeffs = np.linalg.solve(A, x)
print('a = {}, b = {}, c = {}, d = {}'.format(*coeffs))
```

```
a = 1.0, b = 12.5, c = 8.0, d = 9.0
```

That is,

$$C_8H_{18} + \tfrac{25}{2}O_2 \longrightarrow 8CO_2 + 9H_2O.$$

Or, equivalently,

$$2C_8H_{18} + 25O_2 \longrightarrow 16CO_2 + 18H_2O.$$

Balancing a Redox Reaction

Iron(II) is oxidized to iron(III) by hydrogen peroxide in the presence of acid. The species involved in this redox reaction are therefore Fe^{2+}, H_2O_2, H^+, Fe^{3+} and H_2O:

$$a Fe^{2+} + b H_2O_2 + c H^+ \longrightarrow d Fe^{3+} + e H_2O.$$

In addition to balancing the atom stoichiometries, we must also conserve the charge, which leads to the following four equations:

$$
\begin{aligned}
\text{Fe}: & \quad a = d \\
\text{H}: & \quad 2b + c = 2e \\
\text{O}: & \quad 2b = e \\
\text{charge}: & \quad 2a + c = 3d.
\end{aligned}
$$

Again, the problem is underdetermined but we can set the constraint $a = 1$ since we know we can scale all the coefficients by the same amount and keep the reaction balanced. Therefore, in matrix form:[1]

$$
\begin{pmatrix}
-1 & 0 & 0 & 1 & 0 \\
0 & -2 & -1 & 0 & 2 \\
0 & -2 & 0 & 0 & 1 \\
-2 & 0 & -1 & 3 & 0 \\
1 & 0 & 0 & 0 & 0
\end{pmatrix}
\begin{pmatrix}
a \\ b \\ c \\ d \\ e
\end{pmatrix}
=
\begin{pmatrix}
0 \\ 0 \\ 0 \\ 0 \\ 1
\end{pmatrix}
$$

```
A = np.array([[-1,  0,  0, 1, 0],    # Fe
              [ 0, -2, -1, 0, 2],    # H
              [ 0, -2,  0, 0, 1],    # O
              [-2,  0, -1, 3, 0],    # charge
              [ 1,  0,  0, 0, 0]])   # constraint: a = 1

x = np.array([0, 0, 0, 0, 1])
coeffs = np.linalg.solve(A, x)
print('a, b, c, d, e =', coeffs)
```

```
a, b, c, d, e = [1.   0.5 1.   1.   1. ]
```

---

[1] Note that if we want the solved coefficients, a, b, etc. to all have the same sign, we can set up the matrix with reactant and product stoichiometric numbers to have opposite signs. It doesn't matter whether the reactants' entries in this matrix are negative and the products' entries are positive or vice versa: The balancing algorithm has no sense of the direction of the reaction.

Balancing the equation with integer coefficients then gives

$$2Fe^{2+} + H_2O_2 + 2H^+ \longrightarrow 2Fe^{3+} + 2H_2O.$$

Note: There are some reactions that cannot be balanced. It is also possible for a reaction to be balanced in more than one way; if this is the case, then there are an infinite number of ways to balance the reaction since any linear combination of two validly balanced equations is also balanced.

To determine if a reaction equation can be balanced uniquely, one can examine the *nullity* of the chemical composition matrix. The nullity of a matrix is the dimension of its so-called *null space*: in this case, the difference between the number of chemical species (columns), n, and the rank of the chemical composition matrix (the number of linearly-independent columns), r. The rank of a matrix is returned by the method np.linalg.matrix_rank. So, in the combustion reaction above:

```
M = np.array([[ 8, 0, -1,  0],
              [18, 0,  0, -2],
              [ 0, 2, -2, -1]])
n = M.shape[1]                          # Number of columns of M
r = np.linalg.matrix_rank(M)
print(f'n = {n}, r = {r}, nullity = {n-r}')
```

```
n = 4, r = 3, nullity = 1
```

We can expect to balance the reaction uniquely (to within a global scaling factor, as discussed above). However, consider the following reaction:

$$(NH_4)_2SO_4 \longrightarrow NH_4OH + SO_2 \quad \text{[unbalanced]}.$$

```
M = np.array([[-2, 1, 0], [-8, 5, 0], [-1, 0, 1], [-4, 1, 2]])
n = M.shape[1]
r = np.linalg.matrix_rank(M)
print(f'n = {n}, r = {r}, nullity = {n-r}')
```

```
n = 3, r = 3, nullity = 0
```

The conclusion is that the set of simultaneous equations defining the chemical balance has no solution.

Conversely, the reaction

$$CO + CO_2 + H_2 \longrightarrow CH_4 + H_2O$$

has the following chemical composition matrix:

$$\begin{pmatrix} -1 & -1 & 0 & 1 & 0 \\ -1 & -2 & 0 & 0 & 1 \\ 0 & 0 & -2 & 4 & 2 \end{pmatrix}$$

```
M = np.array([[-1, -1, 0, 1, 0], [-1, -2, 0, 0, 1], [0, 0, -2, 4, 2]])
n = M.shape[1]
r = np.linalg.matrix_rank(M)
print(f'n = {n}, r = {r}, nullity = {n-r}')
```

```
n = 5, r = 3, nullity = 2
```

In this case, there is no unique way to balance the equation: Even after fixing a particular stoichiometric coefficient, say that of CO, the problem is still underdetermined. In particular, the reactions

$$CO_2 + H_2 \longrightarrow CO + H_2O$$

and

$$CO_2 + CH_4 \longrightarrow 2\,CO + 2\,H_2$$

together involve all the chemical species present, are linearly independent, and are individually balanced. Therefore, any linear combination of them is also a balanced chemical reaction, for example, 3 times the first equation minus 2 times the second:

$$CO + CO_2 + 7\,H_2 \longrightarrow 2\,CH_4 + 3\,H_2O.$$

Alternatively, 4 times the first minus 3 times the second gives a completely different balancing (i.e., one that is not obtained simply by multiplying the stoichiometries of the previous one by a constant):

$$2\,CO + CO_2 + 10\,H_2 \longrightarrow 3\,CH_4 + 4\,H_2O.$$

There are an infinite number of such balancings, and choosing the relevant one(s) requires knowledge of the mechanism involved.

E18.2

The Moment of Inertia of a Molecule

The moment of inertia of a molecule, considered as a rigid collection of point masses, m_i, can be represented by the tensor:

$$\mathbf{I} = \begin{pmatrix} I_{xx} & I_{xy} & I_{xz} \\ I_{xy} & I_{yy} & I_{yz} \\ I_{xz} & I_{yz} & I_{zz} \end{pmatrix},$$

where

$$I_{xx} = \sum_i m_i(y_i^2 + z_i^2), \qquad I_{yy} = \sum_i m_i(x_i^2 + z_i^2), \qquad I_{zz} = \sum_i m_i(x_i^2 + y_i^2),$$

$$I_{xy} = -\sum_i m_i x_i y_i, \qquad I_{yz} = -\sum_i m_i y_i z_i, \qquad I_{xz} = -\sum_i m_i x_i z_i.$$

The atomic positions, (x_i, y_i, z_i), are measured relative to the molecular center of mass in an arbitrary coordinate frame.

There exists a transformation of the coordinate frame such that this matrix is diagonal: the axes of this transformed frame are called the *principal* axes and the diagonal inertia matrix elements, $I_a \leq I_b \leq I_c$, are the *principal moments of inertia*. It is with respect to these axes that the molecular moments of inertia are usually reported.

A molecule may be classified as follows according to the relative values of I_a, I_b and I_c:

- $I_a = I_b = I_c$: spherical top;
- $I_a = I_b < I_c$: oblate symmetric top;
- $I_a < I_b = I_c$: prolate symmetric top;
- $I_a < I_b < I_c$: asymmetric top.

Furthermore, in spectroscopy, it is conventional to define the *rotational constants*,

$$A = \frac{h}{8\pi^2 c I_a}, \quad B = \frac{h}{8\pi^2 c I_b}, \quad C = \frac{h}{8\pi^2 c I_c},$$

which are reported in wavenumber units (cm^{-1}).

Question

The file CHC13.dat contains the positions of the atoms in the molecule $CHCl_3$ in XYZ format.

```
# CHC13
# Columns are: atom mass (atomic units), x, y, z (Angstroms)
12          0.0000      0.0000      0.4563      # C
1.0079      0.0000      0.0000      1.5353      # H
34.9689     0.0000      1.6845     -0.0838      # Cl
34.9689     1.4588     -0.8423     -0.0838      # Cl
34.9689    -1.4588     -0.8423     -0.0838      # Cl
```

Determine the rotational constants for this molecule and classify it as a spherical, oblate, prolate, or asymmetric top.

Repeat this exercise for hydrogen peroxide using the file H2O2.dat:

```
# H2O2
# Columns are: atom mass (atomic units), x, y, z (Angstroms)
15.9994     0.0000      0.7375     -0.0528      # O
15.9994     0.0000     -0.7375     -0.0528      # O
1.0079      0.8190      0.8170      0.4220      # H
1.0079     -0.8190     -0.8170      0.4220      # H
```

Solution

The four columns of the provided data file are mass (in Da) and (x, y, z) coordinates (in Å).

```
import numpy as np
from scipy.constants import u, h, c
m, x, y, z = np.genfromtxt('CHCl3.dat', unpack=True)
```

To ensure the atomic coordinates are stored relative to the molecular center of mass, we must shift their origin to this position. In the provided coordinates, the center of mass is at

$$r_{CM} = \frac{1}{M} \sum_i m_i r_i, \quad \text{where } M = \sum_i m_i.$$

```
def translate_to_cofm(m, x, y, z):
    """Translate the atom positions to be relative to the CofM."""

    # Total molecular mass.
    M = np.sum(m)
    # Position of centre of mass in original coordinates.
    xCM = np.sum(m * x) / M
    yCM = np.sum(m * y) / M
    zCM = np.sum(m * z) / M
    # Transform to CofM coordinates and return them.
    return x - xCM, y - yCM, z - zCM
```

This function is used in the code below to construct and diagonalize the moment of inertia matrix.

```
def get_inertia_matrix(m, x, y, z):
    """Return the moment of inertia tensor."""

    x, y, z = translate_to_cofm(m, x, y, z)
    Ixx = np.sum(m * (y**2 + z**2))
    Iyy = np.sum(m * (x**2 + z**2))
    Izz = np.sum(m * (x**2 + y**2))
    Ixy = -np.sum(m * x * y)
    Iyz = -np.sum(m * y * z)
    Ixz = -np.sum(m * x * z)
    I = np.array([[Ixx, Ixy, Ixz],
                  [Ixy, Iyy, Iyz],
                  [Ixz, Iyz, Izz]
                 ])
    return I

def get_principal_moi(I):
    """Determine the principal moments of inertia."""

    # The principal moments of inertia are the eigenvalues of the moment
    # of inertia tensor.
    Ip = np.linalg.eigvals(I)
```

```
    # Sort and convert principal moments of inertia to kg.m2 before returning.
    Ip.sort()
    return Ip * u / 1.e20

def get_rotational_constants(filename):
    """Return the rotational constants, A, B and C (in cm-1) for a molecule.

    The atomic coordinates are retrieved from filename which should have
    four columns of data: mass (in Da), and x, y, z coordinates (in Angstroms).

    """

    m, x, y, z = np.genfromtxt(filename, unpack=True)
    I = get_inertia_matrix(m, x, y, z)
    Ip = get_principal_moi(I)
    A, B, C = h / 8 / np.pi**2 / c / 100 / Ip
    return A, B, C
```

```
A, B, C = get_rotational_constants('CHCl3.dat')
print(f'CHCl3: A = {A:.3f} cm-1, B = {B:.3f} cm-1, C = {C:.3f} cm-1.')
```

```
CHCl3: A = 0.109 cm-1, B = 0.109 cm-1, C = 0.057 cm-1.
```

Since we have $A = B > C$, it must be that $I_a = I_b < I_c$, and $CHCl_3$ is an *oblate* symmetric top.

For hydrogen peroxide:

```
A, B, C = get_rotational_constants('H2O2.dat')
print(f'H2O2: A = {A:.3f} cm-1, B = {B:.3f} cm-1, C = {C:.3f} cm-1.')
```

```
H2O2: A = 10.060 cm-1, B = 0.874 cm-1, C = 0.839 cm-1.
```

For this species, $A > B > C$ so $I_a < I_b < I_c$ and H_2O_2 is an *asymmetric* top. However, $B \approx C$ so it is close to a *prolate* symmetric top.

E18.3

Constructing the Cayley Table for a Permutation Group

The distinct rearrangements of three objects, [1,2,3], can be described by six distinct permutations, written in cycle notation as

$$e, (12), (13), (23), (123), (132).$$

Here, e represents the identity permutation (which leaves the objects undisturbed) and, for example, (12) swaps objects 1 and 2; (123) is the cyclic permutation ($1 \rightarrow 2, 2 \rightarrow 3, 3 \rightarrow 1$).

If the objects are labeled within a column array as above,

$$\begin{pmatrix} 1 \\ 2 \\ 3 \end{pmatrix},$$

then these permutations can be represented as 3×3 matrices, for example:

$$\begin{pmatrix} 0 & 1 & 0 \\ 0 & 0 & 1 \\ 1 & 0 & 0 \end{pmatrix} \cdot \begin{pmatrix} 1 \\ 2 \\ 3 \end{pmatrix} = \begin{pmatrix} 2 \\ 3 \\ 1 \end{pmatrix}$$

```
import numpy as np
s = [None] * 6
names = ['e', '(12)', '(13)', '(23)', '(123)', '(132)']
s[0] = np.eye(3, dtype=int)   # e
s[1] = np.array([[0, 1, 0], [1, 0, 0], [0, 0, 1]], dtype=int) # (12)
s[2] = np.array([[0, 0, 1], [0, 1, 0], [1, 0, 0]], dtype=int) # (13)
s[3] = np.array([[1, 0, 0], [0, 0, 1], [0, 1, 0]], dtype=int) # (23)
s[4] = np.array([[0, 1, 0], [0, 0, 1], [1, 0, 0]], dtype=int) # (123)
s[5] = np.array([[0, 0, 1], [1, 0, 0], [0, 1, 0]], dtype=int) # (132)
```

The effect of each of the $3! = 6$ permutations is demonstrated below:

```
A = np.array([[1, 2, 3]]).T
n = len(s)
for i in range(n):
    print(f'{names[i]:>5s} . {A.T} = {(s[i] @ A).T}')
```

```
    e . [[1 2 3]] = [[1 2 3]]
 (12) . [[1 2 3]] = [[2 1 3]]
 (13) . [[1 2 3]] = [[3 2 1]]
 (23) . [[1 2 3]] = [[1 3 2]]
(123) . [[1 2 3]] = [[2 3 1]]
(132) . [[1 2 3]] = [[3 1 2]]
```

We will demonstrate below that the permutation matrices form a *group* under multiplication by considering all of their possible products in a Cayley table.

This requires us to calculate a matrix $m = a \cdot b$ for each pair of matrices, a, b and find the index, k, of m in the list s.

```
def locate_matrix(m):
    """Return the index of matrix m in the list s."""
    for k, p in enumerate(s):
        if np.all(p==m):
            return k
    # NB if we're feeling smart, we could just use
    # return np.where((m==s).all(axis=(1,2)))[0][0]

# Make a Cayley table for the symmetric group, S3.
print('   S3   |' + ('{:^7s}'*n).format(*[names[j] for j in range(n)]))
```

```
print('-'*8 + '+' + '-'*(7*n))
for i in range(n):
    print(f'{names[i]:>7s} |', end='')
    for j in range(n):
        m = s[i] @ s[j]
        k = locate_matrix(m)
        print(f'{names[k]:^7s}', end='')
    print()
```

```
S3   |    e     (12)    (13)    (23)   (123)   (132)
-----+-------------------------------------------------
   e |    e     (12)    (13)    (23)   (123)   (132)
(12) |  (12)     e     (123)   (132)   (13)    (23)
(13) |  (13)   (132)     e     (123)   (23)    (12)
(23) |  (23)   (123)   (132)     e     (12)    (13)
(123)| (123)   (23)    (12)    (13)   (132)     e
(132)| (132)   (13)    (23)    (12)     e     (123)
```

From the above Cayley table, it is clear that the six matrices defined in s do indeed form a group under multiplication:

- Associativity: $(ab)c = a(bc)$ is true because matrix multiplication is associative.
- Closure: For all a, b in the group, $ab = c$ is also in the group.
- Identity element: There is an element e such that $ea = ae = a$ for all a.
- Inverse element: For each a, there is a unique element, b, such that $ab = ba = e$.

The S_3 group is non-Abelian because it is not true that $ab = ba$ for all a, b; in fact, S_3 is the smallest possible non-abelian group.

Note: The S_3 group is isomorphic to the dihedral group of order 6, the group of reflection and rotation symmetries of an equilateral triangle, and hence also the C_{3v} point group. This is the point group to which NH_3 (considered as a rigid molecule) belongs: The reflections permute pairs of hydrogen atoms, the $\pm 120°$ rotations cyclicly permute all three hydrogen atoms clockwise and anticlockwise.

18.5 Exercises

P18.1

(a) Balancing a redox reaction

Balance the equation for the reaction of permanganate and iodide ions in basic solution:

$$MnO_4^-(aq) + I^-(aq) \longrightarrow MnO_2(s) + I_2(aq) \quad \text{[unbalanced]}$$

Hint: You will need to add OH^- ions and H_2O molecules to the reaction in stoichiometric amounts to be determined.

(b) Balancing a complex reaction

Balance the following equation [R. J. Stout, *J. Chem. Educ.* **72**, 1125 (1995)]:

$$Cr_7N_{66}H_{96}C_{42}O_{24} + MnO_4^- + H^+ \longrightarrow Cr_2O_7^{2-} + Mn^{2+} + CO_2 + NO_3^- + H_2O$$
[unbalanced]

(c) The reaction between copper and nitric acid

Copper metal may be thought of as reacting with nitric acid according to the following reaction:

$$aCu(s) + bHNO_3(aq) \longrightarrow cCu(NO_3)_2(aq) + dNO(g) + eNO_2(g) + fH_2O(l)$$

Show that this reaction cannot be balanced uniquely.

Now, when the reaction is carried out with concentrated nitric acid the favored gaseous product is nitrogen dioxide instead of nitric oxide ($d = 0$); conversely, in dilute nitric acid nitric oxide is produced instead of NO_2 ($e = 0$). Write balanced equations for these two cases.

P18.2 With reference to Example E18.3:

(a) Demonstrate that the inverse of each 3×3 permutation matrix is its transpose (i.e., the permutation matrices are *orthogonal*).
(b) Show that there is some power of each permutation matrix that is the identity matrix.
(c) The *parity* of a permutation is -1 or $+1$ according to whether the number of distinct pairwise swaps it can be broken down into is odd or even, respectively. Show that the parity of each permutation is equal to the determinant of its corresponding permutation matrix.

19

Linear Least Squares Fitting I

19.1 Background

In scientific research, it is common to analyze data with a model function that depends linearly on a set of parameters. In the simplest case, this involves fitting a "line of best fit" through the data. That is, finding the intercept, a, and gradient b to the linear model $y = a + bx$, where x is the independent variable (e.g., time or temperature) and y is the model output to be compared with a measured value. The goal of an experimental test of this model is then to measure values y_i for different values of x_i in order to determine the parameters a and b (which presumably contain some physical significance). For example consider the Arrhenius equation in chemical kinetics:

$$k = A \exp\left(-\frac{E_a}{RT}\right),$$

(19.1)

We can conduct experiments to determine the rate constant, k, for different temperatures, T, and want to retrieve the values of the physically meaningful parameters A and E_a. If the experiments were perfect, with no measurement error or uncertainties, and if the Arrhenius equation holds exactly, then we could simply measure k at two different values of T and write a pair of simultaneous equations:

$$k_1 = A \exp\left(-\frac{E_a}{RT_1}\right) \quad \text{and} \quad k_2 = A \exp\left(-\frac{E_a}{RT_2}\right).$$

Solving this pair of equations would then give the unique and precise values of A and E_a directly:

$$E_a = \frac{R \ln(k_1/k_2)}{\frac{1}{T_2} - \frac{1}{T_1}},$$

$$A = k_2 \exp\left(\frac{T_2}{T_2 - T_1}\right).$$

However, in real life, the measurement of k_i (and, indeed, that of the temperatures, T_i, at which they were obtained) is subject to a finite experimental error, and we would expect to get a much better estimate of the parameters A and E_a by making several measurements of k_i at a range of different temperatures. Furthermore, the Arrhenius equation is based on a physical model which breaks down in some circumstances (e.g., where quantum mechanical tunneling is important).[1]

Now we have the problem of deciding which values of the parameters best fit the data; in general, in the presence of noise and error, there will be no choice of A and E_a that is consistent with all the measurements (T_i, k_i): this is an *overdetermined* system. Instead, we seek values for these parameters that lead to the best approximation in the sense that the model predicts values for k that lie "as close as possible, on average" to those measured at the corresponding temperatures.

In practice, it is convenient to algebraically manipulate the model until the data can be plotted on a graph through which a straight line of best fit can be plotted. Frequently (and almost universally in undergraduate chemistry courses), this is done with pen and paper, the best-fit line being judged by eye. In the example of the Arrhenius model for the temperature-dependence of the rate coefficient of a reaction, Equation 19.1 is rearranged as:

$$\ln k = \ln A - \frac{E_a}{RT},$$

and the data points $y_i = \ln k_i$ are plotted against $x_i = \frac{1}{T_i}$; if the model holds, then the points will lie roughly on a straight line (with some scatter); the parameters $\ln A$ and $-E_a/R$ can be read off as the intercept and gradient of this line, respectively.

The problem of determining the parameters that best fit a model which is linear in those parameters is treated more formally by the statistical technique known as *linear regression*. This is also more general than the problem of fitting a straight line through some data: The model itself does not have to be linear, only the parameters that are to be determined for that model. For example, linear regression can be used to fit any polynomial to a set of data points: The model function $y = ax^2 + bx + c$ is linear in its parameters (a, b and c) even though it is obviously not a linear function of x (see also Section 16.1.2). However, the function $y = a\sin(bx + c)$ cannot be fit using linear regression, since it cannot be rearranged into a linear function of its parameters.

Sticking, for now, with the problem of determining the "best" parameters a and b for a straight line model function, $y = a + bx$, what is meant by "best" here? A very common definition that works well in practice is that the best parameters are those that lead to the smallest value of the sum of the squared *residuals* (the differences between the measured and modeled data):

---

[1] We have also treated A as independent of temperature, which, since it is related to the collision rate, it is not.

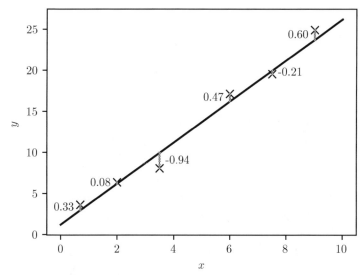

Figure 19.1 An example line of best fit that minimizes the sum of the squared residuals. Each residual (the difference between the measured y_i and the corresponding point on the linear fit, $a + bx_i$) is indicated as a numerical value.

$$\chi^2 = \sum_i [y_i - (a + bx_i)]^2, \tag{19.2}$$

for a set of experimental measurements (x_i, y_i). That is, take the difference between each measured y_i and the value of y predicted by the model at its x_i, square it, and sum over all measurements. The residuals and their relation to a fitted line for some sample data are illustrated in Figure 19.1.

To find the minimum value of χ^2, and hence the parameters that best fit the model function to the data involves solving the so-called *normal equations*, which are derived in the online resources for this book at `https://scipython.com/chem/normal-equations/`.

19.2 Fitting a Line of Best Fit

For the case of a straight-line fit to n data points (x_i, y_i), we have n linear equations in two unknown coefficients (a and b):

$$
\begin{aligned}
y_1 &= a + bx_1, \\
y_2 &= a + bx_2, \\
\vdots\ &=\ \vdots \\
y_n &= a + bx_n.
\end{aligned}
$$

In matrix form, they may be written

$$\mathbf{y} = \mathbf{X}\beta,$$

where

$$\mathbf{y} = \begin{pmatrix} y_1 \\ y_2 \\ \vdots \\ y_n \end{pmatrix}, \quad \mathbf{X} = \begin{pmatrix} 1 & x_1 \\ 1 & x_2 \\ \vdots \\ 1 & x_n \end{pmatrix}, \quad \beta = \begin{pmatrix} a \\ b \end{pmatrix}, \tag{19.3}$$

As mentioned previously, these equations (in general) have no solution, but the particular value of the coefficient vector, $\hat{\beta}$, that minimizes the sum of squared differences between $\mathbf{y}$ and $\mathbf{X}\beta$ is given by the normal equations:

$$(\mathbf{X}^T\mathbf{X})\hat{\beta} = \mathbf{X}^T\mathbf{y}.$$

That is,

$$\hat{\beta} = (\mathbf{X}^T\mathbf{X})^{-1}\mathbf{X}^T\mathbf{y}. \tag{19.4}$$

The "best-fit parameters" are contained in the vector $\hat{\beta}$.

19.3 numpy.linalg.lstsq

The NumPy function that performs linear least-squares fitting is numpy.linalg.lstsq. It takes arrays representing $\mathbf{X}$ and $\mathbf{y}$ and returns the following objects:

- A NumPy array representing the coefficient vector, $\hat{\beta}$ giving the least-squares solution to the normal equations;
- The sum of the squared residuals, χ^2, which can be used as a measure of the goodness-of-fit;
- The rank of the matrix $\mathbf{X}$;
- The singular values of $\mathbf{X}$.

If the calculation of $\hat{\beta}$ does not converge (either because of limited numerical precision or because a unique least-squares solution does not exist due to the nature of $\mathbf{X}$ and $\mathbf{y}$), a LinAlgError exception is raised.

As an example, consider the six data points plotted in Figure 19.1:

```
In [x]: x = np.array([0.7,   2,    3.5,   6,    7.5,   9])
In [x]: y = np.array([3.62, 6.36, 8.07, 17.13, 19.53, 24.91])
```

The matrix $\mathbf{X}$ may be constructed by stacking a sequence of ones on top of the array x and taking the transpose as follows:

```
In [x]: X = np.vstack([np.ones(len(x)), x]).T
In [x]: X

Out[x]:
array([[1. , 0.7],
       [1. , 2. ],
       [1. , 3.5],
       [1. , 6. ],
       [1. , 7.5],
       [1. , 9. ]])
```

The least squares solution is given by:

```
In [x]: np.linalg.lstsq(X, y, rcond=None)
Out[x]:
(array([0.91924427, 2.58203953]),
 array([6.08473075]),
 2,
 array([13.94140915,  1.27558262]))
```

The coefficients $a = 0.9192$ and $b = 2.5820$ are given first, followed by the sum of the squared residuals; the matrix rank of **X** is 2, indicating that its columns are linearly independent of each other; and the singular values of **X** have "reasonable" magnitudes (i.e., are not close to zero, within the floating point precision used for the calculation).[2]

In contrast, consider trying to fit the function $y = ax + b + c(x + 2)$ to the data. Clearly, this should give rise to problems since the basis functions x, 1 and $x + 2$ are not linearly independent and there will be an infinite number of ways to choose the coefficients a, b and c which all fit the data equally well. This can be seen from the values returned by `numpy.linalg.lstsq`:

```
In [x]: X = np.vstack([np.ones(len(x)), x, x + 2]).T
In [x]: np.linalg.lstsq(X, y, rcond=None)

Out[x]:
(array([-0.55426509,  1.84528485,  0.73675468]),
 array([], dtype=float64),
 2,
 array([2.28284677e+01, 1.90815712e+00, 7.67869424e-16]))
```

This should be read as a warning: the matrix **X** has three columns but a rank of 2, and the third singular value is close to zero: it would be better to adapt the basis functions to remove this degeneracy (e.g., eliminate the third column of **X** to fit an unambiguous straight line, or use x^2 in its place to fit a quadratic function).

---

[2] The argument rcond=None suppresses a warning about a change, introduced in version 1.14.0 of the NumPy package, to the way that zero singular values of the matrix **X** are identified.

19.4 Examples

E19.1

Question

The rate coefficient for the isomerization reaction of cyclopropane to propene was measured at five temperatures, as given in the table below. Determine the Arrhenius coefficient and activation energy, and comment on their values.

T/K	k/s^{-1}
700	1.5×10^{-5}
750	4.9×10^{-4}
800	5.9×10^{-3}
850	4.7×10^{-2}
900	5.5×10^{-1}

Solution

If the kinetics follow Arrhenius behavior, it should be possible to fit a plot of $y \equiv \ln k$ against $x \equiv 1/T$ to a straight line with the formula

$$\ln k = \ln A - \frac{E_a}{RT}.$$

This function is linear in the parameters $a = \ln A$ and $b = -E_a/R$ and can be fit using NumPy's linear least squares fitting function, $\texttt{np.linalg.lstsq}$.

```python
import numpy as np
import matplotlib.pyplot as plt
from scipy.constants import R

# The experimental data: rate constant k (s-1) as a function of temperature, T (K).
T = np.array([700, 750, 800, 850, 900])
k = np.array([1.5e-5, 4.9e-4, 5.9e-3, 4.7e-2, 5.5e-1])
```

```python
y = np.log(k)
X = np.vstack((np.ones(len(T)), 1/T)).T
ret = np.linalg.lstsq(X, y, rcond=None)
print('Fit parameters: {}\nSquared residuals: {}\n'
      'Rank of X: {}\nSingular values of X:{}'.format(*ret))
```

```
Fit parameters: [    35.23538457  -32334.97480231]
Squared residuals: [0.16337359]
Rank of X: 2
Singular values of X:[2.23606975e+00 2.51075716e-04]
```

```
# Retrieve the line intercept and gradient; convert to A and Ea
a, b = ret[0]
A, Ea = np.exp(a), -b * R / 1000
print(f'A = {A:.2e} s-1')
print(f'Ea = {Ea:.0f} kJ.mol-1')
```

```
A = 2.01e+15 s-1
Ea = 269 kJ.mol-1
```

This activation energy is quite large since a carbon–carbon bond needs to break to open the cyclopropane ring. The pre-exponential Arrhenius factor is of the order of a vibrational frequency, as expected for a unimolecular reaction.

To plot the best-fit line, we can create a function, get_lnk, that returns ln k using the fitted values of A and E_a use it at two temperatures. Below, this is compared with the experimental data, showing a decent enough fit.

```
plt.plot(1/T, np.log(k), 'x')
T1, T2 = 650, 950
def get_lnk(T):
    return np.log(A) - Ea * 1000 / R / T

plt.plot([1/T1, 1/T2], [get_lnk(T1), get_lnk(T2)])
plt.xlabel('$1/T\;/\mathrm{K^{-1}}$')
plt.ylabel('$\ln (k/\mathrm{s^{-1}})$')
```

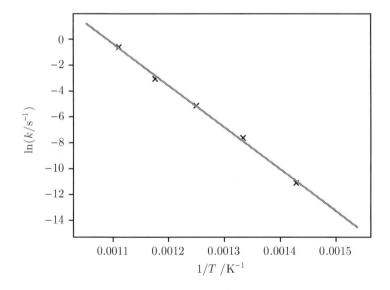

E19.2

Question

The virial and van der Waals equations of state for a gas are given, respectively, by

$$p_{\text{virial}} = \frac{RT}{V_m}\left(1 + \frac{B}{V_m} + \frac{C}{V_m^2} + \cdots\right),$$

$$p_{\text{vdw}} = \frac{RT}{V_m - b} - \frac{a}{V_m^2}.$$

Show that the second and third virial coefficients B and C are related to the van der Waals constants a and b through the relations:

$$B = b - \frac{a}{RT} \quad \text{and} \quad C = b^2.$$

The file `Ar-virial.txt`, available at `https://scipython.com/chem/goc/` contains measurements of the second virial coefficient for argon as a function of temperature, $B(T)$. Fit these data to the above equation to obtain values for a and b. Compare with the literature values $a = 1.355\,L^2\,bar\,mol^{-2}$ and $b = 0.03201\,L\,mol^{-1}$.

Solution

Rearrange the van der Waals equation of state to

$$p_{\text{vdw}} = \frac{RT}{V_m}\left(\frac{1}{1 - \frac{b}{V_m}} - \frac{a}{RTV_m}\right),$$

where $b/V_m \ll 1$ and use the Taylor series expansion $(1-x)^{-1} = 1 + x + x^2 + \cdots$ for $x = b/V_m$ to give

$$p_{\text{vdw}} = \frac{RT}{V_m}\left(1 + \frac{b}{V_m} + \frac{b^2}{V_m^2} + \cdots - \frac{a}{RTV_m}\right).$$

Comparison of the terms in different powers of $1/V_m$ in the virial equation of state,

$$p_{\text{virial}} = \frac{RT}{V_m}\left(1 + \frac{B}{V_m} + \frac{C}{V_m^2} + \cdots\right),$$

gives

$$B = b - \frac{a}{RT} \quad \text{and} \quad C = b^2.$$

The provided data can be plotted using Matplotlib:

```
import numpy as np
from scipy.constants import R
import matplotlib.pyplot as plt
```

```
T, B = np.loadtxt('Ar-virial.txt', skiprows=3, unpack=True)
```

```
plt.plot(T, B, 'x')
plt.xlabel('$T/\mathrm{K}$')
plt.ylabel('$B/\mathrm{cm^3\,mol^{-1}}$')
```

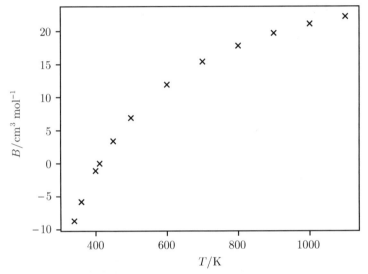

The equation relating $B(T)$ to a and b is linear in the van der Waals parameters, and so can be fitted with ordinary linear least squares (equivalent to finding a best-fit straight line to a plot of B against $1/T$).

The matrix $\mathbf{X}$ is:

$$\mathbf{X} = \begin{pmatrix} 1 & 1/T_1 \\ 1 & 1/T_2 \\ \vdots & \vdots \\ 1 & 1/T_n \end{pmatrix},$$

and given the measured values of $B(T)$,

$$\mathbf{B} = \begin{pmatrix} B(T_1) \\ B(T_2) \\ \vdots \\ B(T_n) \end{pmatrix},$$

we seek the parameter vector, $\hat{\boldsymbol{\beta}}$, that best solves the over-determined system of equations $\boldsymbol{B} = \mathbf{X}\boldsymbol{\beta}$. The elements of $\hat{\boldsymbol{\beta}}$ are then related to the van der Waals constants through

$$\hat{\boldsymbol{\beta}} = \begin{pmatrix} b \\ -\frac{a}{R} \end{pmatrix}.$$

`np.linalg.lstsq` can be used to perform the fit, as before:

```
X = np.vstack((np.ones(len(B)), 1/T)).T
res = np.linalg.lstsq(X, B, rcond=None)
res
```

```
(array([   36.65180953, -15156.14919878]),
 array([2.32226311]),
 2,
 array([3.46410756e+00, 2.38668255e-03]))
```

This fit seems to be well-behaved (the rank of $\mathbf{X}$ is 2, the same as the number of parameters to be fitted). To compare the fit to the data it is sufficient to evaluate $\mathbf{B}_{\text{fit}} = \mathbf{X} \cdot \hat{\boldsymbol{\beta}}$:

```
plt.plot(T, B, 'x', label='observed')
beta_fit = res[0]
Bfit = X @ beta_fit
plt.plot(T, Bfit, label='van der Waals fit')
plt.xlabel('$T/\mathrm{K}$')
plt.ylabel('$B/\mathrm{cm^3\,mol^{-1}}$')
plt.legend()
```

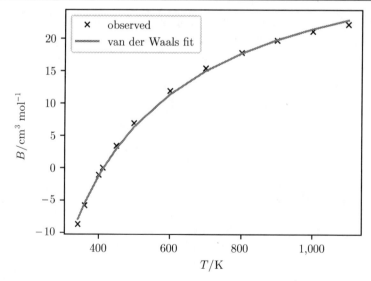

To compare the fitted values of a and b to the provided literature values, it is convenient to convert the fitted parameters, b and $-a/R$, to SI units (e.g. m^3 mol^{-1} from cm^3 mol^{-1}) first.

b is then easily converted into the customary units of L mol^{-1}. Multiplying the second fitted parameter by $-R$ leaves a in SI units: J m^3 $mol^{-2} \equiv (N\,m)\,m^3\,mol^{-2} \equiv Pa\,m^6\,mol^{-2}$ which can be converted into L^2 $bar\,mol^{-2}$ using the factors $1\,bar \equiv 10^5\,Pa$ and $1\,L^2 = 10^6\,m^6$.

```
# Extract the van der Waals constants from the fitted coefficients and convert
# to SI units (a in J.m3.mol-2 and b in m3.mol-1).
b, a = beta_fit[0] / 1.e6, -R * beta_fit[1] / 1.e6
# Now convert to the conventional units, a in L2.bar.mol-2, b in L.mol-1
b = b * 1000
a = a * 1.e6 / 1.e5
print(f'a = {a:2f} L2.bar.mol-2')
print(f'b = {b:4f} L.mol-1')
```

```
a = 1.260152 L2.bar.mol-2
b = 0.036652 L.mol-1
```

These values are in reasonable agreement with the literature values $a = 1.355\,L^2$ $bar\,mol^{-2}$ and $b = 0.03201\,L\,mol^{-1}$.

E19.3

Question

The file VE_H2O-EtOH.txt, available at https://scipython.com/chem/gob/, gives the excess molar volume of water–ethanol mixtures (at 298 K) as a function of the mole fraction of ethanol (EtOH), x_1:

```
# Excess molar volume of water-ethanol mixtures.
# Data from Grolier and Wilhelm, Fluid Phase Equilib. 6, 283 (1981).
# x1 = mole fraction of ethanol, x2 = 1 - x1 = mole fraction of water.
x1     VmE /cm3.mol-1
0.0128 -0.0500
0.0213 -0.0880
0.0341 -0.1510
...
0.9373 -0.2220
```

The data can be read in and plotted as follows (Figure 19.2):

```
import numpy as np
import matplotlib.pyplot as plt
x1, VE = np.genfromtxt('VE_H2O-EtOH.txt', unpack=True, skip_header=4)
```

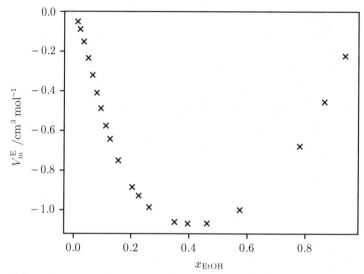

Figure 19.2   Excess molar volume of water–ethanol mixtures plotted as a function of the ethanol mole fraction.

```
plt.plot(x1, VE, 'x')
plt.xlabel('$x_\mathrm{EtOH}$')
plt.ylabel('$V_\mathrm{m}^\mathrm{E} \; /\mathrm{cm^3\,mol^{-1}}$')
```

Fit these data to the Redlich–Kister equation:

$$V_m^E = x_1 x_2 \sum_{k=0}^{k_{max}} a_k (2x_2 - 1)^k$$

using $k_{max} = 6$, and determine the total volume of a mixture of 1 L of pure water with 1 L of pure ethanol. Take molar masses $M(H_2O) = 18$ g mol^{-1} and $M(EtOH) = 46$ g mol^{-1}, and densities $\rho(H_2O) = 1$ g cm^{-3} and $\rho(EtOH) = 0.79$ g cm^{-3}.

Solution

The best fit of the data to this function can be solved using the ordinary linear least-squares approach, since the fit function is linear in the parameters a_k. In matrix form, $\mathbf{V}_m^E = \mathbf{X}\beta$, where $\mathbf{V}_m^E$ is a column vector of the measured excess volumes, and β is a column vector of the coefficients to be determined:

$$\mathbf{V}_m^E = \begin{pmatrix} V_{m,1}^E \\ V_{m,2}^E \\ \vdots \\ V_{m,n}^E \end{pmatrix} = \begin{pmatrix} -0.0500 \\ -0.0880 \\ \vdots \\ -0.2220 \end{pmatrix}, \quad \beta = \begin{pmatrix} a_0 \\ a_1 \\ \vdots \\ a_{k_{max}} \end{pmatrix}.$$

The matrix, **X**, sometimes called the *design matrix*, is

$$
\begin{pmatrix}
x_{1,1}x_{2,1} & x_{1,1}x_{2,1}(2x_{2,1}-1) & x_{1,1}x_{2,1}(2x_{2,1}-1)^2 & \cdots & x_{1,1}x_{2,1}(2x_{2,1}-1)^{k_{max}} \\
x_{1,2}x_{2,2} & x_{1,2}x_{2,2}(2x_{2,2}-1) & x_{1,2}x_{2,2}(2x_{2,2}-1)^2 & \cdots & x_{1,2}x_{2,2}(2x_{2,2}-1)^{k_{max}} \\
\vdots & \vdots & \vdots & \cdots & \vdots \\
x_{1,n}x_{2,n} & x_{1,n}x_{2,n}(2x_{2,n}-1) & x_{1,n}x_{2,n}(2x_{2,n}-1)^2 & \cdots & x_{1,n}x_{2,n}(2x_{2,n}-1)^{k_{max}}
\end{pmatrix}
$$

There are $k_{max}+1$ columns, corresponding to the terms in the Redlich–Kister equation, and n rows in this matrix, one for each of the measured mole fraction pairs, (x_1, x_2); the mole fractions are labeled as $x_{1,1}, x_{1,2}, \cdots, x_{1,n}$ for ethanol and $x_{2,1}, x_{2,2}, \cdots, x_{2,n}$ for water for the n data points.

x_2 can be calculated from x_1, since this is a binary mixture and $x_1 + x_2 = 1$. It makes sense to define a function to create the design matrix:

```
def get_design_matrix(kmax, x1):
    x2 = 1 - x1
    X = np.empty((len(x1), kmax+1))
    for k in range(0, kmax+1):
        X[:, k] = x1 * x2 * (2 * x2 - 1)**k
    return X

kmax = 6
X = get_design_matrix(kmax, x1)
```

The linear least square fit is easily achieved with np.linalg.lstsq:

```
res = np.linalg.lstsq(X, VE, rcond=None)
res
```

```
(array([-4.22206349, -0.93662852, -0.63128236, -1.86290358, -4.18863044,
         2.54097835,  5.9447701 ]),
 array([0.00014473]),
 7,
 array([0.6974695 , 0.30050993, 0.16334442, 0.06713332, 0.03285552,
        0.01020795, 0.0036322 ]))
```

This is a reasonable fit, as can be seen by plotting the data points and the fit function (Figure 19.3):

```
# Get the fitted parameters, a_k, from the fit result.
a = res[0]
# Create a fine grid of water mole fraction values from 0 to 1.
x1_grid = np.linspace(0, 1, 100)
# Get the matrix X for this grid and retrieve the corresponding fitted
# excess volumes.
X = get_design_matrix(kmax, x1_grid)

VE_fit = np.sum(a * X, axis=1)
```

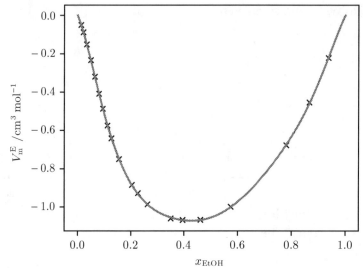

Figure 19.3   Excess molar volume of water–ethanol mixtures as a function of the ethanol mole fraction fitted to the Redlich–Kister equation.

```
plt.plot(x1, VE, 'x')
plt.plot(x1_grid, VE_fit)
plt.xlabel('$x_\mathrm{EtOH}$')
plt.ylabel('$V_\mathrm{m}^\mathrm{E}$')
```

To use the fitted coefficients, it might be more convenient to define a function that takes either a single ethanol mole fraction or an array of mole fraction values and returns the corresponding excess volume:

```
def get_VE_fit(x1):
    x1 = np.atleast_1d(x1)
    X = get_design_matrix(kmax, x1)
    VE_fit = np.sum(a * X, axis=1)
    if len(x1) == 1:
        return VE_fit[0]
    return VE_fit
```

For example, when 1 L of water is mixed with 1 L of pure ethanol:

```
# Molar masses (g.mol-1) and densities (g.cm-3) of water and ethanol.
M_H2O, M_EtOH = 18, 46
rho_H2O, rho_EtOH = 1, 0.79

# Volumes of pure liquids (cm3)
V_H2O, V_EtOH = 1e3, 1e3
# Number of moles of water and ethanol.
n_H2O = rho_H2O * V_H2O / M_H2O
n_EtOH = rho_EtOH * V_EtOH / M_EtOH
# Mole fraction of ethanol.
x_EtOH = n_EtOH / (n_H2O + n_EtOH)
```

```
VE = (n_H2O + n_EtOH) * get_VE_fit(x_EtOH)
print('VmE(1 L water + 1 L ethanol) = {:.2f} cm3'.format(VE))
Vtot = V_H2O + V_EtOH + VE
print('Total volume of mixture = {:.2f} cm3'.format(Vtot))
```

```
VmE(1 L water + 1 L ethanol) = -68.98 cm3
Total volume of mixture = 1931.02 cm3
```

The prediction is that when equal volumes of pure water and ethanol are mixed, the total volume decreases.

19.5 Exercises

P19.1 The *Michaelis–Menten equation* can be used describe the rate, v, of an enzyme-catalyzed reaction. In this model, an enzyme, E, binds to a substrate, S, to form a complex, ES, which either releases a product, P (and regenerates E) or falls apart again without reaction:

$$E + S \underset{k_{-1}}{\overset{k_1}{\rightleftharpoons}} ES \overset{k_2}{\longrightarrow} P + E$$

For many such catalyzed reactions, it is found that

$$v = \frac{d[P]}{dt} = v_{max} \frac{[S]}{K_M + [S]},$$

where $v_{max} = k_2[E]_0 = k_2([E] + [ES])$ is the maximum rate attainable and the Michaelis constant, $K_M = (k_{-1} + k_2)/k_1$.

Pepsin is one of the main enzymes in the digestive systems of mammals, where it helps to break down proteins into smaller peptides which can be absorbed by the small intestine. In a study of the enzymatic kinetics of pepsin with the protein bovine serum albumin (S), the rate of reaction, v, was measured as a function of substrate concentration, [S]; these data are given in the file pepsin-rates.txt:

```
# Reaction rates, v (in mM.s-1), for enzymatic kinetics of pepsin with
# bovine serum albumin (S, in mM).)
S /mM   v /mM.s-1
 0.1 0.00339
 0.2 0.00549
 0.5 0.00875
 1.0 0.01090
 5.0 0.01358
10.0 0.01401
20.0 0.01423
```

The experiments were carried out at 35°C and a pH of 2 with a total concentration of pepsin of $[E]_0 = 0.028$ mM.

Using the *Lineweaver–Burk* method, the Michaelis–Menten equation can be transformed into linear form by plotting $1/v$ against $1/[S]$. Use `np.linalg.lstsq` to fit the data and obtain values for K_M, v_{max} and k_2.

P19.2 Show that Equation 19.4 can be written, in the case of a straight-line regression ("simple linear regression") on n points using the model function $y = a + bx$ as follows:

$$\hat{\beta} = \begin{pmatrix} a \\ b \end{pmatrix} \text{ where } a = \frac{S_y S_{xx} - S_{xy} S_x}{n S_{xx} - S_x^2} \text{ and } b = \frac{n S_{xy} - S_y S_x}{n S_{xx} - S_x^2},$$

and the summary statistics are defined as:

$$S_x = \sum_{i=1}^{n} x_i, \quad S_y = \sum_{i=1}^{n} y_i, \quad S_{xx} = \sum_{i=1}^{n} x_i^2, \quad S_{xy} = \sum_{i=1}^{n} x_i y_i.$$

20

Linear Least Squares Fitting II

20.1 Parameter Uncertainties

It is often useful to be able to estimate the uncertainty in the parameters fitted to
a model function. In general, such estimation is difficult and even controversial
to define, but in the common case of a straight-line best fit to the model function
$y = a + bx$, with independent errors in the dependent variable y only it is (relatively)
straightforward. Unfortunately, the NumPy `linalg.lstsq` method does not provide this information, and we have to return to the explicit solution of the Normal
Equations (see Question P19.2):

$$a = \frac{S_y S_{xx} - S_{xy} S_x}{n S_{xx} - S_x^2},$$

$$b = \frac{n S_{xy} - S_y S_x}{n S_{xx} - S_x^2},$$

where

$$S_x = \sum_{i=1}^n x_i, \quad S_y = \sum_{i=1}^n y_i, \quad S_{xx} = \sum_{i=1}^n x_i^2, \quad S_{xy} = \sum_{i=1}^n x_i y_i.$$

It can be shown that:

$$\sigma_a = \sigma_y \sqrt{\frac{S_{xx}}{n S_{xx} - S_x^2}} \quad \text{and} \quad \sigma_b = \sigma_y \sqrt{\frac{n}{n S_{xx} - S_x^2}},$$

where σ_y is the uncertainty in the values y_i. In a real experiment, we may only
be able to estimate σ_y, for example, from knowledge about the methodology and
instrument used to measure y_i. In the simplest case, where y is measured directly,
it may be possible to provide a value of σ_y without too much trouble. For example,
if y is a measured length, it might be reasonable to assign the value $\sigma_y = 1$ mm if a
ruler is used; for times measured with a stopwatch, $\sigma_y = 0.5$ s would be about right.

However, frequently y is formed from a combination of measured quantities and its uncertainty may be less easy to determine. If the uncertainties (random errors) in the measured quantities, p, q, r, ... can be assumed to be *uncorrelated*, a common approach is to propagate their uncertainties into an uncertainty in the function $y = y(p, q, r, ...)$ with the formula:

$$\sigma_y = \sqrt{\left(\frac{\partial y}{\partial p}\right)^2 \sigma_p^2 + \left(\frac{\partial y}{\partial q}\right)^2 \sigma_q^2 + \left(\frac{\partial y}{\partial r}\right)^2 \sigma_r^2 + ...} \qquad (20.1)$$

This is effectively the first non-zero term in a multivariate Taylor series expansion of the variances, σ_i^2, and is therefore only approximate: For highly non-linear functions y it might be expected to give flawed results.

In the absence of this information, the best that can be done might be to estimate σ_y from the measurements of y_i themselves, and the only thing we have to compare them to is the model function:

$$\sigma_a \approx \sqrt{\frac{1}{n} \sum_{i=1}^{n} (y_i - a - bx_i)^2}.$$

Unfortunately, we don't know the true values of the model parameters, a and b; we only have estimates of them from the fit. In practice, this reduces the number of degrees of freedom in the problem (after all, a straight line can always be fit exactly between two points) and the above equation should be modified as follows:

$$\sigma_a \approx \sqrt{\frac{1}{n-2} \sum_{i=1}^{n} (y_i - a - bx_i)^2}.$$

20.2 Example

E20.1

Question

The temperature of n_g moles of a gas in degrees centigrade, θ, as a function of its pressure, p, at constant volume, V, may be expected to follow the ideal gas law:

$$pV = n_g RT = n_g R(\theta - \theta_0) \quad \Rightarrow \quad \theta = \frac{V}{n_g R} p + \theta_0,$$

where θ_0 is the value of absolute zero in degrees centigrade.

The following data were measured using an accurate barometer (such that there is negligible uncertainty in the pressure) but a rather less accurate thermometer.

p /bar	θ /°C
0.85	−16
0.90	3
0.95	19.5
1.00	35
1.05	52.5
1.10	63

Use linear least-squares regression to estimate the value of θ_0, with an estimate of its uncertainty.

Solution

The model function is of the form $y = a + bx$ with $x \equiv p$, $y \equiv \theta$, $a \equiv \theta_0$ and $b \equiv V/(n_g R)$. The best-fit values for the parameters a and b can be found by linear regression:

$$a = \frac{S_y S_{xx} - S_{xy} S_x}{\Delta},$$

$$b = \frac{n S_{xy} - S_y S_x}{\Delta},$$

where $\Delta = n S_{xx} - S_x^2$ and the following shorthand expressions are used:

$$S_x = \sum_{i=1}^{n} x_i, \quad S_y = \sum_{i=1}^{n} y_i, \quad S_{xx} = \sum_{i=1}^{n} x_i^2, \quad S_{xy} = \sum_{i=1}^{n} x_i y_i.$$

```
import numpy as np
import matplotlib.pyplot as plt
```

```
x = np.array([0.85, 0.90, 0.95, 1.00, 1.05, 1.10])
y = np.array([-15.5, 3, 20, 35, 52.5, 63])
```

```
n = len(x)
Sx, Sy, Sxx, Sxy = np.sum(x), np.sum(y), np.sum(x*x), np.sum(x*y)
Delta = n * Sxx - Sx**2
```

```
a = (Sy * Sxx - Sxy * Sx) / Delta
b = (n*Sxy - Sy * Sx) / Delta
a, b
```

```
(-283.4380952380927, 317.7142857142826)
```

Our estimate of absolute zero is, therefore, $\theta_0 = a = -283.4$ K. The uncertainty in the parameter a is

$$\sigma_a = \sigma_y \sqrt{\frac{S_{xx}}{\Delta}},$$

but we don't know what the uncertainty in the measured temperatures, σ_y, is. The best we can hope to do is to estimate σ_y from the fit, assuming it to be the same for each measurement:

$$\sigma_y \approx \sqrt{\frac{1}{n-2} \sum_{i=1}^{n} (y_i - a - bx_i)^2}.$$

```
sigy = np.sqrt(np.sum((y - a - b*x)**2) / (n-2))
siga = sigy * np.sqrt(Sxx / Delta)
sigy, siga
```

```
(2.3772232701600573, 11.123585797652261)
```

We could report the derived value of absolute zero as $-280 \pm 11°$C. Note that the uncertainty in the parameter a is much greater than that in the temperatures, $y_i \equiv \theta_i$, since these values have to be extrapolated a long way back to the intercept and are therefore sensitive to the fitted slope, b.

On the other hand, $\sigma_\theta = 2.4\,°$C is quite a bit larger than the uncertainty we might expect in the measurement of the temperature by a regular thermometer, suggesting that either the experiment was not carried out very carefully or that the ideal gas approximation does not hold well at the experimental pressures for this gas.

```
# Make a figure with two subplots, side by side.
WIDTH, HEIGHT, DPI = 800, 400, 100
fig_kw = dict(figsize=(WIDTH/DPI, HEIGHT/DPI), dpi=DPI)
fig, ax = plt.subplots(nrows=1, ncols=2, **fig_kw)

# Left-hand plot: the fitted data with inferred error bars.
ax[0].errorbar(x, y, sigy, fmt='x', c='k', capsize=4)
xfit = np.array([x[0], x[-1]])
ax[0].plot(xfit, a + b*xfit, 'k')
ax[0].set_xlabel(r'$p/\mathrm{bar}$')
ax[0].set_ylabel(r'$\theta/\mathrm{^\circ\!C}$')

# Right-hand plot: the extrapolation back to absolute zero.
ax[1].plot(x, y, 'x', c='k')
xfit = np.array([0, x[-1]])
ax[1].plot(xfit, a + b*xfit, 'k--')
ax[1].set_xlabel(r'$p/\mathrm{bar}$')
ax[1].set_ylabel(r'$\theta/\mathrm{^\circ\!C}$')
ax[1].set_xlim(0)
plt.tight_layout()
```

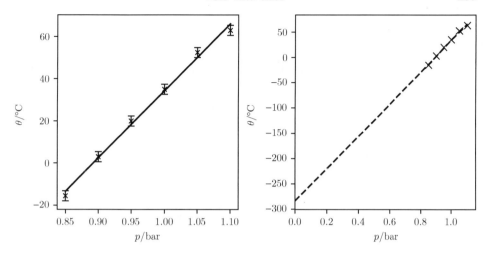

20.3 Exercises

P20.1 Use Equation 20.1 to show that the uncorrelated random uncertainties, σ_p, σ_q and σ_x in the parameters p, q and x propagate into an uncertainty in the following functions of y as follows (the quantities a, b and c may be assumed to be precisely known):

1. $y = ap \pm bq \implies \sigma_y = \sqrt{a^2\sigma_p^2 + b^2\sigma_q^2}$

2. $y = cpq \implies \sigma_y = |y|\sqrt{\left(\dfrac{\sigma_p}{p}\right)^2 + \left(\dfrac{\sigma_q}{q}\right)^2}$

3. $y = a\ln(bx) \implies \sigma_y = \left|\dfrac{a}{x}\right|\sigma_x$

4. $y = ae^{bx} \implies \sigma_y = |by|\sigma_x$

P20.2 A novel way to measure the gravitational acceleration, g, involves an indirect measurement of the pressure inside a submerged, inverted test tube containing a small amount of water. The experimental set-up proposed by Quiroga et al.[1] is shown in the Figure 20.1 below.

Let the ambient pressure be p_0 and the length and cross-sectional area of the test tube be l_0 and A, respectively, so that it has a volume $V_0 = Al_0$. Treating the n moles of air inside it as an ideal gas at temperature T, we have $p_0V_0 = nRT$.

When the tube is submerged to a distance h, water rises inside it and the air it contains is compressed to a pressure p_1 and volume V_1, where $p_1V_1 = p_0V_0$ if the temperature is constant. The new pressure is $p_1 = p_0 + \rho gl_1$, where $\rho = 1\,\mathrm{g\,cm^{-1}}$ is the water density, g is the gravitational acceleration (to be determined) and l_1 is

[1] Quiroga et al., *Phys. Teach.* **48**, 386 (2010).

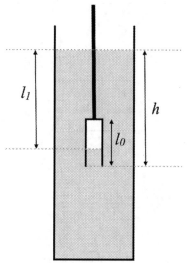

Figure 20.1   A diagram of the "diving bell" apparatus used to measure the gravitational acceleration, g.

the measured distance from the bulk water surface to the top of the water level in
the test tube.

From the above figure, it is clear that $V_1 = A[l_0 - (h - l_1)]$, and so

$$p_1 = \frac{l_0}{l_0 - (h - l_1)} p_0.$$

Therefore,

$$\left[\frac{l_0}{l_0 - (h - l_1)} - 1 \right] p_0 = \rho g l_1.$$

That is, if measurements of l_1 are made for different submersion depths, h, a plot of
the lefthand side of this equation against l_1 should yield a straight line with gradient
ρg, from which g can be deduced.

Assuming that σ_h and σ_{l_1} (the uncertainties in the measurements of h and l_1
respectively) dominate and that they are uncorrelated, show that the uncertainty in
the quantity

$$y = p_1 - p_0 = \left[\frac{l_0}{l_0 - (h - l_1)} - 1 \right] p_0 = \rho g l_1$$

is

$$\sigma_y = \frac{l_0 p_0}{[l_0 - (h - l_1)]^2} \sqrt{\sigma_h^2 + \sigma_{l_1}^2}.$$

Use the following data to estimate g from a linear least squares fit and the uncertainty in this estimate. Take an average value for σ_y, calculated using the above formula.

These data were collected on a day with an ambient air pressure of $p_0 = 1037$ mbar and using a test tube of length 20 cm. Take $\sigma_h = \sigma_{l_0} = 1$ mm.

```
# Measured distances, h and L1, in cm.
h = [29.9, 35. , 39.9, 45.1, 50. , 55. , 59.9, 65. , 70.1, 74.9, 80.1,
      85.1, 90.1, 95. ]
L1 = [29.3, 34.3, 39.2, 44.5, 49.1, 54. , 58.7, 64. , 68.9, 73.5, 78.6,
      83.7, 88.6, 93.4]
```

21

Numerical Integration

The `scipy.integrate` package contains methods for evaluating definite integrals. It can handle both proper integrals (with finite limits) and improper integrals (those with one or more infinite limits). These methods use numerical algorithms (adaptive quadrature) – for symbolic integration, see the SymPy package (Chapter 35).

21.1 Integrals of a Single Variable

The `scipy.integrate.quad` method takes a function, `func`, and upper and lower limits, a and b, and returns the integral of `func` with respect to its first argument between these limits. For example, to evaluate:

$$\int_0^2 x^2 \, dx = \frac{8}{3}$$

```
In [x]: from scipy.integrate import quad

In [x]: def func(x):
   ...:     return x**2

In [x]: quad(func, 0, 2)
Out[x]: (2.666666666666667, 2.960594732333751e-14)
```

`quad` returns a `tuple` containing two values: the value of the integral and an estimate of the absolute error. For simple functions, an anonymous (`lambda`) function is often used:

```
In [x]: quad(lambda x: x**2, 0, 2)
Out[x]: (2.666666666666667, 2.960594732333751e-14)
```

Functions with singularities or discontinuities can cause problems in numerical integration. Consider the integral:

$$\int_{-1}^{1} \frac{\mathrm{d}x}{\sqrt{|1 - 4x^2|}}.$$

```
In [x]: f = lambda x: 1 / np.sqrt(np.abs(1 - 4*x**2))
In [x]: quad(f, -1, 1)
<ipython-input-59-8c58327d07ca>:1: RuntimeWarning: divide by zero encountered
in double_scalars
  f = lambda x: 1 / np.sqrt(np.abs(1 - 4*x**2))
Out[x]: (inf, inf)
```

When this happens, a sequence of *break points* can be specified as the argument `points`. In this case, the trouble lies with the singularities at $x = \pm\frac{1}{2}$ so the integral can be evaluated as follows,

```
In [x]: quad(f, -1, 1, points=[-0.5, 0.5])
Out[x]: (2.8877542237184186, 3.632489864457966e-10)
```

To specify an infinite limit, use `np.inf`:

$$\int_{-\infty}^{\infty} \mathrm{e}^{-x^2}\, \mathrm{d}x = \sqrt{\pi} \approx 1.772.$$

```
In [x]: quad(lambda x: np.exp(-(x**2)), -np.inf, np.inf)
Out[x]: (1.7724538509055159, 1.4202636780944923e-08)
```

More complicated functions may take additional parameters: These should take the form of arguments after the variable to be integrated over, and specified in the tuple, `args`, passed to `quad`. For example, to evaluate the integral

$$\int_{0}^{\infty} \mathrm{e}^{-at} \cos kt\, \mathrm{d}t = \frac{a}{a^2 + k^2}$$

for parameters $a = \frac{5}{2}$ and $k = \frac{\pi}{8}$:

```
In [x]: def func(t, a, k):
   ...:     return np.exp(-a * t) * np.cos(k * t)

In [x]: a, k = 2.5, np.pi / 8

In [x]: quad(func, 0, np.inf, args=(a, k))
Out[x]: (0.39036805433229327, 7.550213632675355e-11)
```

21.2 Integrals of Two and Three Variables

The `scipy.integrate` methods `dblquad` and `tplquad` evaluate double and triple integrals, respectively:

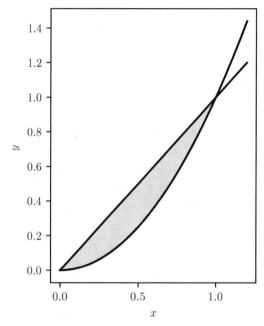

Figure 21.1   The region bounded by the line $y = x$ and the curve $y = x^2$.

$$\int_a^b \int_{g(x)}^{h(x)} f(x, y)\, dy\, dx,$$

$$\int_a^b \int_{g(x)}^{h(x)} \int_{q(x,y)}^{r(x,y)} f(x, y, z)\, dz\, dy\, dx.$$

Note that the limits on all but the outer integral must be provided as callable objects (e.g., a Python function) since they may depend on other variables. As an example, take the integral of the function $f(x, y) = xy^3$ over the region enclosed by the line $y = x$ and the parabola $y = x^2$. The line and curve meet at $x = 0$ and $x = 1$, and since $y = x^2$ lies below $y = x$ for $0 < x < 1$, the integral is

$$I = \int_0^1 \int_{x^2}^x xy^3\, dy\, dx.$$

The region is illustrated in Figure 21.1, which can be generated using Matplotlib:

```
x = np.linspace(0, 1.2, 1000)
h = x**2
g, h = x**2, x

plt.plot(x, h, c="k")
plt.plot(x, g, c="k")
plt.fill_between(x[x <= 1], g[x <= 1], h[x <= 1])
plt.xlabel('$x$')
plt.ylabel('$y$')
plt.show()
```

This particular integral can be evaluated analytically:

$$I = \int_0^1 \left[\frac{1}{4}xy^4 \right]_{y=x^2}^{y=x} dx = \int_0^1 \frac{1}{4}x^5 - \frac{1}{4}x^9 \, dx$$

$$= \left[\frac{1}{24}x^6 - \frac{1}{40}x^{10} \right]_0^1 = \frac{1}{60}.$$

Numerically, using `dblquad`, the function must be defined with the innermost variable (y) given as the first argument, and its limits themselves defined as functions the outermost variable (x):

```
In [x]: from scipy.integrate import dblquad

In [x]: def func(y, x):
   ...:     return x * y**3

In [x]: g = lambda x: x**2      # lower limit of y-integral

In [x]: h = lambda x: x         # upper limit of y-integral

In [x]: a, b = 0, 1             # limits of the x-integral

In [x]: dblquad(func, a, b, g, h)
Out[x]: (0.016666666666666666, 5.826435989325954e-16)
```

which is $\frac{1}{60}$ within the numerical error in the quadrature.

To calculate the *area* of the region, one can simply integrate the function $f(x, y) = 1$ over it:

```
In [x]: dblquad(lambda y, x: 1, 0, 1, g, h)
Out[x]: (0.16666666666666666, 2.7755575615628914e-15)
```

and, indeed:

$$\int_0^1 \int_{x^2}^x dy \, dx = \int_0^1 x - x^2 \, dx = \frac{1}{2} - \frac{1}{3} = \frac{1}{6}.$$

A similar approach can be taken with triple integrals, such as those over three-dimensional space. The volume of the unit sphere can be expressed in Cartesian coordinates as eight times the integral over the positive octant:

$$8 \int_0^1 \int_0^{\sqrt{1-x^2}} \int_0^{\sqrt{1-x^2-y^2}} dz \, dy \, dx,$$

Using `tplquad`, with the inner limits again expressed as anonymous functions:

```
In [x]: A, _ = tplquad(lambda z, y, x: 1,
                       0, 1,
                       lambda x: 0, lambda x: np.sqrt(1 - x**2),
```

```
                          lambda x, y: 0, lambda x, y: np.sqrt(1 - x**2 - y**2))
In [x]: 8*A
Out[x]: 4.188790204786391
```

which is $4\pi/3$ within numerical error.

The same volume expressed in spherical polar coordinates becomes

$$\int_0^{2\pi} \int_0^{\pi} \int_0^1 r^2 \sin\theta \, drd\theta d\phi.$$

```
In [x]: from scipy.integrate import tplquad
In [x]: tplquad(lambda phi, theta, r: r**2 * np.sin(theta),
                0, 1,
                lambda theta: 0, lambda theta: np.pi,
                lambda theta, phi: 0, lambda theta, phi: 2*np.pi)
Out[x]: (4.18879020478639, 4.650491330678174e-14)
```

21.3 Examples

E21.1

Some Integrals from Quantum Mechanics

```
import numpy as np
from scipy.integrate import quad, dblquad, tplquad
```

(a) The one-dimensional harmonic oscillator ground state wavefunction is

$$\psi_0(q) = Ne^{-q^2/2}$$

where the displacement from equilibrium, x, has been scaled as the dimensionless quantity, $q = (\mu k/\hbar^2)^{1/4}x$, μ is the oscillator mass and k the force constant. Using this scaling for displacement, the classical oscillator turning points are $q_{\pm} = \pm 1$, and so the probability of tunneling (i.e., for the oscillator to be found outside the classical limits) is

$$P(q > q_+) + P(q < q_-) = 2P(q > q_+) = 2\int_{q_+}^{\infty} |\psi_0(q)|^2 \, dq.$$

This integral can only be evaluated numerically. First, determine the normalization constant from the requirement that

$$\langle \psi_0 | \psi_0 \rangle = 1 \quad \Rightarrow N^2 \int_{-\infty}^{\infty} |\psi_0(q)|^2 \, dq = 1$$

```
def func(q):
    return np.exp(-q**2)
```

```
I, err = quad(func, -np.inf, np.inf)
N = 1 / np.sqrt(I)
N
```

```
0.7511255444649425
```

In fact, this integral can be evaluated analytically and has the value $I = \sqrt{\pi}$, so $N = \pi^{-1/4}$:

```
np.pi**-0.25
```

```
0.7511255444649425
```

The tunneling probability is then:

```
I, err = quad(func, 1, np.inf)
Ptunnelling = 2 * N**2 * I
Ptunnelling
```

```
0.1572992070502851
```

or about 16%.

(b) The two-dimensional particle in a rectangular box of has wavefunctions given by:

$$\psi_{n,m}(x, y) = N \sin \left(\frac{n\pi x}{L_x} \right) \sin \left(\frac{m\pi y}{L_y} \right),$$

where the particle is confined to a region of zero potential for $0 < x < L_x$ and $0 < y < L_y$; and $n = 1, 2, 3, \ldots$ and $m = 1, 2, 3, \ldots$ are quantum numbers. The normalization constant, N, ensures that $\langle \psi_{n,m} | \psi_{n,m} \rangle = 1$. To find it, evaluate the integral in the following expression and rearrange for N:

$$N^2 \int_0^{L_y} \int_0^{L_x} \sin^2 \left(\frac{n\pi x}{L_x} \right) \sin^2 \left(\frac{m\pi y}{L_y} \right) \, dx \, dy = 1$$

Although this integral is not hard to evaluate analytically and gives $N = 2/\sqrt{L_x L_y}$, we can also use dblquad:

```
def func(x, y, n, m, Lx, Ly):
    """Return the square of the 2D particle-in-a-box wavefunction."""
    return (np.sin(n * np.pi * x / Lx) * np.sin(m * np.pi * y / Ly))**2

# Example parameters: box dimensions (length units) and quantum numbers.
Lx, Ly = 1.5, 2.5
n, m = 2, 1
# Evaluate the integral of the square of the unnormalized wavefunction.
```

```
I, unc = dblquad(func, 0, Ly, lambda y: 0, lambda y: Lx, args=(n, m, Lx, Ly))
# The normalization constant is the reciprocal of the square root of this integral.
N = 1 / np.sqrt(I)
N
```

```
1.0327955589886444
```

(the units are those of inverse length). This is the same as the analytical result:

```
2 / np.sqrt(Lx*Ly)
```

```
1.0327955589886444
```

(c) The following electron–electron repulsion integral appears in an approximate treatment of the wavefunction of a helium atom:

$$I = \langle 11|11 \rangle = \langle \varphi_1(\mathbf{r}_1)\varphi_1(\mathbf{r}_2)| \frac{1}{r_{12}} |\varphi_1(\mathbf{r}_1)\varphi_1(\mathbf{r}_2)\rangle,$$

where

$$\varphi_1(\mathbf{r}_i) = \sqrt{\frac{Z^3}{\pi}} e^{-Zr_i}$$

is the hydrogenic 1s orbital occupied by electron i and the electron–electron distance, $r_{12} = \sqrt{r_1^2 - 2r_1r_2\cos\theta + r_2^2}$ with θ the angular separation of the position vectors $\mathbf{r}_1$ and $\mathbf{r}_2$.

This integral can be evaluated analytically and is found to have the value $5Z/8$, but here is a numerical approach.

First note that the integral is over six coordinates: in spherical polar coordinates, $r_1, \theta_1, \phi_1, r_2, \theta_2, \phi_2$:

$$I = \left(\frac{Z^3}{\pi}\right)^2 \int_0^{2\pi} \int_0^{\pi} \int_0^{\infty} \int_0^{2\pi} \int_0^{\pi} \int_0^{\infty} \frac{e^{-2Zr_1}e^{-2Zr_2}}{r_{12}} r_1^2 \sin\theta_1 r_2^2 \sin\theta_2$$
$$dr_1 \, d\theta_1 \, d\phi_1 \, dr_2 \, d\theta_2 \, d\phi_2$$

The symmetry lets us integrate over both ϕ_1 and ϕ_2, and also θ_2 (by letting $\theta = \theta_1$: We need the angular separation of the electrons, so we can fix one and vary the other to obtain all the needed values of θ). This introduces a factor of $2\pi \cdot 2\pi \cdot 2 = 8\pi^2$ and reduces the integral to one over three coordinates:

$$I = \left(\frac{Z^3}{\pi}\right)^2 8\pi^2 \int_0^{\infty} \int_0^{\infty} \int_0^{\pi} \frac{e^{-2Zr_1}e^{-2Zr_2}}{r_{12}} r_1^2 r_2^2 \sin\theta_1 \, d\theta_1 \, dr_1 \, dr_2$$

```
def psi1(r, Z=1):
    return np.sqrt(Z**3 / np.pi) * np.exp(-Z * r)
```

```
def func(theta1, r2, r1, Z=1):
    return ( (psi1(r1, Z) * psi1(r2, Z) * r1 * r2)**2 * np.sin(theta1)
        / np.sqrt(r1**2 + r2**2 - 2*r1*r2*np.cos(theta1))
        )
```

```
Z = 1
I, unc = tplquad(func, 0, np.inf,
                        lambda r1: 0, lambda r1: np.inf,
                        lambda r1, theta1: 0, lambda r1, theta1: np.pi,
                args=(Z,))
I *= 8 * np.pi**2
I, unc
```

```
(0.6250000692531371, 1.489297783794502e-08)
```

This is, within numerical error, equal to 5/8.

E21.2

Question

How much work is done in compressing gaseous nitrogen, initially at 1 atm, isothermally at 298 K to 200 bar into a 15 L scuba diving cylinder? Take N_2 to obey the van der Waals equation of state formula,

$$p = \frac{nRT}{V - nb} - \frac{n^2 a}{V^2}$$

with parameters $a = 1.370\,L^2\,bar\,mol^{-2}$ and $b = 0.0387\,L\,mol^{-1}$.

Obtain V_1 and then use the following expression for the work done:

$$w = -\int_{V_1}^{V_2} p\,dV,$$

to compare the value for w obtained by analytical integration and by with numerical integration using `scipy.integrate.quad`. Also, compare the result with the same analysis carried out under the assumption that N_2 behaves as an ideal gas.

Solution

If the process is isothermal then the temperature, $T = 298$ K is constant. Let the initial pressure be $p_1 = 101325$ Pa and the volume V_1 (to be determined). The final pressure and volume are $p_2 = 2 \times 10^7$ Pa and $V_2 = 15\,L = 1.5 \times 10^{-2}\,m^3$.

In SI units, the van der Waals parameters are $a = 0.137 \text{ m}^6 \text{ Pa mol}^{-2}$ and $b = 3.87 \times 10^{-5} \text{ m}^3 \text{ mol}^{-1}$.

```
import numpy as np
from numpy.polynomial import Polynomial
from scipy.integrate import quad

# Define some quantities
R = 8.314                   # Gas constant, J.K-1.mol-1
T = 298                     # Temperature (K)
p1 = 101325                 # Initial pressure (Pa)
p2, V2 = 2.e7, 1.5e-2       # Final pressure (Pa) and volume (m3)
a, b = 0.137, 3.87e-5       # van der Waals parameters (m6.Pa.mol-2 and m3.mol-1)
```

First, we need to know n, the number of moles of gas compressed, and hence V_1, the initial volume of the gas. Rearranging the van der Waals equation gives:

$$pV^2(V - nb) = nRTV^2 - n^2a(V - nb)$$
$$\Rightarrow \quad abn^3 - aVn^2 + V^2(bp + RT)n - pV^3 = 0.$$

This is a cubic equation in n which could be solved analytically, but is easier to handle numerically:

```
# A NumPy Polynomial representing the cubic equation in n.
poly = Polynomial([-p2 * V2**3, V2**2 * (b*p2 + R*T), -a * V2, a * b])
roots = poly.roots()
roots
```

```
array([133.6509615 +296.59957559j, 133.6509615 -296.59957559j,
       120.29497623  +0.j          ])
```

Of the three roots, only one is real so we are still in the region of phase space where the van der Waals equation is valid. Extracting this value and explicitly dropping its (zero) imaginary part gives n:

```
np.isreal(roots)
```

```
array([False, False,  True])
```

```
n = roots[np.isreal(roots)][0].real
n
```

```
120.29497623236388
```

Comparing this with the value given by assuming ideal gas behavior, it can be seen that there is a small difference:

```
nideal = p2 * V2 / R / T
nideal
```

```
121.08628931873625
```

The volume of this amount of nitrogen gas at 1 atm pressure can be approximated well by the ideal gas law, however:

```
V1 = n * R * T / p1
V1
```

```
2.9414208226397256
```

The work integral can be evaluated analytically for a van der Waals gas,

$$w = -\int_{V_1}^{V_2} p \, dV = -\int_{V_1}^{V_2} \frac{nRT}{V - nb} - \frac{n^2 a}{V^2} \, dV$$

$$= -nRT \ln\left(\frac{V_2 - nb}{V_1 - nb}\right) - n^2 a \left(\frac{1}{V_2} - \frac{1}{V_1}\right).$$

```
term1 = -n   * R * T * np.log((V2 - n*b) / (V1 - n * b))
term2 = -n**2 * a * (1/V2 - 1/V1)
work = term1 + term2
print(work)
```

```
1552012.608838411
```

This is about 1.55 MJ. For completeness, compare with the work required assuming the gas to be ideal:

```
work_ideal = -n   * R * T * np.log(V2 / V1)
work_ideal
```

```
1573230.4681104904
```

or 1.57 MJ: a little greater because an ideal gas has no attractive forces between its molecules.

To take the numerical approach:

```
def p(V, n, R, T, a, b):
    return n * R * T / (V - n * b) - n**2 * a / V**2

wp, err = quad(p, V1, V2, args=(n, R, T, a, b))
print(-wp, err)
```

```
1552012.6088384124 0.004105310628183645
```

Unsurprisingly, this is close to the analytical value found above.

21.4 Exercise

P21.1   Calculate the work required, per kg, to compress H_2 at 298 K to 200 bar. What proportion of the combustion enthalpy does this correspond to? Take $m(H_2) = 2 \, \text{g mol}^{-1}$ and treat hydrogen as a van der Waals gas with parameters $a = 0.2476 \, \text{L}^2 \, \text{bar mol}^{-2}$ and $b = 0.02661 \, \text{L mol}^{-1}$. The molar enthalpy of combustion is $\Delta_c H_m^{\ominus}(H_2) = -286 \, \text{kJ mol}^{-1}$.

Repeat the exercise for methane, using the values $m(CH_4) = 16 \, \text{g mol}^{-1}$, $a = 2.300 \, \text{L}^2 \, \text{bar mol}^{-2}$, $b = 0.04301 \, \text{L mol}^{-1}$, $\Delta_c H_m^{\ominus}(CH_4) = -891 \, \text{kJ mol}^{-1}$. Comment on the difference between the two gases.

22

Optimization with `scipy.optimize`

22.1 Multivariate Minimization and Maximization

A common numerical task is to find the maximum or minimum of a function. These are really the same problem, since the minimum of $f(x)$ is the maximum of $-f(x)$, and so we will deal only with minimization in this chapter.

The `scipy.optimize` package contains the useful `minimize` routine which implements several different algorithms for minimization. This will be described below, but you should be aware that, for an arbitrary function, there is no guarantee that a particular algorithm will find the desired minimum (or any minimum at all, in some cases). Usually, the output of the `scipy.optimize.minimize` call will provide helpful information to help analyze any problems.

Some of the minimization algorithms do not need any more than the function itself (which may be a function of more than one variable), $f(x_1, x_2, \ldots)$, implemented as the Python function, `fun`, and an initial guess for the parameters that minimize this function, $x_0 = (x_1^{(0)}, x_2^{(0)}, \ldots)$, provided as a `tuple`, `x0`, of suitable length:

```
scipy.optimize.minimize(fun, x0, ...)
```

To take a simple, one-dimensional example, consider the function $f(x) = x^2 - 4x + 8$. This quadratic function has a single minimum at $x = 2$, as can be readily verified by differentiation. The default minimization algorithm will find this point using pretty much any initial guess for it:

```
In [x]: def fun(x):
   ...:         return x**2 - 4 * x + 8

In [x]: minimize(fun, (0,))          # Initial guess, x0 = 0
Out[x]:
        fun: 4.0
  hess_inv: array([[0.5]])
        jac: array([5.96046448e-08])
```

```
  message: 'Optimization terminated successfully.'
     nfev: 6
      nit: 2
     njev: 3
   status: 0
  success: True
        x: array([2.00000002])
```

The returned object summarizes how the algorithm proceeded. The most important information in this object is as follows:

- fun: 4.0 – the value of the function at the minimum is 4;
- success: True and message: 'Optimization terminated success- fully.' – the optimization was successful;
- nfev: 6 – there were six function evaluations;
- nit: 2 – the algorithm took two iterations;
- x: array([2.00000002]) – the converged value of the parameter that minimizes $f(x)$ is indeed 2.

The other entries in the returned object relate to the first and second derivatives of the function (the *Jacobian* and the *Hessian*) and will not be described here.

Here is an example of where things can go wrong: Consider the function $f(\theta) = \cos\theta$.

```
In [x]: def f(theta):
   ...:         return np.cos(theta)

In [x]: minimize(f, (0,))          # Initial guess, theta0 = 0
Out[x]:
        fun: 1.0
  hess_inv: array([[1]])
        jac: array([-7.4505806e-09])
   message: 'Optimization terminated successfully.'
      nfev: 2
       nit: 0
      njev: 1
    status: 0
   success: True
         x: array([0.])
```

Here, we started the minimization with a guess ($\theta^{(0)} = 0$) that happened to fall on a maximum of the function and the default algorithm didn't manage to find its way downhill from there. Note that no warning or error was reported in this case: It is up to the user to check that the returned value for the minimum makes sense.

One solution to this problem is to start with a better guess, for example, $\theta^{(0)} = 1$ (the more you know about your function, the better):

```
In [x]: minimize(f, (1,))
Out[x]:
```

```
      fun: -0.9999999999887697
 hess_inv: array([[1.00540073]])
      jac: array([4.74601984e-06])
  message: 'Optimization terminated successfully.'
     nfev: 14
      nit: 3
     njev: 7
   status: 0
  success: True
        x: array([3.14159739])
```

Now, $\cos\theta$ has an infinite number of minima, and the one returned will depend on the initial guess: here, starting at $\theta^{(0)} = 1$ has led to the closest minimum, at $\theta = \pi$. Had we started with a guess of, say, $\theta^{(0)} = 10$, the returned minimum would have been the one at $\theta = 3\pi$.

Alternatively, a different minimization algorithm can be chosen with the `method` argument.[1] The direct-search Nelder–Meld method (which does not need the function derivatives and so is less likely to get confused if initialized on a stationary point of the function) would be a good choice:

```
In [x]: minimize(f, (0,), method="Nelder-Mead")    # Initial guess, theta0 = 0
Out[x]:
 final_simplex: (array([[3.1415625],
       [3.141625 ]]), array([-1., -1.]))
           fun: -0.9999999995453805
       message: 'Optimization terminated successfully.'
          nfev: 56
           nit: 28
        status: 0
       success: True
             x: array([3.1415625])
```

For a multivariate example, consider the two-dimensional function

$$f(x, y) = x^2 - ax - xy + by + 4y^2,$$

where a and b are constants.

```
In [x]: def f(X, a, b):
   ...:     x, y = X
   ...:     return x**2 - a*x - x*y + b*y + 4*y**2
```

Note that the Python function should receive the variables on which the minimization is to be performed (x and y) as a *sequence* (`list`, `tuple`, NumPy `array`, etc.). This sequence is given the name `X` here, and its contents are unpacked into the variables `x` and `y` for use. Any additional arguments (the constants a and b) follow this

---

[1] See the documentation at `https://docs.scipy.org/doc/scipy/reference/optimize.html` for details of the available algorithms.

object and are passed to the `minimize` method in the `args` argument. For example, to use $a = 4$, $b = 10$:

```
In [x]: X0 = (0, 0)      # Initial guesses for x and y

In [x]: minimize(f, X0, args=(4, 10))   # use constants a=4, b=10
Out[x]:
        fun: -8.266666666662827
   hess_inv: array([[0.54192027, 0.06855076],
          [0.06855076, 0.13374671]])
        jac: array([3.57627869e-06, 1.43051147e-06])
    message: 'Optimization terminated successfully.'
       nfev: 21
        nit: 6
       njev: 7
     status: 0
    success: True
          x: array([ 1.46666866, -1.06666624])
```

That is, the minimum lies at $(x, y) = (\frac{44}{30}, -\frac{16}{15})$.

22.2 Univariate Minimization and Maximization

A more convenient way to minimize a function of a single variable is provided by the method `scipy.optimize.minimize_scalar`. In its simplest use, an initial guess is not required:

```
In [x]: def f(x):
   ...:      return x**2 - 4 * x + 8

In [x]: minimize_scalar(f)
Out[x]:
        fun: 4.0
    message: '\nOptimization terminated successfully;\nThe returned value
             satisfies the termination criteria\n(using xtol = 1.48e-08 )'
       nfev: 9
        nit: 4
    success: True
          x: 2.0
```

However, if it is possible to *bracket* the required minimum, it can be faster and more effective, particularly for functions with multiple minima. For example, the function

$$f(x) = \left(\frac{x}{10}\right)^2 + \sin x$$

is depicted in Figure 22.1. The points $(a, f(a))$, $(b, f(b))$ and $(c, f(c))$ are said to bracket a minimum in a continuous scalar function $f(x)$ if, for $a < b < c$, $f(b) <$

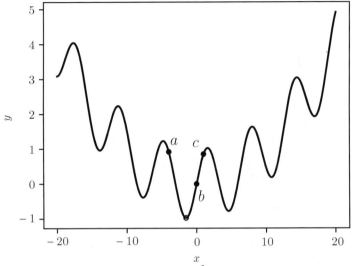

Figure 22.1   A scalar function, $f(x) = \left(\frac{x}{10}\right)^2 + \sin x$, with several minima; the minimum close to $x = 0$ is bracketed by the points a, b and c.

$f(a)$ and $f(b) < f(c)$. Loosely speaking, if $f(b)$ is less than $f(a)$ *and* $f(c)$, then there must be a minimum somewhere between a and c if the function is continuous.

```
In [x]: def f(x):
   ...:     return (x/10)**2 + np.sin(x)

In [x]: minimize_scalar(f, bracket=(-4, 0, 1))
Out[x]:
      fun: -0.9758098306412542
  message: '\nOptimization terminated successfully;\nThe returned value satisfies
               the termination criteria\n(using xtol = 1.48e-08 )'
     nfev: 13
      nit: 9
  success: True
        x: -1.5399916226762196
```

Choosing a different bracket leads to a different minimum:

```
In [x]: minimize_scalar(f, bracket=(15, 16, 19))
Out[x]:
      fun: 1.9264465683873035
  message: '\nOptimization terminated successfully;\nThe returned value satisfies
               the termination criteria\n(using xtol = 1.48e-08 )'
     nfev: 14
      nit: 10
  success: True
        x: 16.933261653884006
```

22.3 Example

E22.1

The Variational Principle Applied to the Particle in a Box

The one-dimensional particle-in-a-box system is one of the simplest in quantum mechanics and can be solved exactly. A particle of mass M is confined to some finite region where its potential energy is zero (i.e., the potential barrier outside this region is infinite). The Schrödinger equation is therefore

$$\hat{H}\psi = -\frac{\hbar^2}{2M}\frac{d^2\psi}{dx^2} + V(x) = E\psi,$$

where here we will take the box to be centered on the origin:

$$V(x) = \begin{cases} 0 & -\frac{L}{2} \leq x \leq \frac{L}{2} \\ \infty & x < -\frac{L}{2} \text{ and } x > \frac{L}{2} \end{cases}$$

The solution of this equation is a standard exercise, and gives the following wave-functions and energies:

$$\psi(x) = \begin{cases} \sqrt{\frac{2}{L}}\sin\left(\frac{\pi n x}{L}\right) & n \text{ even} \\ \sqrt{\frac{2}{L}}\cos\left(\frac{\pi n x}{L}\right) & n \text{ odd} \end{cases}$$

$$\text{and} \quad E_n = \frac{n^2\hbar^2\pi^2}{2ML^2} = \frac{n^2 h^2}{8ML^2},$$

where the quantum number $n = 1, 2, 3, \ldots$.

The Variational Principle implies that an approximate wavefunction, $\phi(x)$, will have an expected energy greater than or equal to the ground state energy, E_1.

$$\langle E \rangle = \frac{\langle \phi | \hat{H} | \phi \rangle}{\langle \phi | \phi \rangle} \geq E_1.$$

In the variational approach to determining the wavefunctions and energies of a system, an approximate wavefunction is constructed that depends on some parameters $a_1, a_2, \ldots, a_m$, and the expectation value of its energy minimized with respect to these parameters. This example will demonstrate the approach using the particle in a box system, since we can compare the approximation with the known exact values.

The approximate wavefunction chosen will be a polynomial in x. For convenience, we will work with in the *Hartree atomic unit system* in which $\hbar = 1$ and take $M = 1$ and $L = 2$, so that the "box" lies between $-1 \leq x \leq 1$. This is equivalent to the situation of an electron in a box of length $2a_0$ where $a_0 = 52.9$ pm is the Bohr radius. The ground state energy is then $\pi^2/8 = 1.2337\,E_h$ in units of the hartree, $E_h = 4.3597 \times 10^{-18}$ J.

```python
import numpy as np
from numpy.polynomial import Polynomial
import matplotlib.pyplot as plt
```

```python
# Particle mass, box length.
mass, L = 1, 2
# Grid of points on the x-axis, for plotting purposes.
x = np.linspace(-1, 1, 1000)

def psi(n):
    """Return the exact particle-in-a-box wavefunction for quantum number n."""
    if n % 2:
        return np.cos(np.pi * n * x / L)
    return np.sin(np.pi * n * x / L)

def E(n):
    return (n * np.pi)**2/ 2 / mass / L**2
```

```python
def plot_wavefunction(n):
    plt.plot(x, psi(n))
    plt.xlabel(r'$x\;/a_0$')
    plt.ylabel(r'$\psi(x)\;/a_0^{-1/2}$')

# We're interested in the ground state, n=1.
n = 1
plot_wavefunction(n)
```

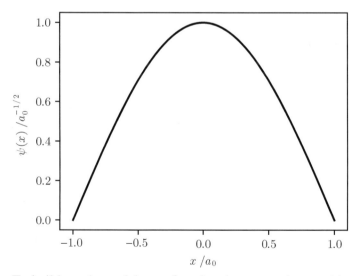

To build a polynomial wavefunction that approximates this ground state, note that ψ_1 is an *even* function: $\psi_1(-x) = \psi(x)$, so we only need to include even powers of x. Furthermore, the first order of approximation is a quartic polynomial, since the coefficients of a quadratic are constrained by the requirements that $\phi(-1) =$

$\phi(1) = 0$ and the normalization condition. Concretely, this quadratic can be written as

$$\phi_0 = N(x^2 + b),$$

where $\phi_0(1) = 0$ implies $b = -1$. The normalization condition $\langle \phi_0 | \phi_0 \rangle = 1$ then constrains the constant $N = \pm\sqrt{15/16}$. Both $\pm\phi_0(x)$ are valid approximations with the same energy: Since we have chosen $\psi(x)$ to be positive above, the choice

$$\phi_0 = \sqrt{\frac{15}{16}}(1 - x^2)$$

provides a direct comparison. The expectation value of the energy of this state is

$$\langle \phi_0 | \hat{H} | \phi_0 \rangle = -\frac{\hbar^2}{2M} \int_{-1}^{1} \phi_0 \frac{d^2\phi_0}{dx^2} \, dx = -\frac{1}{2}\frac{15}{16} \int_{-1}^{1} (1 - x^2)(-2) \, dx$$

$$= \frac{15}{16}\left[x - \frac{x^3}{3} \right]_{-1}^{1} = \frac{15}{8}\frac{2}{3} = \frac{5}{4}.$$

This energy, $1.25E_h$, is indeed greater than the true ground state energy, $1.2337E_h$ (and pretty close, given that we haven't actually optimized anything).

Now consider the general case of a polynomial of even order $2m \geq 4$. The same two constraints as before apply, and we can write the unnormalized trial wavefunction as:

$$\phi_m = -\left(1 + \sum_{k=1}^{m} a_k \right) + \sum_{k=1}^{m-1} a_k x^{2k} + x^{2m},$$

which has been constructed to satisfy the condition $\phi_m(-1) = \phi_m(1) = 0$.

We need functions to set up the polynomial from the coefficient parameters, a_k, and to evaluate the normalization integral, and the Rayleigh–Ritz ratio, $\langle E \rangle = \langle \phi | \hat{H} | \phi \rangle / \langle \phi | \phi \rangle$:

```
def phi_t(a):
    """
    Return the unnormalized trial wavefunction approximated as a polynomial
    with coefficients a[k] (ordered by smallest to largest powers to x).

    """

    ncoeffs = len(a) * 2 + 3
    coeffs = np.zeros(ncoeffs)
    coeffs[0] = -(1 + sum(a))
    coeffs[2:ncoeffs-1:2] = a
    coeffs[-1] = 1
    return Polynomial(coeffs)
```

```
def get_N2(phi):
    """Return the square of the normalization constant, <phi|phi>."""
    den = (phi * phi).integ()
    return den(1) - den(-1)

def rayleigh_ritz(a):
    """Return the Rayleigh-Ritz ratio, <phi|H|phi> / <phi|phi>."""
    phi = phi_t(a)
    phipp = phi.deriv(2)
    num = -(phi * phipp).integ() / 2

    N2 = get_N2(phi)
    return (num(1) - num(-1)) / N2
```

Next, we can write a function to minimize $\langle E \rangle$ with respect to a_k for a given level of approximation using `scipy.optimize.minimize`:

```
from scipy.optimize import minimize

def get_approx(m):
    """Get the optimum parameters, a_k, by minimizing <phi|H|phi> / <phi|phi>."""

    # Initial guess for the parameters: just set them all equal to unity.
    a0 = [1] * m
    res = minimize(rayleigh_ritz, a0)
    return res
```

How well does this work?

```
E1 = E(1)
mmax = 7
Eapprox = [None] * (mmax+1)
Eapprox[1] = 5 / 4
a = {}

for m in range(2, 7):
    res = get_approx(m)
    Eapprox[m] = res['fun']
    a[m] = res['x']

print('m          <E>/Eh     error')
error_ppm = [None] * (mmax+1)
for m in range(1, 7):
    # The approximation error, in parts per million.
    error_ppm[m] = (Eapprox[m] - E1) / E1 * 1.e6
    print(f'{m:.7f} {Eapprox[m]:.7f} {error_ppm[m]:>9.3f} ppm')
```

```
m          <E>/Eh     error
1.0000000 1.2500000 13211.836 ppm
2.0000000 1.2337224    17.676 ppm
3.0000000 1.2337066     4.888 ppm
4.0000000 1.2337043     3.015 ppm
5.0000000 1.2337025     1.559 ppm
6.0000000 1.2337013     0.577 ppm
```

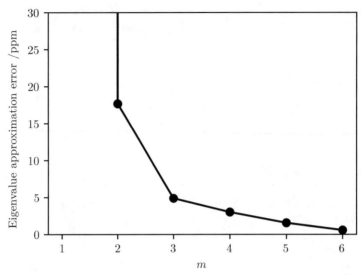

Figure 22.2   The improvement in approximation error to the ground state energy
of the particle-in-a-box system using a polynomial function with an increasing
number of terms.

Even one variational term improves the approximation a lot (from 13,221 ppm =
0.13%) to 18 ppm as can be seen from the output above and the plot produced by
the following code (Figure 22.2).

```
plt.plot(range(1, mmax+1), error_ppm[1:], 'o-')
plt.ylim(0, 30)
plt.xlabel(r'$m$')
plt.ylabel(r'Eigenvalue approximation error /ppm')
```

```
# The exact ground-state wavefunction.
plt.plot(x, psi(n), label=r'$\psi$')

def get_phi_approx(m):
    """Return the normalized approximate wavefunction, phi_1, to order m."""
    if m == 1:
        return np.sqrt(15/16) * (1 - x**2)
    phi = phi_t(a[m])
    if phi(0) < 0:
        phi = - phi
    return phi(x) / np.sqrt(get_N2(phi))

for m in range(1, mmax):
    plt.plot(x, get_phi_approx(m), label=f'$\phi_{m}$')
plt.xlabel(r'$x\;/a_0$')
plt.ylabel(r'$\psi(x)\;/a_0^{-1/2}$')
plt.legend()
```

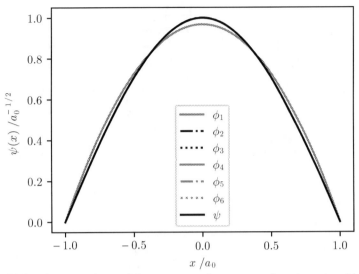

Figure 22.3 A comparison of the true ground state wavefunction, ψ, with polynomial approximations, ϕ_i. On this scale, all wavefunctions apart from ϕ_1 overlap with the exact solution.

All the approximate wavefunctions apart from the quadratic one overlap with the true ground state wavefunction, ψ, on this scale (Figure 22.3). It might be better, therefore, to plot the *difference* between the approximations, ϕ_m and ψ (Figure 22.4).

```python
# Plot a baseline, zero-error.
plt.plot([-1, 1], [0, 0], c='k', lw=1)
for m in range(1, mmax):
    plt.plot(x, psi(n) - get_phi_approx(m), label=f'$\Delta\phi_{m}$')
plt.ylim(-2.e-3, 2.e-3)
plt.xlabel(r'$x\;/a_0$')
plt.ylabel(r'$(\psi - \phi_m)\;/a_0^{-1/2}$')
plt.legend()
```

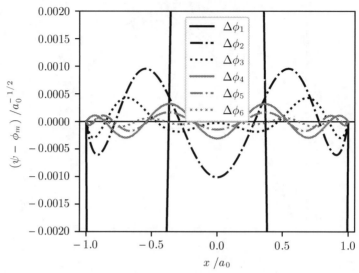

Figure 22.4   A comparison of the difference between the true ground state wave-function, ψ, and successive polynomial approximations, ϕ_i. The relatively large error in the first approximate wavefunction, $\Delta\phi_1$, is off the scale of the plot.

22.4 Exercises

P22.1   The Planck distribution describes the wavelength-dependency of the emitted energy density from a *black body*,

$$u(\lambda, T) = \frac{8\pi^2 hc}{\lambda^5} \frac{1}{e^{hc/\lambda k_B T} - 1},$$

Determine the wavelength of the maximum of this distribution for the temperatures 300 K (room temperature), 1,800 K (the melting point of iron), 5800 K (the surface temperature of the sun) and 11,000 K (the temperature of the blue supergiant star, Rigel).

Wien's displacement law predicts that this wavelength of maximum emission is proportional to $1/T$:

$$\lambda_{\max} T = b,$$

where b is a constant known as *Wien's displacement constant*.

Use the above calculation to determine a value for b and compare it with the "exact" value of b, which is available within the `scipy.constants` module.

P22.2   The one-dimensional quartic oscillator is one characterized by a potential energy proportional to the fourth power of the displacement. Taking the Hamiltonian for a quantum mechanical quartic oscillator to be

$$\hat{H} = -\frac{\hbar^2}{2\mu}\frac{d^2}{dx^2} + \frac{1}{2}kx^4,$$

minimize the expectation energy, $\langle E \rangle$, of the trial wavefunction $\psi = \exp(-\alpha x^2/2)$ with respect to its parameter, α (a) numerically, using `scipy.optimize.minimize` or `scipy.optimize.minimize_scalar`; (b) analytically by differentiation of $\langle E \rangle$ with respect to α.

Hints: first define the dimensionless quantity, $q = (k\mu/\hbar^2)^{1/6}x$. This simplifies the Schrödinger equation to

$$-\frac{1}{2}\frac{d^2\psi}{dq^2} + \frac{1}{2}q^4\psi = E'\psi,$$

where the energy, E', is measured in units of $(k\hbar^4/\mu^2)^{1/3}$.
For the analytical solution, define

$$I_n = \int_{-\infty}^{\infty} q^{2n}e^{-\alpha q^2}\, dq,$$

integrate by parts to derive the recursion relation

$$I_n = \frac{2n-1}{2\alpha}I_{n-1},$$

and note that $I_0 = \sqrt{\pi/\alpha}$.

Vibrational Spectroscopy

23.1 The Harmonic Oscillator Model

The simplest model for the vibration of a diatomic molecule is the one-dimensional harmonic oscillator, for which the potential energy is proportional to the square of the atoms' displacement from equilibrium, x, along the internuclear axis: $V(x) = \frac{1}{2}kx^2$, where $x = r - r_e$ and r is the instantaneous interatomic separation and r_e is the equilibrium bond length. The bond *force constant*, k, is a measure of how "stiff" the bond is: stronger bonds, with a higher k, require more force to stretch them and store more potential energy for a given displacement from equilibrium. The harmonic oscillator Hamiltonian is therefore

$$\hat{H} = -\frac{\hbar^2}{2\mu}\frac{d^2}{dx^2} + \frac{1}{2}kx^2,$$

where $\mu = m_1 m_2/(m_1 + m_2)$ is the reduced mass of the two atoms.

The harmonic oscillator is important because it can be used to approximate any real potential in the vicinity of its equilibrium point, and its corresponding Schrödinger equation, $\hat{H}\psi = E\psi$, can be solved exactly. The first step in the solution to this equation,

$$-\frac{\hbar^2}{2\mu}\frac{d^2\psi}{dx^2} + \frac{1}{2}kx^2\psi = E\psi$$

is a change of variable to the dimensionless quantity $q = (\mu k/\hbar^2)^{1/4}x$:

$$-\frac{1}{2}\frac{d^2\psi}{dq^2} + \frac{1}{2}q^2\psi = \frac{E}{\hbar\omega}\psi, \tag{23.1}$$

where $\omega = \sqrt{k/\mu}$ is the classical oscillator frequency. The solutions to Equation 23.1 are the one-dimensional harmonic oscillator wavefunctions and energies defined by a vibrational quantum number $v = 0, 1, 2, \ldots$:

$$\psi_v(q) = N_v H_v(q) \exp(-q^2/2) \quad \text{and } E_v = \hbar\omega(v + \tfrac{1}{2}),$$

where $N_v = (\sqrt{\pi}2^v v!)^{-1/2}$ is a normalization constant and $H_v(q)$ is the Hermite polynomial of order v, defined by

$$H_v(q) = (-1)^v e^{q^2} \frac{d^v}{dq^v}\left(e^{-q^2}\right).$$

The first few Hermite polynomials are listed below:

$$H_0(q) = 1$$
$$H_1(q) = 2q$$
$$H_2(q) = 4q^2 - 2$$
$$H_3(q) = 8q^3 - 12q$$
$$H_4(q) = 16q^4 - 48q^2 + 12$$

The Hermite polynomials follow a useful recurrence relation:

$$H_{v+1}(q) = 2qH_v(q) - 2vH_{v-1}(q),$$

so that, given the first two polynomials, $H_0 = 1$ and $H_1 = 2q$, all the others can be calculated.

23.2 Example

E23.1

The `scipy.special` package contains the function, `hermite`, which returns a Hermite polynomial object. For example:

```
from scipy.special import hermite

H2 = hermite(2)        # H_2(q) = 4q^2 - 2
print(H2(3))           # 4.3^2 - 2
```

```
34.00000000000001
```

The harmonic oscillator wavefunctions for any vibrational quantum number, v, can therefore be calculated and plotted using NumPy and Matplotlib.

```
import numpy as np

def psi_v(q, v):
    """Return the harmonic oscillator wavefunction, psi_v, at q."""
    Nv = 1 / np.sqrt( np.sqrt(np.pi) * 2**v * np.math.factorial(v))
    Hv = hermite(v)
    return Nv * Hv(q) * np.exp(-q**2 / 2)
```

```
import matplotlib.pyplot as plt

q = np.linspace(-4, 4, 500)
def plot_psi(q, v):
    psi = psi_v(q, v)
    plt.plot(q, psi, label=f'$v={v}$')
    plt.xlabel(r'$q$')
    plt.ylabel(r'$\psi_v$')
plot_psi(q, 0)
plot_psi(q, 1)
plt.legend()
```

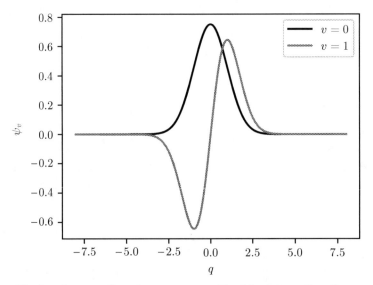

We can verify that the wavefunctions are normalized by integrating the corresponding probability distribution:

$$\int_{-\infty}^{\infty} |\psi_v(q)|^2 \, dq = 1$$

Using `scipy.integrate.quad`:

```
from scipy.integrate import quad

def P_v(q, v):
    """Return the probability distribution function for psi_v."""
    return psi_v(q, v)**2

v = 0
area, abserr = quad(P_v, -np.inf, np.inf, args=(v,))
print('area =', area)
```

```
area = 1.0
```

An important feature of the harmonic oscillator probability distribution is that there is a finite chance of finding the oscillator outside its classical turning points, a phenomenon known as *quantum mechanical tunneling* (see also Example E21.1).

The turning points occur when the classical oscillator's kinetic energy is zero and its total energy is potential energy. In our dimensionless units,

$$\frac{E}{\hbar\omega} = \tfrac{1}{2}q_\pm^2 \Rightarrow v + \tfrac{1}{2} = \tfrac{1}{2}q_\pm^2,$$

and hence

$$q_\pm = \pm\sqrt{2v+1}$$

This can be indicated on a plot of the harmonic oscillator probability density functions, and the probability of tunneling can be calculated from the integral of $|\psi_v|^2$ in the regions $q > q_+$ and $q < q_-$.

```python
def get_A(v):
    """Return the classical oscillator amplitude for state v."""
    return np.sqrt(2*v + 1)


def plot_tunnelling(v):
    """Plot the |psi_v(q)|^2 and illustrate the tunnelling effect."""

    P = P_v(q, v)

    # Plot the probability density distribution and store the line colour.
    line, = plt.plot(q, P, label=f'$v={v}$')
    c = line.get_color()

    # Get the classical oscillator amplitude; q_+ = A, q_- = -A.
    A = get_A(v)
    # The probability of tunnelling is the area outside -A < q < +A.
    P_tunnel = 2 * quad(P_v, A, np.infty, args=(v,))[0]
    print(f'v={v} tunnelling probability = {P_tunnel:.4f} ({P_tunnel*100:.2f}%)')

    # Indicate the tunnelling as a shaded region.
    plt.vlines([-A, A], 0, P_v(A, v), color=c)
    plt.fill_between(q[q < -A], P[q < -A], fc=c, alpha=0.2)
    plt.fill_between(q[q > +A], P[q > +A], fc=c, alpha=0.2)

    # Some tidying: remove the top and right frame borders
    ax = plt.gca()
    ax.spines['top'].set_visible(False)
    ax.spines['right'].set_visible(False)
    # Label the axes and ensure the plot y-axis starts at 0.
    plt.xlabel(r'$q$')
    plt.ylabel(r'$|\psi_v(q)|^2$')
    plt.ylim(0)

plot_tunnelling(0)
plot_tunnelling(4)
plt.legend()
```

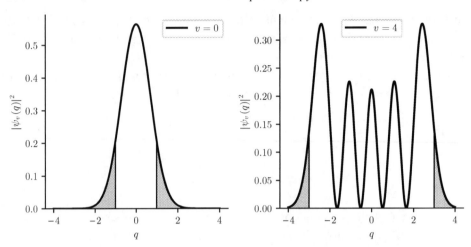

It is also helpful to compare the quantum mechanical probability distribution with the classical one. To derive this, consider the (classical) equation of motion for the potential $V(x) = \frac{1}{2}kx^2$:

$$F = -\frac{dV}{dx} = ma \Rightarrow -kx = \mu\frac{d^2x}{dt^2} \Rightarrow \frac{d^2x}{dt^2} = -\omega^2 x,$$

where $\omega = \sqrt{k/\mu}$. Solving this differential equation gives the classical trajectory:

$$x(t) = A\sin(\omega t + \phi),$$

where ϕ is a phase factor determined by the initial conditions and the amplitude is determined by $V(A) = E$ (all the energy is potential energy at the turning points), and hence $A = \sqrt{2E/k}$. The motion is oscillatory, with period $T = 2\pi/\omega$.

The classical probability density distribution, $P_{cl}\,dx$, can be thought of as proportional to the amount of time, dt, the oscillator spends at each infinitesimal region of its motion, dx:

$$P_{cl}\,dx = \frac{dt}{T}$$

Therefore,

$$P_{cl} = \frac{N\,dt}{T\,dx}, \quad \text{where } \frac{dt}{dx} = \left(\frac{dx}{dt}\right)^{-1} = [A\omega\cos(\omega t + \phi)]^{-1}$$

$$P_{cl} = \frac{N}{2\pi}\frac{1}{A\cos(\omega t + \phi)} = \frac{N}{2\pi}\frac{1}{\sqrt{A^2 - x^2}},$$

where the normalization constant, N, must be chosen so that

$$\int_{-A}^{A} P_{cl}\,dx = 1,$$

with the result that $N = 2$. The normalized classical probability density function is therefore

$$P_{cl} = \frac{1}{\pi} \frac{1}{\sqrt{A^2 - x^2}}.$$

Note that this function tends to infinity at the turning points, $\pm A$ (though the area under P_{cl} is finite). It has a minimum at $x = 0$: At the equilibrium point, it is moving fastest (all its energy is kinetic energy at this point).

```python
def P_cl(q, A):
    return 1/np.pi / np.sqrt(A**2 - q**2)

def plot_QM_and_classical_probabilities(q, v):
    P_qm_v = P_v(q, v)
    A = get_A(v)
    P_cl_v = P_cl(q, A)

    plt.plot(q, P_cl_v, label='classical')
    plt.plot(q, P_qm_v, label=f'quantum, $v={v}$')
    ymax = 1
    # As an alternative to pyplot.plot, pyplot.vlines([x1, x2, ...], y1, y2)
    # plots vertical lines between y-coordinates y1 and y2 at x = x1, x2, ...
    plt.vlines([-A, A], 0, ymax, ls='--')
    plt.ylim(0, ymax)
    plt.xlabel('$q$')
    plt.ylabel('$P(q)$')
    plt.legend()
```

The ground state ($v = 0$) probability density function of the quantum harmonic oscillator couldn't be more different from its classical counterpart:

```python
q = np.linspace(-3, 3, 1000)
plot_QM_and_classical_probabilities(q, 0)
```

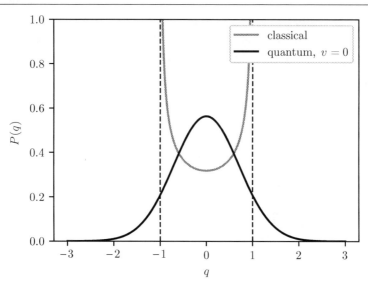

However, as the energy increases there is a closer agreement between the quantum and classical models (a phenomenon known as the *correspondence principle*).

```
q = np.linspace(-8, 8, 1000)
P_plot_QM_and_classical_probabilities(q, 20)
```

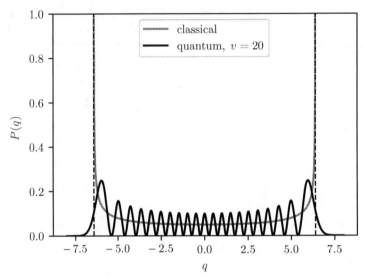

As long as one is looking for the oscillator in a not-too-small interval of space, $a \leq q \leq b$, the average of the quantum probability distribution in this region is close to the classical one.

23.3 Exercise

P23.1 By taking an average of $|\psi_v(q)|^2$ over a series of intervals of suitable constant width to smooth out its peaks, compare the classical and quantum mechanical probability density distributions for the $v = 20$ excited vibrational state of the harmonic oscillator.

24

The Morse Oscillator

The harmonic approximation does not provide a good model for the behavior of a real diatomic molecule in higher vibrational states. The true interatomic potential is not a parabola: real molecules dissociate at high excitation. Furthermore, the vibrational states of a real molecule converge with increasing vibrational quantum number, v, whereas the harmonic approximation predicts an infinite ladder of equally spaced levels given by $E_v = \hbar\omega(v + \frac{1}{2})$ (where ω is the oscillator angular frequency). The potential and energy levels for $^1H^{35}Cl$ are depicted in Figure 24.1 and compared with the harmonic approximation.

A better model for a real molecule is the *Morse oscillator*:

$$V(x) = D_e \left[1 - e^{-ax}\right]^2,$$

where D_e is the molecule's dissociation energy (measured from the bottom of the potential), the parameter $a = \sqrt{k_e/2D_e}$ and $k_e = (d^2V/dx^2)_e$ is the bond force constant at the bottom of the potential well.

The Morse oscillator Schrödinger equation can be solved exactly, and the corresponding energies are usually written (in wavenumber units) as

$$G(v) = \omega_e(v + \frac{1}{2}) - \omega_e x_e(v + \frac{1}{2})^2,$$

where

$$\omega_e = \frac{a}{2\pi c}\sqrt{\frac{2D_e}{\mu}} \quad \text{and} \quad \omega_e x_e = \frac{\omega_e^2}{4D_e}$$

are the equilibrium harmonic frequency and the anharmonicity constant, respectively. For most diatomic molecules, the energy gap, $G(1) - G(0)$, between the $v = 0$ ground state and the $v = 1$ first excited state is of the order of 1000 cm^{-1}. The ratio of the populations of these states is

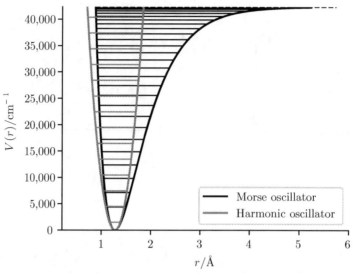

Figure 24.1   A comparison of the interatomic potential and vibrational energy levels for the $^1\text{H}^{35}\text{Cl}$ molecule and their harmonic approximations.

$$\frac{n_1}{n_0} = \exp\left(-\frac{\Delta E}{k_B T}\right) = \exp\left(-\frac{[G(1) - G(0)]hc}{k_B T}\right),$$

which evaluates to less than 1% at room temperature: Almost all of the molecules are in the $v = 0$ ground state.

The fundamental vibrational band, corresponding to the transition $v = 0 \to 1$, occurs at a wavenumber given by

$$\tilde{v}_{0 \to 1} = G(1) - G(0) = \omega_e - 2\omega_e x_e$$

Overtone bands, for which $\Delta v > 1$ are predicted to be observed at wavenumbers

$$\tilde{v}_{0 \to v} = G(v) - G(0) = \omega_e v - \omega_e x_e v(v + 1).$$

24.1 Example

E24.1

Inferring Morse Parameters from a Vibrational Spectrum

The vibrational spectrum of the CO molecule shows a fundamental band at $2143\,\text{cm}^{-1}$ and the first overtone band at $4260\,\text{cm}^{-1}$.

This information can be used to estimate the Morse parameters, D_e and a, and hence the bond strength and maximum vibrational level.

The observed bands can be used to estimate the parameters ω_e and $\omega_e x_e$:

$$\tilde{\nu}_{0\rightarrow 1} = \omega_e - 2\omega_e x_e = 2143\ \text{cm}^{-1},$$

$$\tilde{\nu}_{0\rightarrow 2} = 2\omega_e - 6\omega_e x_e = 4260\ \text{cm}^{-1}.$$

Therefore,

$$\omega_e x_e = \tfrac{1}{2}(2\tilde{\nu}_{0\rightarrow 1} - \tilde{\nu}_{0\rightarrow 2}),$$

$$\omega_e = \tilde{\nu}_{0\rightarrow 1} + 2\omega_e x_e.$$

```
nu_0_1, nu_0_2 = 2143, 4260
wexe = (2*nu_0_1 - nu_0_2)/2
we = nu_0_1 + 2*wexe
print(f'we = {we} cm-1, wexe = {wexe} cm-1.')
```

```
we = 2169.0 cm-1, wexe = 13.0 cm-1.
```

From the Morse oscillator model,

$$D_e = \frac{\omega_e^2}{4\omega_e x_e}\ \text{and}$$

$$a = 2\pi c\omega_e\sqrt{\frac{\mu}{2D_e}},$$

where μ is the reduced mass of the molecule.

```
import numpy as np
from scipy.constants import h, c, N_A, u
import matplotlib.pyplot as plt

# Dissociation energy (measured from the potential minimum, cm-1).
De = we**2 / 4 / wexe
# Reduced mass of (12C)(16O) (kg).
mu = 12 * 16 / (12 + 16) * u
# Convert everything to SI units in calculating a: De /J = hc.De /cm-1
a = 2 * np.pi * (c * 100) * we * np.sqrt(mu / 2 / (De * h * (c * 100)))
# Convert a from m-1 to A-1.
a *= 1.e-10
print(f'De = {De:.0f} cm-1, a = {a:.1f} A-1')
```

```
De = 90472 cm-1, a = 2.3 A-1
```

The Morse parameter, D_e, is the dissociation energy measured from the potential minimum. The bond strength is the energy required to dissociate the molecule from its ground state ($v = 0$, with energy $G(0) = \tfrac{1}{2}\omega_e - \tfrac{1}{4}\omega_e x_e$).

```
E0 = we / 2 - wexe / 4
D0 = De - E0
# Convert dissociation energy, D0, to kJ.mol-1.
```

```
bond_strength = D0 * h * (c * 100) * N_A / 1000
print(f'CO bond strength = {bond_strength:.0f} kJ.mol-1')
```

```
CO bond strength = 1069 kJ.mol-1
```

The vibrational energy levels converge with increasing quantum number, v; the highest energy level, v_{max} is given by the condition

$$\frac{dG(v)}{dv} = 0$$

$$\Rightarrow \omega_e - 2\omega_e x_e (v_{max} + \tfrac{1}{2}) = 0$$

$$\Rightarrow v_{max} = \frac{\omega_e}{2\omega_e x_e} - \frac{1}{2}$$

```
vmax = int(we / 2 / wexe - 0.5)
print(f'vmax = {vmax}')
```

```
vmax = 82
```

To plot the potential energy curve and energy levels, create an array of bond displacements, $x = r - r_e$ and use the Morse oscillator formula for $V(x)$:

```
# Grid of bond displacement values, x = r - re, in Angstrom.
x = np.linspace(-0.6, 2, 1000)
# Morse oscillator potential.
V = De * (1 - np.exp(-a*x))**2
```

We will draw the energy levels as horizontal lines between the classical turning points, which lie on the potential curve. Rearranging the above formula for $V(x)$ for an arbitrary energy level, E_v, these turning points are found to be at

$$x_{\pm} = -\frac{1}{a} \ln \left(1 \mp \sqrt{\frac{E_v}{D_e}} \right).$$

```
# Plot the Morse oscillator potential.
plt.plot(x, V)
plt.xlabel(r'$(r - r_\mathrm{e})/\mathrm{\AA}$')
plt.ylabel(r'$D_\mathrm{e}/\mathrm{cm^{-1}}$')
plt.ylim(0, De)

# Arrays of vibrational quantum number and corresponding energy.
v = np.arange(0, vmax+1, 1, dtype=int)
Ev = np.array([we * (v+0.5) - wexe * (v+0.5)**2])
# Classical turning points.
xm = - np.log(1 + np.sqrt(Ev / De)) / a
xp = - np.log(1 - np.sqrt(Ev / De)) / a
# Plot the energy levels as horizontal lines between xm and xp at Ev
plt.hlines(Ev, xm, xp)
```

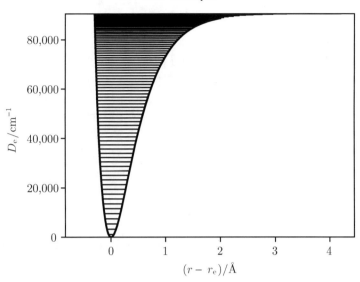

25

Solving Ordinary Differential Equations

Ordinary differential equations (ODEs) can be solved numerically with `scipy.integrate.solve_ivp` ("solve an initial value problem"). This method solves *first order* differential equations; to solve a higher-order equation, it must be decomposed into a system of first order equations, as described below.

25.1 A Single First-Order ODE

The general form of a first-order ODE is

$$\frac{dy}{dt} = f(t, y),$$

where $f(t, y)$ is some function of the independent variable (e.g., time, t) and y itself; the solution that is sought, $y(t)$, in general, varies with time in some way determined by an *initial condition*: a known value of y at some particular time (e.g., $t = 0$). Concretely, consider the kinetics of a first-order reaction, $A \rightarrow P$ with a rate constant, k. The independent variable is time, t; the solution we look for is $y(t) = [A](t)$, the reactant concentration as a function of time; and this solution depends on the initial concentration, $y(0) = [A]_0$, which we know. The first order kinetics are summarized by the differential equation:

$$\frac{d[A]}{dt} = -k[A],$$

or, denoting the reactant concentration as y, the equation $dy/dt = -ky$: in this case, the right-hand side function, $f(t, y) = -ky$, does not depend on t (though we expect y to, of course).

The analytical solution to this differential equation is easy to obtain:

$$[A](t) = [A]_0 e^{-kt},$$

but `solve_ivp` can also be used to give a numerical solution. We need to provide:

- a Python function that returns dy/dt as $f(t, y)$, given t and y;
- the initial and final time values for which the solution is desired, as a tuple, (t0, tf); and
- the initial condition, y0.

```
def dydt(t, y):
    return -k * y

soln = solve_ivp(dydt, (t0, tf), [y0])
```

(Note that the initial conditions must be provided in a `list` (or similar sequence), even if there is just one such condition, as here. The returned object, `soln`, contains information about the solution, as described in the following example.)

Example: The decomposition of N_2O_5

The decomposition of N_2O_5 is first order with a rate constant, $k = 5 \times 10^{-4}\,\text{s}^{-1}$. Given an initial concentration of $[N_2O_5] = 0.02\,\text{mol dm}^{-3}$, what is the concentration of N_2O_5 after 2 hours?

```
import numpy as np
from scipy.integrate import solve_ivp
import matplotlib.pyplot as plt

# First-order reaction rate constant, s-1.
k = 5.0e-4
# Initial concentration of N2O5, mol.dm-3.
y0 = 0.02
# Initial and final time points for the integration (s).
t0, tf = 0, 2 * 60 * 60

def dydt(t, y):
    """Return dy/dt = f(t, y) at time t."""
    return -k * y

# Integrate the differential equation.
soln = solve_ivp(dydt, (t0, tf), [y0])
print(soln)
```

The returned object is printed as:

```
 message: 'The solver successfully reached the end of the integration interval.'
    nfev: 50
    njev: 0
     nlu: 0
     sol: None
  status: 0
 success: True
```

```
     t: array([0.00000000e+00, 4.61789293e-01, 5.07968222e+00, 5.12586115e+01,
        5.13047904e+02, 2.37586665e+03, 4.14917646e+03, 5.96008698e+03,
        7.20000000e+03])
t_events: None
     y: array([[0.02      , 0.01999538, 0.01994927, 0.01949393, 0.01547473,
        0.00610149, 0.0025153 , 0.00101769, 0.00054751]])
y_events: None
```

The solution was obtained successfully in 50 function evaluations and the initial, intermediate and final values of t and y are reported. The concentration of N_2O_5 after 2 hours (7,200 secs) is $5.48 \times 10^{-4} \, \text{mol dm}^{-3}$, as confirmed by the analytical solution:

```
In [x]: y0 * np.exp(-k * tf)
Out[x]: 0.0005464744489458512        # mol.dm-3
```

The nine time points, given as the array t in the returned solution, that were used to determine the value of $y(t_f)$ were chosen by the ODE solver algorithm. To follow the reactant concentration as a function of time in higher resolution, provide a suitable sequence of time points as the argument t_eval:

```
In [x]: # Integrate the differential equation, report at 21 time points.
In [x]: t_eval = np.linspace(t0, tf, 21)
   ...: soln = solve_ivp(dydt, (t0, tf), [y0], t_eval=t_eval)

In [x]: t, y = soln.t, soln.y[0]

In [x]: t
Out[x]:
array([   0.,  360.,  720., 1080., 1440., 1800., 2160., 2520., 2880.,
       3240., 3600., 3960., 4320., 4680., 5040., 5400., 5760., 6120.,
       6480., 6840., 7200.])

In [x]: y
Out[x]:
array([0.02      , 0.0167054 , 0.01395191, 0.0116484 , 0.00972759,
       0.00812852, 0.00679511, 0.00567696, 0.0047402 , 0.00395862,
       0.00330752, 0.00276439, 0.00230921, 0.00192808, 0.00161016,
       0.00134539, 0.00112455, 0.00093947, 0.00078467, 0.00065544,
       0.00054751])
```

Better still, setting the dense_output argument to True defines an OdeSolution object named sol as one of the returned objects. This can be called to generate interpolated values of the solution for arbitrary intermediate values of the time points:

```
# Solve the ODE
soln = solve_ivp(dydt, (t0, tf), [y0], dense_output=True)
# Evaluate the solution, y(t), at 41 time points.
t = np.linspace(t0, tf, 41)
y = soln.sol(t)[0]
```

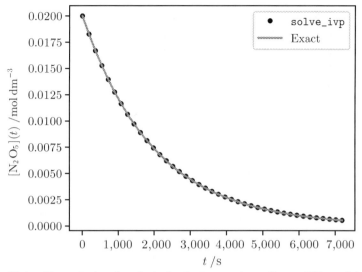

Figure 25.1   Numerical and analytical solutions to the ordinary differential equation governing the decomposition of $[N_2O_5]$.

```
# Plot and compare the numerical and exact solutions.
plt.plot(t, y, 'o', color='k', label=r'solve_ivp')
plt.plot(t, y0 * np.exp(-k*t), color='gray', label='Exact')
plt.xlabel('t /s')
plt.ylabel('[N_20_5](t) /mol.dm-3')
plt.legend()
plt.show()
```

The resulting plot, shown in Figure 25.1 demonstrates that the numerical algorithm was able to follow the true solution accurately in this case.

25.2 Coupled First-Order ODEs

Given a set of coupled first-order ODEs in more than one dependent variable, $y_1(t), y_2(t), \ldots, y_n(t)$:

$$\frac{dy_1}{dt} = f_1(y_1, y_2, \ldots, y_n; t),$$

$$\frac{dy_2}{dt} = f_2(y_1, y_2, \ldots, y_n; t),$$

$$\cdots$$

$$\frac{dy_n}{dt} = f_n(y_1, y_2, \ldots, y_n; t).$$

A numerical solution to each of them can be obtained using `solve_ivp`. In this case, the function provided must receive the instantaneous values of y_i in a sequence such as a `list` and return another sequence of their derivatives in the same order. The form of this function is therefore:

```
def deriv(t, y):
    # y = [y1, y2, y3, ...] is a sequence of dependent variables.
    dy1dt = f1(y, t)    # calculate dy1/dt as f1(y1, y2, ..., yn; t)
    dy2dt = f2(y, t)    # calculate dy2/dt as f2(y1, y2, ..., yn; t)
    # ... etc
    # Return the derivatives in a sequence such as a tuple:
    return dy1dt, dy2dt, ..., dyndt
```

As a simple example, consider the sequence of first-order reactions:

$$A \rightarrow B \qquad\qquad\qquad k_1$$
$$B \rightarrow P \qquad\qquad\qquad k_2$$

The equations governing the rate of change of A and B are

$$\frac{d[A]}{dt} = -k_1[A],$$
$$\frac{d[B]}{dt} = k_1[A] - k_2[B].$$

This pair of coupled equations can again be solved analytically (see Chapter 11), but in our numerical solution, let $y_1 \equiv [A]$ and $y_2 \equiv [B]$:

$$\frac{dy_1}{dt} = -k_1 y_1,$$
$$\frac{dy_2}{dt} = k_1 y_1 - k_2 y_2.$$

The derivative function would then be defined as:

```
def deriv(t, y):
    """Return dy_i/dt for each y_i at time t."""
    y1, y2 = y
    dy1dt = -k1 * y1
    dy2dt = k1 * y1 - k2 * y2
    return dy1dt, dy2dt
```

Here, the values of `k1` and `k2` are assumed to be picked up in the enclosing scope (typically they might be defined in the global scope of the script); it is probably better to pass them explicitly as arguments to the `deriv` function, after the `y` array. In this case, `solve_ivp` needs to know what they are and they should be packed into its argument, `args`, as in the following complete example:

```
# First-order reaction rate constants, s-1.
k1, k2 = 1, 0.1
# Initial concentrations of species A and B (mol.dm-3).
y0 = A0, B0 = 1, 0
# Initial and final time points for the integration (s).
```

```
t0, tf = 0, 40

def deriv(t, y, k1, k2):
    """Return dy_i/dt for each y_i at time t."""
    y1, y2 = y
    dy1dt = -k1 * y1
    dy2dt = k1 * y1 - k2 * y2
    return dy1dt , dy2dt

# Integrate the differential equation.
soln = solve_ivp(deriv, (t0, tf), y0, dense_output=True, args=(k1, k2))
print(soln.message)

t = np.linspace(t0, tf, 200)
A, B = soln.sol(t)
# The concentration, [P], is determined by conservation.
P = A0 - A - B

plt.plot(t, A, label=r'$\mathrm{[A]}$')
plt.plot(t, B, label=r'$\mathrm{[B]}$')
plt.plot(t, P, label=r'$\mathrm{[P]}$')
plt.xlabel(r'$t\;/\mathrm{s}$')
plt.ylabel(r'$\mathrm{conc.\;/mol\,dm^{-3}}$')
plt.legend()
plt.show()
```

The solution produced, illustrated in Figure 25.2 agrees well with the analytical solution from Chapter 11.

The default algorithm used by `solve_ivp` is `'RK45'`, an explicit Runge–Kutta method, which is a good general-purpose approach for many problems. There is, however, a common class of differential equation, including many of those that arise in chemical kinetics, known as *stiff*. A problem is said to be stiff if a numerical method is required to take excessively small steps in its intervals of integration in relation to the smoothness of the exact underlying solution. Stiff problems frequently occur when terms in the ODE represent a variable changing in magnitude with very different timescales (e.g., when a chemical reaction mechanism involves reaction steps with very different rate constants). The methods `'Radau'`, `'BDF'` and `'LSODA'` are worth trying if you suspect your ODE is stiff. The method is identified by passing a string to the argument `method` in the `solve_ivp` call, as in Example E25.1 at the end of this chapter.

25.3 A Single Second-Order ODE

A differential equation with a single dependent variable of order n can be written as a system of n first-order differential equations in n dependent variables; in this form, `solve_ivp` can be used to obtain a solution to second- and higher-order ODEs, as in the following example.

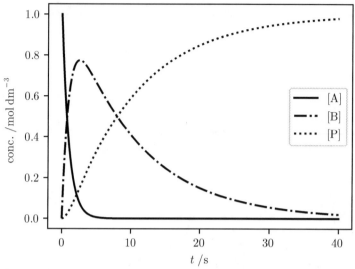

Figure 25.2   Numerical solution to the ODEs governing the reaction sequence $A \rightarrow B \rightarrow P$ for rate constants $k_1 = 1\,s^{-1}$ and $k_2 = 0.1\,s^{-1}$ and taking $[A]_0 = 1\,mol\,dm^{-3}$, $[B]_0 = [P]_0 = 0$.

Consider the equation of motion for a harmonic oscillator characterized by a mass, m and spring (force) constant, k: The restoring force upon displacement from equilibrium by a distance, x, is $-kx$ and so

$$F = m\frac{d^2x}{dt^2} \quad \Rightarrow \quad \frac{d^2x}{dt^2} = -\omega^2 x,$$

where $\omega = \sqrt{k/m}$. This equation may be decomposed into two first-order equations as follows:

$$\frac{dx_1}{dt} = x_2,$$

$$\frac{dx_2}{dt} = -\omega^2 x_1,$$

where x_1 is identified with x and x_2 with dx/dt.

Once again, there is a well-known analytical solution to this differential equation, so comparison with the numerical approach is straightforward (Figure 25.3):

```
# Mass (kg), force constant (N.m-1)
m, k = 1, 2
omega = np.sqrt(k / m)

# Oscillator initial conditions (position, m, and velocity, m.s-1).
x0 = 0.01, 0
# Initial and final time points for the integration (s).
t0, tf = 0, 10
```

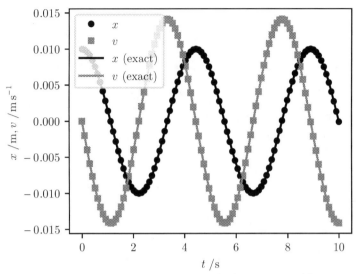

Figure 25.3   Numerical and exact solutions to the harmonic oscillator ODE for a mass, $m = 1$ kg and force constant $k = 2$ N m^{-1} with initial conditions $x(0) = 1$ cm and $v(0) = 0$.

```
def deriv(t, x, omega):
    """Return dx_1/dt and dx_2/dt = dx_1^2/dt^2 time t."""
    x1, x2 = x
    dx1dt = x2
    dx2dt = -omega**2 * x1
    return dx1dt , dx2dt


# Integrate the differential equation.
soln = solve_ivp(deriv, (t0, tf), x0, dense_output=True, args=(omega,))
print(soln.message)

t = np.linspace(t0, tf, 200)
x, v = soln.sol(t)

plt.plot(t, x, label=r'$x$')
plt.plot(t, v, label=r'$v$')
plt.xlabel(r'$t\;/\mathrm{s}$')
plt.ylabel(r'$x \;/\mathrm{m}, v \;/\mathrm{m\,s^{-1}}$')
plt.legend()
plt.show()
```

25.4 Example

E25.1

A simple mechanism for the formation of ozone in the stratosphere consists of the following four reactions (known as the *Chapman cycle*):

$$O_2 + h\nu \rightarrow 2O \qquad\qquad k_1 = 3 \times 10^{-12}\,\mathrm{s}^{-1}$$

$$O_2 + O + M \rightarrow O_3 + M \qquad k_2 = 1.2 \times 10^{-33}\,\mathrm{cm^6\,molec^{-2}\,s^{-1}}$$

$$O_3 + h\nu' \rightarrow O + O_2 \qquad\qquad k_3 = 5.5 \times 10^{-4}\,\mathrm{s}^{-1}$$

$$O + O_3 \rightarrow 2O_2 \qquad\qquad k_4 = 6.9 \times 10^{-16}\,\mathrm{cm^3\,molec^{-1}\,s^{-1}}$$

where M is a nonreacting third body taken to be at the total air molecule concentration for the altitude being considered. These reactions lead to the following rate equations for $[O]$, $[O_3]$ and $[O_2]$:

$$\frac{d[O_2]}{dt} = -k_1[O_2] - k_2[O_2][O][M] + k_3[O_3] + 2k_4[O][O_3],$$

$$\frac{d[O]}{dt} = 2k_1[O_2] - k_2[O_2][O][M] + k_3[O_3] - k_4[O][O_3],$$

$$\frac{d[O_3]}{dt} = k_2[O_2][O][M] - k_3[O_3] - k_4[O][O_3].$$

The rate constants apply at an altitude of 25 km, where $[M] = 9 \times 10^{17}$ molec cm^{-3}.

This mechanism can be modeled using a numerical differential equation solver, though it is more sensible to invoke the steady-state approximation, which predicts equilibrium concentrations of ozone and atomic oxygen given by:

$$[O_3] = \sqrt{\frac{k_1 k_2}{k_3 k_4}}[O_2][M]^{\frac{1}{2}}, \qquad \frac{[O]}{[O_3]} = \frac{k_3}{k_2[O_2][M]}.$$

The following code performs the numerical integration of the above ODEs using `scipy.integrate.solve_ivp` and compares the result with the steady-state values above.

```python
import numpy as np
from scipy.integrate import solve_ivp
import matplotlib.pyplot as plt

# Reaction rate constants.
k1 = 3.e-12          #      O2 + hv -> 2O       (s-1)
k2 = 1.2e-33         # O2 + O + M -> O3 + M  (cm6.molec-2.s-1)
k3 = 5.5e-4          #    O3 + hv' -> O + O2   (s-1)
k4 = 6.9e-16         #     O + O3 -> 2O2       (cm3.molec-1.s-1)

def deriv(t, c, M):
    """ Return the d[X]/dt for each species."""
    O2, O, O3 = c
    dO2dt = -k1*O2 - k2*M*O*O2 + k3*O3 + 2*k4*O*O3
    dOdt = 2*k1*O2 - k2*M*O*O2 + k3*O3 - k4*O*O3
    dO3dt = k2*M*O*O2 - k3*O3 - k4*O*O3
    return dO2dt, dOdt, dO3dt

# Total molecule concentration, M, and O2 concentration, cO2.
M = 9.e17
cO2 = 0.21*M
```

```
# Initial conditions for [O2], [O], [O3]
c0 = [cO2, 0, 0]
# Integrate the differential equations over a suitable time grid (s).
ti, tf = 0, 5.e7
# NB We need a solver that is robust to stiff ODEs.
soln = solve_ivp(deriv, (ti, tf), c0, args=(M,), method='LSODA')
t, c = soln.t, soln.y

# Steady-state approximation solution for comparison.
cO3ss = np.sqrt(k1 * k2 / k3 / k4 * M) * cO2
cOss = k3 * cO3ss / k2 / cO2 / M
print('Numerical values:\n[O] = {:g} molec/cm3, [O3] = {:g} molec/cm3'
          .format(*c[1:,-1]))
print('Steady-state values:\n[O]ss = {:g} molec/cm3, [O3]ss = {:g} molec/cm3'
          .format(cOss, cO3ss))

# Plot the evolution of [O3] and [O] with time
plt.plot(t,c[2], label=r'$\mathrm{[O_3]}$')
plt.plot(t,c[1], label=r'$\mathrm{[O]}$')
plt.yscale('log')
plt.ylim(1.e5, 1.e14)
plt.xlabel(r'$t\;/\mathrm{s}$')
plt.ylabel(r'$[\cdot\,]\;/\mathrm{molec\,cm^{-3}}$')
plt.legend()
plt.show()
```

```
Numerical values:
[O] = 4.69124e+07 molec/cm3, [O3] = 1.7407e+13 molec/cm3
Steady-state values:
[O]ss = 4.7055e+07 molec/cm3, [O3]ss = 1.74634e+13 molec/cm3
```

Clearly, the steady-state approximation works well for this system (Figure 25.4).

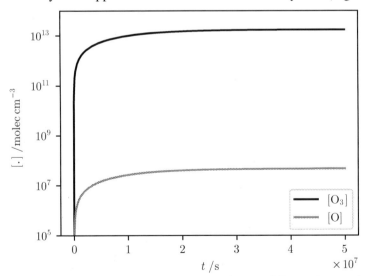

Figure 25.4   Ozone and oxygen atom concentrations rapidly reach a steady state under the Chapman Cycle reaction mechanism.

26

The Oregonator

26.1 Theory and Analysis

The Belousov–Zhabotinsky (BZ) reaction is a famous example of a chemical oscillator: a complex mechanism in which one of its component species undergoes a periodic change in concentration. There are different variants, but a commonly used composition is a mixture of potassium bromate ($KBrO_3^-$), cerium (IV) sulfate ($Ce(SO_4)_2$) and propanedioic acid ($CH_2(CO_2H)_2$) in dilute sulfuric acid.

The oscillating concentrations of Ce^{3+} and Ce^{4+} are seen as a color change according to which of two principle processes dominate. At high bromide ion concentration, bromate is reduced:

$$2Br^- + BrO_3^- + 3H^+ + 3CH_2(CO_2H)_2 \rightarrow 3BrCH(CO_2H)_2 + 3H_2O$$

The Br^- is consumed by this reaction and at some point its concentration is low enough that a second process takes over: the oxidation of Ce^{3+} in an autocatalytic step involving bromous acid, $HBrO_2$:

$$2Ce^{3+} + BrO_3^- + HBrO_2 + 3H^+ \rightarrow 2Ce^{4+} + H_2O + 2HBrO_2$$

However, at some point the bromide ion concentration will begin to rise again through the reduction of Ce^{4+}:

$$4Ce^{4+} + BrCH(CO_2H)_2 + 2H_2O \rightarrow 4Ce^{3+} + HCO_2H + 2CO_2 + 5H^+ + Br^-$$

and subsequent reactions consuming $HBrO_2$. Then the first process dominates again.

The BZ reaction has a complex mechanism with many intermediate steps, but its essential features of its chemical dynamics may be represented by the following simple five-step model:

$$k_1: \quad A + Y \quad \rightarrow X + P$$
$$k_2: \quad X + Y \quad \rightarrow 2P$$
$$k_3: \quad A + X \quad \rightarrow 2X + 2Z$$
$$k_4: \quad\quad 2X \quad \rightarrow A + P$$
$$k_5: \quad B + Z \quad \rightarrow \tfrac{1}{2}fY$$

Here, X is identified with $HBrO_2$, Y is Br^- and Z is Ce^{4+}. The reactants, $A = BrO_3^-$ and $B = CH_2(CO_2H)_2$, and P represents the product species, $BrCH(CO_2H)_2$ and HOBr, which are assumed to be present at much higher concentrations than [X], [Y] and [Z] and therefore effectively constant for the purposes of analyzing the oscillations in these intermediates. f is an adjustable parameter in the model which generalizes it to some extent (essentially by controlling the concentration of Ce^{4+}).

This model is known as the *Oregonator* and was created in the 1970s by Field and Noyes at the University of Oregon.[1] According to the acknowledgments in this paper, the existence, form and significance of the Oregonator occurred to one of the authors, "during an exceedingly dull sermon."

The rate equations for the intermediate species may be written:

$$\frac{d[X]}{dt} = k_1[A][Y] - k_2[X][Y] + k_3[A][X] - 2k_4[X]^2,$$

$$\frac{d[Y]}{dt} = -k_1[A][Y] - k_2[X][Y] + \tfrac{1}{2}fk_5[B][Z],$$

$$\frac{d[Z]}{dt} = 2k_3[A][X] - k_5[B][Z],$$

where the adjustable parameter, f, takes values in the range 0–3. Although we could go ahead and integrate these equations numerically, it is helpful to scale them so as to make them dimensionless and reduce the number of constants:

$$\alpha \frac{dx}{d\tau} = \gamma y - xy + x(1 - x)$$

$$\beta \frac{dy}{d\tau} = -\gamma y - xy + fz$$

$$\frac{dz}{d\tau} = x - z$$

where

$$x = \frac{2k_4[X]}{k_3[A]}, \quad y = \frac{k_2[Y]}{k_3[A]}, \quad z = \frac{k_4k_5[B][Z]}{(k_3[A])^2}; \quad \text{and}$$

$$\alpha = \frac{k_5[B]}{k_3[A]}, \quad \beta = \frac{2k_4k_5[B]}{k_2k_3[A]}, \quad \gamma = \frac{2k_1k_4}{k_2k_3},$$

as derived at the end of this chapter.

[1] R. J. Field, E. Koros, R. M. Noyes, *J. Am. Chem. Soc.* **94**, 8649 (1972).

The rate constants, k_i, used in the above model relate to the rates of particular reaction steps in the full mechanism and depend on the solution pH through:

$$k_1 = (2.1 \, \text{M}^{-3} \, \text{s}^{-1})[\text{H}^+]^2$$

$$k_2 = (3 \times 10^6 \, \text{M}^{-2} \, \text{s}^{-1})[\text{H}^+]$$

$$k_3 = (42 \, \text{M}^{-2} \, \text{s}^{-1})[\text{H}^+]$$

$$k_4 = (3000 \, \text{M}^{-2} \, \text{s}^{-1})[\text{H}^+]$$

The value of k_5 may be taken to be $1 \, \text{M}^{-1} \, \text{s}^{-1}$ and the stoichiometric factor,[2] $f = 1$.

SciPy's solve_ivp function can be used to integrate the differential equations of the Oregonator model, given a set of initial conditions, which here we take to be $x(0) = y(0) = z(0) = 1$. The reactant concentrations (which we assume to be constant over the few cycles of the reaction studied here) are taken to be $[A] = 0.06 \, \text{M}$ and $[B] = 0.02 \, \text{M}$.

```python
import numpy as np
from scipy.integrate import solve_ivp
from matplotlib import pyplot as plt
```

```python
def deriv(tau, X, alpha, beta, gamma, f):
    """Return the derivatives dx/dtau, dy/dtau and dz/dtau."""

    x, y, z = X
    dxdtau = (gamma*y - x*y + x*(1-x)) / alpha
    dydtau = (-gamma*y - x*y + f*z) / beta
    dzdtau = x - z
    return dxdtau, dydtau, dzdtau
```

```python
def solve_oregonator(Hp=0.8, f=1, A=0.06, B=0.02):
    """Integegrate the Oregonator differential equations.

    Hp is the H+ concentration in M, f the stoichiometric parameter and
    A and B the constant reactant concentrations.

    """

    kp1, kp2, kp3, kp4 = 2.1, 3e6, 42, 3e3
    k1, k2, k3, k4 = kp1 * Hp**2, kp2 * Hp, kp3 * Hp, kp4 * Hp
    k5 = 1
    alpha = k5 * B / k3 / A
    beta = 2 * k4 * k5 * B / k2 / k3 / A
    gamma = 2 * k1 * k4 / k2 / k3

    # Initial and final (scaled) times for the integration.
    tau_i, tau_f = 0, 40
    # Initial conditions, x(0), y(0), z(0).
```

---

[2] R. J. Field, "Oregonator". *Scholarpedia*, **2**(5):1386 (2007).

```
X0 = (1, 1, 1)
# Solve the differential equations, using the Radau method (an
# implicit Runge-Kutta algorithm suited to stiff ODEs.
soln = solve_ivp(deriv, (tau_i, tau_f), X0, dense_output=True,
                 args=(alpha, beta, gamma, f), method='Radau')

# Interpolate the solution onto a suitable grid of (scaled) times.
tau = np.linspace(tau_i, tau_f, 10000)
x, y, z = soln.sol(tau)
return tau, x, y, z
```

```
def plot_oregonator(tau, x, y, z):
    """Plot the scaled concentrations, x(tau), y(tau) and z(tau)."""

    fig, axes = plt.subplots(nrows=3, ncols=1)
    axes[0].plot(tau, np.log10(x))
    axes[1].plot(tau, np.log10(y))
    axes[2].plot(tau, np.log10(z))
    axes[0].set_xticklabels([])
    axes[1].set_xticklabels([])
    axes[0].set_ylabel(r'$\log_{10}(x)\;(\mathrm{HBrO_2})$')
    axes[1].set_ylabel(r'$\log_{10}(y)\;(\mathrm{Br^-})$')
    axes[2].set_ylabel(r'$\log_{10}(z)\;(\mathrm{Ce^{4+}})$')
    axes[2].set_xlabel(r'scaled time, $\tau$')
```

```
tau, x, y, z = solve_oregonator(Hp=0.8, f=1)
plot_oregonator(tau, x, y, z)
```

The result is shown in Figure 26.1.

The periodic nature of the reaction composition can also be shown in a plot of its *limit cycle*: the closed trajectory of the concentrations x, y and z (after an initial induction time). For example, the following code generates Figure 26.2:

```
def plot_limit_cycle(x, y, z):
    """Plot the limit cycle as a pair of plots (x-z and y-z)."""

    fig, axes = plt.subplots(nrows=1, ncols=2)
    axes[0].plot(np.log10(x), np.log10(z))
    axes[0].set_xlabel('$\log_{10}(x)\;(\mathrm{HBrO_2})$')
    axes[0].set_ylabel(r'$\log_{10}(z)\;(\mathrm{Ce^{4+}})$')

    axes[1].plot(np.log10(y), np.log10(z))
    axes[1].set_xlabel('$\log_{10}(y)\;(\mathrm{Br^-})$')
    axes[1].set_ylabel(r'$\log_{10}(z)\;(\mathrm{Ce^{4+}})$')
    plt.tight_layout()

plot_limit_cycle(x, y, z)
```

In contrast, no oscillation is observed within this model for values of f greater than about 2.4, as can be seen in the following plots (Figures 26.3 and 26.4):

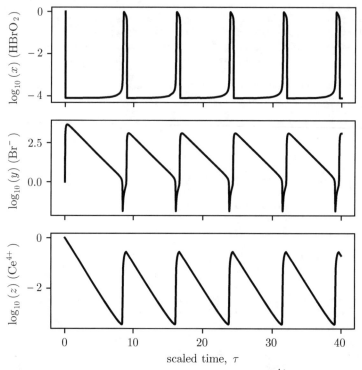

Figure 26.1   Scaled concentrations of HBrO$_2$, Br$^-$ and Ce^{4+} as a function of the scaled time coordinate, τ, for stoichiometric parameter, $f = 1$.

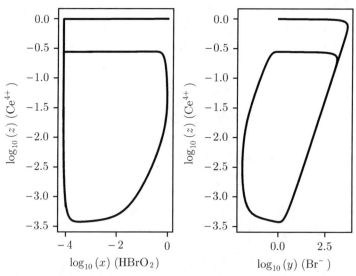

Figure 26.2   Periodic behavior in the concentrations of HBrO$_2$, Br$^-$ and Ce^{4+} plotted as a limit cycle, for stoichiometric parameter, $f = 1$.

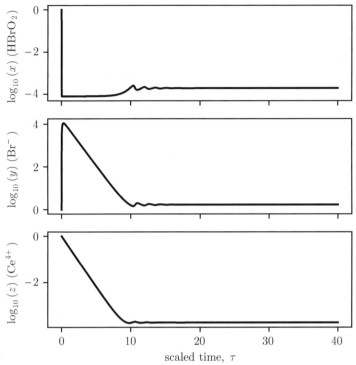

Figure 26.3   Scaled concentrations of $HBrO_2$, Br^- and Ce^{4+} as a function of the scaled time coordinate, τ, for stoichiometric parameter, $f = 2.42$.

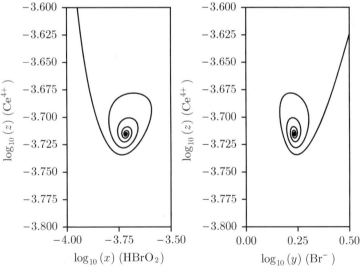

Figure 26.4   Convergence in the concentrations of $HBrO_2$, Br^- and Ce^{4+} plotted as a limit cycle, for stoichiometric parameter, $f = 2.42$.

```
tau, x, y, z = solve_oregonator(Hp=0.8, f=2.42)
plot_oregonator(tau, x, y, z)
```

```
fig, axes = plt.subplots(nrows=1, ncols=2)
axes[0].plot(np.log10(x), np.log10(z))
axes[0].set_xlabel('$\log_{10}(x)\;(\mathrm{HBrO_2})$')
axes[0].set_ylabel(r'$\log_{10}(z)\;(\mathrm{Ce^{4+}})$')
axes[0].set_xlim(-4, -3.5)
axes[0].set_ylim(-3.8, -3.6)

axes[1].plot(np.log10(y), np.log10(z))
axes[1].set_xlabel('$\log_{10}(y)\;(\mathrm{Br^-})$')
axes[1].set_ylabel(r'$\log_{10}(z)\;(\mathrm{Ce^{4+}})$')
axes[1].set_xlim(0, 0.5)
axes[1].set_ylim(-3.8, -3.6)

plt.tight_layout()
```

Derivation of Dimensionless Rate Equations The rate equations governing the concentrations of the intermediate species in the Oregonator mechanism are:

$$\frac{d[X]}{dt} = k_1[A][Y] - k_2[X][Y] + k_3[A][X] - 2k_4[X]^2,$$

$$\frac{d[Y]}{dt} = -k_1[A][Y] - k_2[X][Y] + \tfrac{1}{2}fk_5[B][Z],$$

$$\frac{d[Z]}{dt} = 2k_3[A][X] - k_5[B][Z].$$

Our goal is to scale $[X]$, $[Y]$, $[Z]$ and t so as to make these equations dimensionless. Choose scaling factors p, q, r and s to define new, dimensionless variables: $x = p[X]$, $y = q[Y]$, $z = r[Z]$ and $\tau = st$. Then,

$$\frac{s}{p}\frac{dx}{d\tau} = \frac{k_1[A]}{q}y - \frac{k_2}{pq}xy + \frac{k_3[A]}{p}x - \frac{2k_4}{p^2}x^2,$$

$$\frac{s}{q}\frac{dy}{d\tau} = -\frac{k_1[A]}{q}y - \frac{k_2}{pq}xy + \frac{fk_5[B]}{2r}z,$$

$$\frac{s}{r}\frac{dz}{d\tau} = \frac{2k_3[A]}{p}x - \frac{k_5[B]}{r}z.$$

There is no unique choice of the scaling parameters, but a sensible start is to set $s = k_5[B]$ since then $\tau = k_5[B]t$ is quite simple, and the last equation becomes

$$\frac{dz}{d\tau} = \frac{2k_3[A]r}{k_5[B]p}x - z.$$

We can also choose the relation $r = k_5[B]p/(2k_3[A])$ to simplify further to

$$\frac{dz}{d\tau} = x - z.$$

Next, rearrange the first equation with the goal of setting the coefficients of the terms in x and y to unity.

$$\frac{s}{k_3[A]}\frac{dx}{d\tau} = \frac{k_1 p}{k_3 q}y - \frac{k_2}{k_3[A]q}xy + x - \frac{2k_4}{k_3[A]p}x^2.$$

The last term suggests that we choose $p = 2k_4/(k_3[A])$, and the second term motivates the choice $q = k_2/(k_3[A])$. We are left with

$$\alpha\frac{dx}{d\tau} = \gamma y - xy + x(1 - x),$$

where

$$\alpha = \frac{k_5[B]}{k_3[A]} \text{ and } \gamma = \frac{k_1 p}{k_3 q} = \frac{2k_1 k_4}{k_2 k_3}.$$

Our earlier choice relating r and p leaves $r = k_4 k_5[B]/(k_3[A])^2$ and hence (after rearranging to a form in which the coefficient of xy is -1):

$$\beta\frac{dy}{d\tau} = -\gamma y - xy + fz$$

where

$$\beta = \frac{2k_4 k_5[B]}{k_2 k_3[A]}.$$

26.2 Exercise

P26.1 The *Brusselator* is a model for an autocatalytic chemical mechanism that can show oscillatory behavior under some conditions. It consists of the following four reactions, in which the reactants, A and B, are assumed to be in excess and therefore to have constant concentrations.

$A \rightarrow X$	k_1
$2X + Y \rightarrow 3X$	k_2
$B + X \rightarrow Y + D$	k_3
$X \rightarrow E$	k_4

The products, D and E, are assumed to be unreactive and play no further role in the mechanism once formed, and the analysis of the chemical dynamics is focused on the intermediate species X and Y.

Obtain expressions for the constants p, q and r that scale the time variable and intermediate concentrations into the non-dimensional form: $x = p[X]$, $y = q[Y]$ and $\tau = rt$ to leave the following pair of coupled ordinary differential equations:

$$\frac{dx}{d\tau} = 1 + ax^2y - (b+1)x,$$

$$\frac{dy}{d\tau} = -ax^2y + bx,$$

where $a = k_1^2 k_2 [A]^2 / k_4^3$ and $b = k_3[B]/k_4$.

The system has a single *fixed point*: the pair of values $(x_\star, y_\star)$ that remain unchanged in time under these differential equations. show that $x_\star = 1$ and $y_\star = b/a$.

By solving the above pair of differential equations, plot x and y as a function of τ using the initial conditions $(x_0, y_0) = (0, 0)$ for reactant concentrations (a) $(a, b) = (1, 0.7)$, (b) $(a, b) = (1, 1.8)$ and (c) $(a, b) = (1, 2.05)$. For each solution, also plot the phase space plot (the curve of y against x).

27

Root-Finding with `scipy.optimize`

27.1 Root-Finding Algorithms

`scipy.optimize` package contains, in addition to the minimization routines described in Chapter 22, implementations of algorithms to find the roots of arbitrary univariate and multivariate functions. This chapter will give a brief introduction to the use of `scipy.optimization.root_scalar` to find the roots of a univariate (scalar) function, $f(x)$. In principle, this function can be any continuous function of its single variable, x, including those generated by some algorithm (as opposed to, say, one that can be written in algebraic form). Here, we will demonstrate three commonly used root-finding methods to determine the positions of the radial nodes in the 4s atomic orbital ($n = 4$, $l = 0$, $m_l = 0$), which may be written:

$$\psi_{400}(r, \theta, \phi) = R_{40}(r)Y_{00}(\theta, \phi) = N(24 - 36\rho + 12\rho^2 - \rho^3)e^{-\rho/2},$$

where $\rho = 2Zr/n = Zr/2$ and $N = (Z^3/\pi)^{1/2}/192$ combines the spherical harmonic Y_{00} and a normalization constant. Z is the nuclear charge in multiples of the electron charge, e, and the units of r are bohr, $a_0 \approx 52.9$ pm; for simplicity, we have assumed the nucleus to be infinitely heavy.

Of course, in this case, the radial nodes (the values of r for which $\psi = 0$) are simply the roots of the pre-exponential cubic polynomial and could be found by other means. A less-trivial use of `root_scalar` is given in the worked example at the end of the chapter.

First, plot the radial part of the wavefunction to approximately locate the nodes (Figure 27.1).

```
import numpy as np
import matplotlib.pyplot as plt

# A grid of distance from the nucleus, in atomic units (bohr).
r = np.linspace(0, 60, 1000)
# Nuclear charge.
Z = 1
```

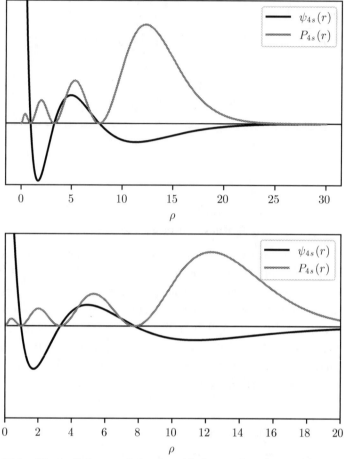

Figure 27.1   The radial part of the 4s orbital wavefunction and its probability distribution on two scales.

```
# This factor applies to the 4s orbital only.
N = np.sqrt(Z**1.5 / np.pi) / 192

rho = Z * r / 2
def psi(rho):
    return N * (24 - 36*rho + 12*rho**2 - rho**3) * np.exp(-rho / 2)

def make_plot(ax):
    ax.plot(rho, psi(rho), label=r'$\psi_{4s}(r)$')
    ax.axhline(0, c='k', lw=1)
    ax.set_yticks([])
    ax.set_xlabel(r'$\rho$')
    ax.plot(rho, 4 * np.pi * psi(rho)**2 * rho**2, label=r'$P_{4s}(r)$')
    ax.legend()

fig, axes = plt.subplots(nrows=2, ncols=1)
make_plot(axes[0])
```

```
axes[0].set_ylim(-0.01,0.02)

make_plot(axes[1])
axes[1].set_xlim(0, 20)
axes[1].set_ylim(-0.02,0.02)
```

The root_scalar method requires a callable function, f, (the one to determine a root of) any additional arguments it might need, args, and optionally, a specification of the algorithm to use, method. Some methods require additional information (described below), for example, a way to calculate the first derivative of the function or a bracketing interval to locate the specific root desired.

```
from scipy.optimize import root_scalar
```

Bisection

The bisection method requires a bracketing interval: values of $x = a$ and $x = b$ such that the desired root is known to lie between a and b. This means that the signs of $f(a)$ and $f(b)$ must be different. It proceeds by evaluating the function at $x = c$, the midpoint between a and b, dividing the bracketing interval in two. If the signs of $f(a)$ and $f(c)$ are different, the new bracket is (a, c); conversely, if $f(c)$ and $f(b)$ differ in sign, the new bracket is (c, b). The algorithm then repeats until the bracket becomes smaller than some tolerance.

For example, from the plots, it can be seen that one of the nodes lies between $\rho = 6$ and $\rho = 8$:

```
root_scalar(psi, method='bisect', bracket=(6, 8))
```

```
      converged: True
           flag: 'converged'
 function_calls: 42
     iterations: 40
           root: 7.758770483143962
```

The returned object has a flag, converged, indicating the algorithm was successful, some statistics on how the algorithm proceeded, and the value of the root: here $\rho = 7.759$.

The bisection method can be slow (it required 42 function calls here), but it has the merit that for well-behaved functions, it cannot fail.

Brent's Algorithm

The Brent algorithm is another reliable choice that is generally faster than bisection. Without going into detail, it too requires a bracket for the desired root. The basic version is referred to with method='brentq':

```
root_scalar(psi, method='brentq', bracket=(6, 8))
```

```
      converged: True
           flag: 'converged'
 function_calls: 7
     iterations: 6
           root: 7.758770483143634
```

This has returned the same root with only seven function calls.

Newton–Raphson

Instead of bracketing the root, the Newton–Raphson method starts with an initial guess, x0, and iteratively approximates the root's position by evaluating:

$$x_{i+1} = x_i - \frac{f(x_i)}{f'(x_i)},$$

where $f'(x)$ is the first derivative of the function. That is, at each iteration, the root is approximated as x_{i+1}, the x-axis intercept of the tangent to the curve at $f(x_i)$.

In this case, it is easy to code a function to return the first derivative. In fact, since $e^{-\rho/2}$ is not zero at the nodes, we can divide ψ by this factor and find the roots of its polynomial pre-exponential function.

```python
def func(rho):
    """4s orbital wavefunction: pre-exponential function."""
    return 24 - 36*rho + 12*rho**2 - rho**3

def funcp(rho):
    """First derivative of 4s pre-exponential function."""
    return -36 + 24*rho - 3*rho**2
```

To determine the root of interest, a good starting guess might be $\rho = 10$:

```python
root_scalar(func, method='newton', fprime=funcp, x0=10)
```

```
      converged: True
           flag: 'converged'
 function_calls: 12
     iterations: 6
           root: 7.758770483143631
```

It is important to be aware of the circumstances under which the Newton–Raphson method can fail to find its root. Consider the starting guess, $\rho = 6$:

```python
root_scalar(func, method='newton', fprime=funcp, x0=6)
```

```
/Users/christian/.../site-packages/scipy/optimize/_zeros_py.py:302:
RuntimeWarning: Derivative was zero.
  warnings.warn(msg, RuntimeWarning)
```

```
     converged: False
          flag: 'convergence error'
function_calls: 2
    iterations: 1
          root: 6.0
```

As indicated in the warning report, this failed because the function's first derivative was zero at the initial guess: We started the algorithm on a maximum in $f(x)$. Without any notion of how to construct an extrapolation from this point back to $f = 0$, no improvement in the estimate of the root could be made.

To find all three nodes, we can inspect the plotted function to construct a sequence of brackets. In the below, the obtained value of ρ (in bohr) is converted into a value of r in pm.

```python
import scipy.constants as pc
# Bohr radius in pm.
a0 = pc.physical_constants['Bohr radius'][0] * 1.e12

brackets = ((0, 2), (3, 5), (6, 10))
for i, bracket in enumerate(brackets, start=1):
    ret = root_scalar(func, method='brentq', bracket=bracket)
    if ret.converged:
        rho0 = ret.root
        r0 = rho0 * 2 / Z       # NB only applies to 4s orbital!
        print(f'Node #{i}: r = {r0 * a0:>7.3f} pm')
```

```
Node #1: r =   99.043 pm
Node #2: r =  349.829 pm
Node #3: r =  821.153 pm
```

27.2 Example

E27.1

The Quartic Oscillator: Numerical Solution

An approximate wavefunction for the quartic oscillator ground state was found using the variational method in Exercise P22.2, using a trial function of with the form of the harmonic oscillator ground state: $\psi_{approx}(q) = \exp\left(-\alpha q^2/2\right)$.

Here, we will solve the quartic oscillator Schrödinger equation numerically, using the *shooting method*. This involves treating the differential equation,

$$-\frac{1}{2}\frac{d^2\psi}{dq^2} + \frac{1}{2}q^4\psi = E\psi$$

as an initial value problem: integrating this equation for a guessed value of E and adjusting this parameter until the boundary conditions of the problem are met.

Concretely, knowing that the wavefunction must decay to zero smoothly as $q \to \pm\infty$, we might start at $q = -q_{max}$ (q_{max} being some large-enough finite value standing in for infinity), where ψ can be taken to have some fixed value (any constant value for $\psi(-q_{max})$ will do, since we can normalize the wavefunction later) and its gradient is zero. The goal is to integrate as a function of q for different values of E, seeking the value that results in the boundary condition $\psi(q_{max}) = 0$ being met.

First, we need to rewrite the second-order differential equation as a pair of first order differential equations with

$$\psi = y_1 \quad \text{and} \quad \frac{d\psi}{dq} = y_2,$$

so that

$$\frac{d^2\psi}{dq^2} = \frac{dy_2}{dq} = (q^4 - 2E)y.$$

To illustrate the process, start with the approximate variational solution, $E = 0.541$ (see Exercise P22.2).

```python
from scipy.integrate import solve_ivp
import matplotlib.pyplot as plt

def diffSE(q, Y, E):
    """Return the derivatives dy1/dq and dy2/dq."""
    y, dydq = Y
    d2ydq2 = (q**4 - 2*E) * y
    return dydq, d2ydq2
```

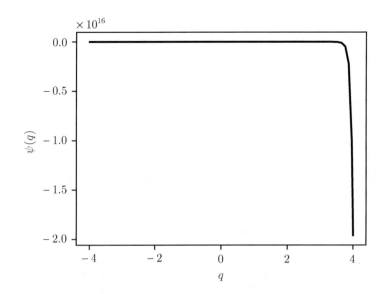

```
# Integrate from q=-qmax to q=qmax.
qmax = 4

Eapprox = 0.541
# At q=-qmax, take psi = 1, dpsi/dq = 0.
y0 = (1, 0)
q_span = (-qmax, qmax)
ret = solve_ivp(diffSE, q_span, y0, args=(Eapprox,))
# Plot the integrated wavefunction as a function of q
plt.plot(ret.t, ret.y[0])
plt.xlabel(r'$q$')
plt.ylabel(r'$\psi(q)$')
```

We know that this energy is greater than the true ground state energy, and in this case the value predicted for $\psi(q_{max})$ is large and negative. Trying a value for E which is too small overshoots our goal of $\psi(q_{max}) = 0$ in the other direction:

```
Eapprox = 0.5
ret = solve_ivp(diffSE, q_span, y0, args=(Eapprox,))
# Plot the integrated wavefunction as a function of q
plt.plot(ret.t, ret.y[0])
plt.xlabel(r'$q$')
plt.ylabel(r'$\psi(q)$')
```

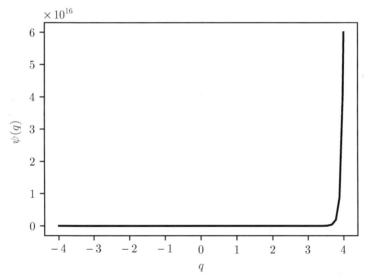

We can treat our predicted value of $\psi(q_{max})$ as a function of E and seek its root, which we now know is bracketed between 0.5 and 0.541.

```
import numpy as np
from scipy.integrate import quad
from scipy.optimize import root_scalar

def func(E):
```

```
    res = solve_ivp(diffSE, q_span, y0, args=(E,))
    # Return the last data point of the wavefunction array.
    return res.y[0][-1]
ret = root_scalar(func, method='brentq', bracket=(0.5, 0.541))
ret
```

```
    converged: True
         flag: 'converged'
function_calls: 8
    iterations: 7
         root: 0.5301075503022544
```

The numerical solution to the Schrödinger equation predicts an energy[1] of 0.530 which, as expected, is lower than the variational energy, 0.541. We can plot both wavefunctions on the same graph to compare them (Figure 27.2).

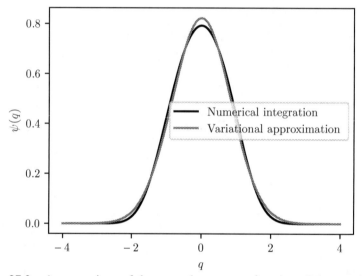

Figure 27.2   A comparison of the ground state wavefunction of the quartic oscillator calculated by numerical integration of the Schrödinger equation and by applying the Variational approximation.

```
E = ret.root
# Repeat the solution with the accurate energy and dense_output=True so
# we can evaluate the wavefunction at different q.
soln = solve_ivp(diffSE, q_span, y0, args=(E,), dense_output=True)
qgrid, dq = np.linspace(-qmax, qmax, 1000, retstep=True)
psi_num = soln.sol(qgrid)[0]
# Normalize the numerical wavefunction.
N, _ = quad(lambda q: soln.sol(q)[0]**2, -qmax, qmax)
psi_num /= np.sqrt(N)
```

[1] The energy is given in units of $(k\hbar^4/\mu^2)^{1/3}$: see Exercise P22.2.

```
plt.plot(qgrid, psi_num, label='Numerical integration')

alpha = 3**(1/3)
psi_var = (np.pi / alpha)**-0.25 * np.exp(-alpha / 2 * qgrid**2)
plt.plot(qgrid, psi_var, label='Variational approximation')
plt.xlabel(r'$q$')
plt.ylabel(r'$\psi(q)$')
plt.legend()
```

28

Rotational Spectroscopy

28.1 Diatomic Molecules: The Rigid Rotor

28.1.1 Energy Levels of the Rigid Rotor

The simplest model representing the rotation of a diatomic molecule treats it as two point masses (m_1 and m_2, the nuclei) separated by a constant distance (a fixed bond length, r). The corresponding Schrödinger equation, $\hat{H}\psi = E\psi$, can be written with the Hamiltonian operator in the form

$$\hat{H} = -\frac{\hbar^2}{2\mu}\nabla^2 = -\frac{\hbar^2}{2I}\left[\frac{1}{\sin\theta}\frac{\partial}{\partial\theta}\left(\sin\theta\frac{\partial}{\partial\theta}\right) + \frac{1}{\sin^2\theta}\frac{\partial^2}{\partial\phi^2}\right],$$

where $I = \mu r^2$ is the molecule's moment of inertia in terms of its reduced mass, $\mu = m_1 m_2/(m_1 + m_2)$. This is identical in form to the angular part of the Hamiltonian for the hydrogen atom and the corresponding wavefunctions are the spherical harmonic functions. In atomic physics, these functions are usually written $Y_l^m(\theta, \phi)$ where l and m are the orbital angular momentum and magnetic quantum numbers, respectively. For a rotating molecule, these quantum numbers are labeled J and M_J instead, representing the rotational angular momentum and its projection along the (space-fixed) z-axis.

The energy levels of the rigid rotor are therefore

$$E(J) = \frac{\hbar^2}{2I}J(J+1)$$

and are $2J + 1$-fold degenerate, corresponding to the values $M_J = -J, -J+1, \ldots, J-1, J$.

Spectroscopists often use the wavenumber, cm^{-1}, as their unit of energy, where $1\ cm^{-1} = hc\,J$ when the speed of light, c, is taken in $cm\ s^{-1}$. The energies are then reported as so-called *terms*,

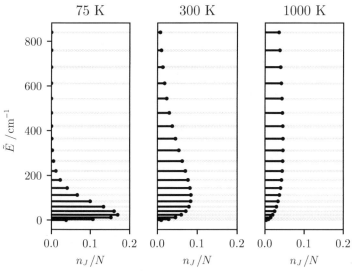

Figure 28.1 Fractional populations of the rotational states of N_2 at three temperatures. A Jupyter notebook with an interactive plot is available at `https://scipython.com/chem/cal/`.

$$F(J) = \frac{E(J)}{hc} = BJ(J+1) \quad \text{where } B = \frac{\hbar^2}{2Ihc} = \frac{h}{8\pi^2 cI}$$

is the molecule's *rotational constant* (also in cm^{-1}).

The fractional population of the Jth rotational level is given by statistical mechanics as:

$$\frac{n_J}{N} = \frac{(2J+1) \exp\left(-\frac{hcBJ(J+1)}{k_B T}\right)}{q_{rot}},$$

where q_{rot} is the rotational partition function. A typical diatomic molecule, such as N_2 has a rotational constant of about $2\,cm^{-1}$; at room temperature, the thermal energy available, $k_B T$, corresponds to about $200\,cm^{-1}$, so we can expect a large number of rotational energy levels to be accessible in most cases (the exception being perhaps the very lightest (small I) diatomics, such as H_2 at low temperatures) – see Figure 28.1.

28.1.2 Rigid Rotor Transitions

A rotating molecule may absorb or emit a photon in making a transition between its different energy levels. The *selection rules* determining which transitions are allowed can be derived from the transition moment integral,

$$\mu_T = \int_\tau \psi_f^* \, \hat{\boldsymbol{\mu}} \, \psi_i \, d\tau,$$

where $\hat{\boldsymbol{\mu}}$ is the electric dipole moment operator and the integral is carried out over the space of all the relevant coordinates. In the context of a rotational transition between states (J'', M_J'') and (J', M_J'),

$$\mu_T = \int_\tau Y_{J'M_J'}^* \, \hat{\boldsymbol{\mu}} \, Y_{J''M_J''} \sin\theta \, d\theta \, d\phi.$$

The properties of spherical harmonics have been extensively studied, and their symmetry is such that this integral is zero unless the following conditions on $\Delta J = J' - J''$ and $\Delta M_J = M_J' - M_J''$ hold:

$$\Delta J = \pm 1 \quad \text{and} \quad \Delta M_J = 0, \pm 1.$$

These are the (electric dipole) selection rules.[1] The intensity of the transition is proportional to the square of the transition dipole moment, $|\mu_T|^2$. In absorption $(J+1 \leftarrow J)$, for a closed-shell diatomic molecule, it is found that:

$$|\mu_T|^2 = \mu_0 \frac{J+1}{2J+1},$$

where μ_0 is the permanent electric dipole of the molecule. Non-polar molecules, such as homonuclear diatomics like N_2, do not show pure rotational spectra.

28.1.3 Carbon Monoxide and the Rigid Rotor Approximation

E28.1

Carbon monoxide is a closed-shell diatomic molecule with an equilibrium rotational constant, $B_e = 1.931$ cm^{-1}. Its pure-rotational (microwave) spectrum is relatively easy to measure, and the transition wavenumbers and intensities are known to high accuracy.

We will compare the experimental absorption spectrum at temperatures 300 K and 1,500 K with the spectrum predicted by the rigid-rotor approximation using the above value of B_e. The intensity of a pure-rotational absorption transition for a closed-shell diatomic molecule is:

$$S \propto \tilde{v} \frac{(J+1)}{q_{\text{rot}}} \exp\left(-\frac{E''}{k_B T}\right) \left[1 - \exp\left(-\frac{\tilde{v}hc}{k_B T}\right)\right],$$

where E'' is the energy of the lower state and $\tilde{v}$ is the transition wavenumber.

Experimental data is available in the files CO-rot-300.txt and CO-rot-1500.txt, which can be downloaded from https://scipython.com/chem/cal/.

---

[1] It is possible to observe transitions due to magnetic dipole and electric quadrupole interactions, but they are very much weaker.

First, some imports and definitions:

```
import numpy as np
import matplotlib.pyplot as plt
from scipy.constants import k as kB, h, c
# For this section we will work with the speed of light, c in cm.s-1.
c *= 100
```

There will be a vibrational dependence on the rotational constant, and even in the ground vibrational state, B_0 is a little smaller than B_e; in the absence of any information about this effect, we will take $B = B_e$.

To deduce how the spectrum intensity changes with temperature, we need the rotational level energies (in J),

$$E(J) = hcBJ(J+1)$$

and the transition wavenumber,

$$\tilde{\nu} = \frac{\Delta E}{hc} \quad \text{where} \quad \Delta E = E(J+1) - E(J) = 2B(J+1).$$

The rotational partition function, q_{rot}, is a *sum over J* so for the relative transition intensities at a fixed temperature it is a constant and we don't need to calculate it explicitly.

```
# Rotational constant of CO, cm-1.
B = 1.93128087
def get_energy(J, B):
    """Return the energy, in J, of the rotational state J."""
    return h * c * B * J * (J+1)

# Set up a suitable grid of values for the rotational constant, J.
J = np.linspace(0, 80, 81, dtype=int)
```

The spectrum, modeled by the rigid-rotor approximation, can be calculated using the above formulas.

```
def get_rigid_rotor_spectrum(T, J, B):
    """Return the line positions and intensities for a diatomic molecule.

    The molecule is treated as a rigid rotor with rotational constant, B (in cm-1).
    J is a one-dimensional array of values of the rotational quantum number, J.
    T is the temperature (in K).

    The intensities are normalized so that the strongest has intensity 1.

    """

    E_lower = get_energy(J, B)
    E_upper = get_energy(J+1, B)
    nu_rr = E_upper - E_lower
    S_rr = (2 * B * (J+1)**2 * np.exp(-E_lower / kB / T)
```

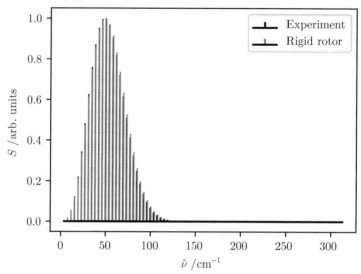

Figure 28.2   A comparison of experimental and theoretical (rigid-rotor) rotational spectra for CO at 300 K.

```
                * (1 - np.exp(-nu_rr / kB / T)) )
    S_rr /= max(S_rr)

    return nu_rr / h / c, S_rr
```

```
def plot_spectra_comparison(T):
    """Plot a comparison of the modelled and experimental rotational spectra."""
    # Rigid rotor spectrum.
    nu_rr, S_rr = get_rigid_rotor_spectrum(T, J, B)
    # Read in and normalize the experimental spectrum.
    nu_expt, S_expt = np.genfromtxt(f'CO-rot-{T}.txt', usecols=(0,1),
                                    skip_header=1, unpack=True)
    S_expt /= max(S_expt)

    plt.stem(nu_expt, S_expt, linefmt='b-', markerfmt='none', basefmt='k',
            label='Experiment')
    plt.stem(nu_rr, S_rr, linefmt='g-', markerfmt='none', basefmt='k',
            label='Rigid rotor')
    plt.legend()
    plt.xlabel(r'$\tilde{\nu}\;/\mathrm{cm^{-1}}$')
    plt.ylabel(r'$S\;/\mathrm{arb.\;units}$')
```

```
T = 300
plot_spectra_comparison(T)
plt.xlim(0, 150)
```

The agreement (Figure 28.2) is pretty good at this temperature, both for the line positions and their intensities.

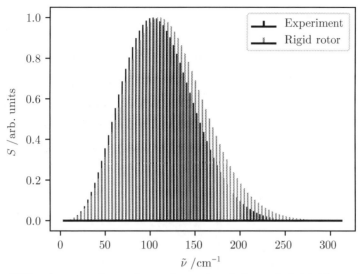

Figure 28.3 A comparison of experimental and theoretical (rigid-rotor) rotational spectra for CO at 1500 K.

```
T = 1500
plot_spectra_comparison(T)
plt.xlim(0, 250)
```

At 1,500 K there is a definite discrepancy at high J (Figure 28.3), mostly due to the neglect of centrifugal distortion.

28.2 Centrifugal Distortion

When a real molecule rotates, its nuclei are subject to a centrifugal force acting to move them away from the axis of rotation. Since real chemical bonds are not rigid, this acts to increase the molecule's moment of inertia and decrease its effective rotational constant, B. The size of the effect is related to the atomic masses, the bond length and the bond force constant, k, (weaker bonds are more easily stretched and lead to greater centrifugal distortion). In practice, the effect can be accounted for by considering an additional term in the energy expression for the rotational energy levels:

$$F(J) = BJ(J+1) - D[J(J+1)]^2,$$

where D is a small (relative to B), positive constant:

$$D = \frac{h^3}{32\pi^4 \mu^2 kcr^6},$$

Additional, smaller terms involving $[J(J+1)]^3$, etc. may also be introduced for high-accuracy spectra extending to large J.

A more accurate model for the transition wavenumbers is therefore

$$\tilde{\nu} = F(J+1) - F(J) = 2B(J+1) - 4D(J+1)^3.$$

28.2.1 Centrifugal Distortion in Hydrogen Iodide

E28.2

Consecutive transitions were observed at the following wavenumbers for the $^1\mathrm{H}^{127}\mathrm{I}$ molecule at 298 K. We will deduce the rotational constant, B_0 and first centrifugal distortion constant, D_0; and estimate the bond length, r, and force constant, k, from these data.

Transition wavenumbers, $\tilde{\nu}/\mathrm{cm}^{-1}$		
76.98849425	89.74487244	102.35117559
115.10755378	127.71385693	140.32016008
152.77638819	165.23261631	177.68884442

The relevant atomic masses are $m(^1\mathrm{H}) = 1.0078u$ and $m(^{127}\mathrm{I}) = 126.9044u$.

Ignoring centrifugal distortion, the transition wavenumbers in a pure rotational spectrum are $\tilde{\nu} \approx 2B_0(J+1)$ and so are spaced by roughly $2B_0$. Clearly, the first transition is not $J = 1 \leftarrow 0$.

```
import numpy as np
import matplotlib.pyplot as plt

nu = np.array([76.98849425,  89.74487244, 102.35117559,
              115.10755378, 127.71385693, 140.32016008,
              152.77638819, 165.23261631, 177.68884442])
```

Estimating $2B_0$ from the spacing of the first two lines, we can use the formula for the transition wavenumbers to obtain $(J+1)$ for the first transition:

```
nu[0] / (nu[1] - nu[0])
```

```
6.035294117444165
```

$(J+1)$ is about 6, so this first transition is $J = 6 \leftarrow 5$. The others are $7 \leftarrow 6$, $8 \leftarrow 7$, etc. The transitions are labeled by their lower-state J value, so it makes sense to set up a NumPy array of these J values:

```
nJ = len(nu)
Jmin = 5
```

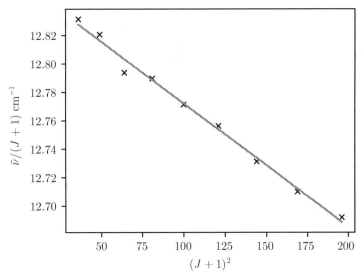

Figure 28.4   Linear least-squares fit to the rotational and centrifugal distortion constants of $^1\mathrm{H}^{127}\mathrm{I}$.

```
J = np.linspace(Jmin, Jmin + nJ - 1, nJ, dtype=int)
J
```

```
array([ 5,  6,  7,  8,  9, 10, 11, 12, 13])
```

The more accurate formula for the wavenumbers, including centrifugal distortion, is

$$\tilde{\nu} = 2B_0(J+1) - 4D_0(J+1)^3$$

Therefore,

$$\frac{\tilde{\nu}}{J+1} = 2B_0 - 4D_0(J+1)^2$$

and the constants B_0 and D_0 may be obtained by a linear regression (Figure 28.4).

```
plt.plot((J+1)**2, nu / (J+1), 'x')
x = np.stack((2*np.ones(nJ), -4*(J+1)**2)).T
xx = np.linalg.lstsq(x, nu/(J+1), rcond=None)
B0, D0 = xx[0]
print(B0, D0)
plt.plot((J+1)**2, 2*B0 - 4*D0*(J+1)**2)
plt.xlabel(r'$(J+1)^2$')
plt.ylabel(r'$\tilde{\nu}/(J+1)\;\mathrm{cm^{-1}}$')
```

The fit seems pretty good and the values obtained are $B_0 = 6.429 \text{ cm}^{-1}$ and $D_0 = 2.152 \times 10^{-4} \text{ cm}^{-1}$.

An estimate of the bond length can be obtained from the definition

$$B_0 = \frac{h}{8\pi^2 c I},$$

where $I = \mu r^2$ (ignoring vibration of the bond).

```
from scipy.constants import u, h, c
# Convert c to cm.s-1.
c *= 100

# Reduced mass of HI
mI, mH = 126.9044, 1.0078
mu = mI * mH / (mI + mH) * u
I = h / 8 / np.pi**2 / c / B0
r = np.sqrt(I / mu)
print('r(HI) = {:.1f} pm'.format(r * 1.e12))
```

```
r(HI) = 161.9 pm
```

The first centrifugal distortion constant is related to the bond force constant through

$$D = \frac{h^3}{32\pi^4 \mu^2 k c r^6},$$

so given our measurement for D_0, we can estimate k:

```
k = h**3 / 32 / np.pi**4 / mu**2 / c / r**6 / D0
print(f'k(HI) = {k:.1f} N.m-1')
```

```
k(HI) = 288.3 N.m-1
```

in pretty good agreement with the literature value of $k = 294 \text{ N m}^{-1}$.

28.3 Polyatomic Molecules

A linear polyatomic molecule (i.e., a molecule with more than two atoms), such as HCN and C_2H_2, in so far as its vibrations can be ignored, can be treated with the same formulas for its rotational motion as above: It has a single rotational constant, B, written in terms of the moment of inertia,

$$I = \sum_i m_i r_i^2,$$

where r_i is the distance of atom i from the center of mass and m_i is the atomic mass. In the case of linear and diatomic molecules, there is only one rotational axis to consider: we can pick any axis perpendicular to the internuclear axis, passing through the centre of mass.[2]

In the more general case of non-linear polyatomic molecules, we can identify three principal rotational axes associated with the principal moments of inertia, conventionally labeled $I_a \leq I_b \leq I_c$ which correspond to three rotational constants, A, B and C, through the relations

$$A = \frac{h}{8\pi^2 cI_a}, \quad B = \frac{h}{8\pi^2 cI_b}, \quad C = \frac{h}{8\pi^2 cI_c}.$$

In some cases, the symmetry of the molecule may make it obvious what these principal rotational axes are; in general, it is necessary to find them by diagonalizing the inertia matrix as in Example E18.2.

It is helpful to classify polyatomic molecules according to the relative values of I_a, I_b and I_c:

- $I_a = I_b = I_c$: spherical top (e.g., CH_4, SF_6);
- $I_a = I_b < I_c$: oblate symmetric top (e.g., BF_3, NH_3, C_6H_6);
- $I_a < I_b = I_c$: prolate symmetric top (e.g., CH_3Cl, CH_3CN);
- $I_a < I_b < I_c$: asymmetric top (e.g., H_2O, C_6H_5OH).

(Viewed this way, a linear molecule is an extreme case of a prolate symmetric top with $I_a = 0$ along the internuclear axis.)

28.3.1 Symmetric Tops

Symmetric tops are molecules that have a symmetry such that they have exactly two equal moments of inertia; this is the case for molecules whose point group has a C_n rotational axis ($n > 2$): one of the principal inertial axes lies along this axis (e.g., colinear with the C–Cl bond in CH_3Cl) and the other two are equal and lie perpendicular to this axis.

The Hamiltonian operator for the rotational energy of a non-linear polyatomic molecule is

$$\hat{H}_{rot} = \frac{\hat{J}_a^2}{2I_a} + \frac{\hat{J}_b^2}{2I_b} + \frac{\hat{J}_c^2}{2I_c}$$

where the total rotational angular momentum operator, $\hat{J}^2 = \hat{J}_a^2 + \hat{J}_b^2 + \hat{J}_c^2$. For a symmetric top, two of I_a, I_b and I_c are equal, so it is possible to rewrite the Hamiltonian as

---

[2] Since the nuclei can be treated as point masses and the electrons are very much lighter, the moment of inertia about the internuclear axis itself can be taken as zero and rotation about this axis does not need to be considered.

$$\hat{H}_{rot} = \frac{\hat{J}^2}{2I_b} + \hat{J}_a^2 \left(\frac{1}{2I_a} - \frac{1}{2I_b} \right) \quad \text{(prolate tops),} \tag{28.1}$$

$$\hat{H}_{rot} = \frac{\hat{J}^2}{2I_b} + \hat{J}_c^2 \left(\frac{1}{2I_c} - \frac{1}{2I_b} \right) \quad \text{(oblate tops).} \tag{28.2}$$

Considering the $\hat{J}_a$ operator (prolate tops) or $\hat{J}_c$ operator (oblate tops) to represent rotation of the molecule about its z-axis (symmetry axis), the corresponding energy levels are therefore

$$F(J, K) = BJ(J + 1) + (A - B)K^2 \quad \text{(prolate tops),} \tag{28.3}$$

$$F(J, K) = BJ(J + 1) + (C - B)K^2 \quad \text{(oblate tops),} \tag{28.4}$$

where $K = -J, -J + 1, \ldots, J - 1, J$ is a quantum number defining the component of the molecule's total rotational angular momentum that lies along the symmetry axis: If $K = 0$, the molecule is tumbling end-over-end; if $K = J$, it is spinning almost entirely about this axis. For rigid molecules, the energy levels are twofold degenerate (apart from the $K = 0$ case): The energy is the same whether the molecule spins clockwise or anticlockwise about its symmetry axis.

The quantization of the angular momentum of a symmetric top about its ("molecule-fixed") symmetry axis, z, should not be confused with its spatial quantization: Each state J, K is *also* $2J + 1$-fold degenerate, corresponding to the projections of the total rotational angular momentum in the space-fixed, Z-axis: there is a further quantum number, $M_J = -J, -J + 1, \ldots, J - 1, J$ describing this projection.

The selection rule on K is $\Delta K = 0$: rotation around the symmetry axis of the molecule does not cause a change in electric dipole, so changes in the rotational state with respect to this axis cannot be induced by interaction with light. Therefore, for a rigid molecule

$$\tilde{\nu} = F(J + 1, K) - F(J, K) = 2B(J + 1),$$

as for linear molecules and diatomics.

28.3.2 *The Rotational Spectrum of Phosphine*

E28.3

The phosphine molecule, PH_3, is a oblate symmetric top with rotational constants, $A = B = 4.524 \, \text{cm}^{-1}$ and $C = 3.919 \, \text{cm}^{-1}$.

The file PH3-spec.txt, available at https://scipython.com/chem/cal/, provides a low-pressure, high-resolution spectrum of phosphine in the region $\tilde{\nu} = 0 - 200 \, \text{cm}^{-1}$. The following code plots this spectrum as Figure 28.5.

```
import numpy as np
from scipy.constants import c
```

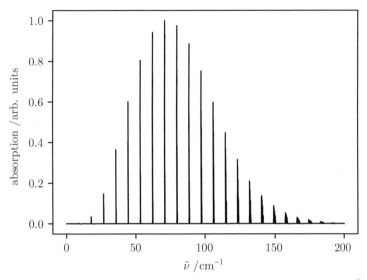

Figure 28.5   Rotational spectrum of PH$_3$.

```
import matplotlib.pyplot as plt
c *= 100
```

```
# Read in the high-resolution spectrum.
nu, spec = np.genfromtxt('PH3-spec.txt', unpack=True, skip_header=2)
```

```
def plot_spec():
    plt.plot(nu, spec)
    plt.xlabel(r'$\tilde{\nu}\;/\mathrm{cm^{-1}}$')
    plt.ylabel(r'absorption /arb. units')
plot_spec()
```

As predicted by Equation 28.3.1, the spectrum resembles that of a diatomic molecule, with lines spaced at approximately $2B \approx 9 \, \mathrm{cm}^{-1}$. However, at larger J some more structure is apparent (Figure 28.6):

```
plot_spec()
# Zoom in on the region between 140 and 160 cm-1.
plt.xlim(140, 160)
plt.ylim(0, 0.2)
# Label the groups of lines by their J+1 <- J.
plt.text(140.2, 0.16, '$J=16\leftarrow 15$')
plt.text(148, 0.1, '$J=17\leftarrow 16$')
plt.text(156, 0.08, '$J=18\leftarrow 17$')
```

The structure is due to centrifugal distortion. For a symmetric top, there are (to first order) three centrifugal distortions corresponding to the following terms in the rotational Hamiltonian operator:

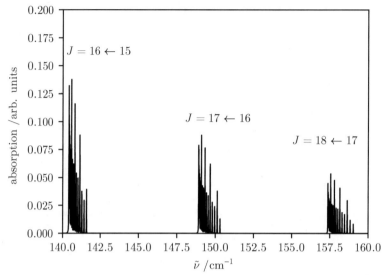

Figure 28.6   The rotational spectrum of PH_3 in the region $140-160\,\text{cm}^{-1}$ showing structure due to centrifugal distortion.

$$\hat{H}_{\text{rot}} = B\hat{J}^2 + (C-B)\hat{J}_z^2 - D_J\hat{J}^4 - D_{JK}\hat{J}^2\hat{J}_z^2 - D_K\hat{J}_z^4$$

$$F(J,K) = BJ(J+1) + (C-B)K^2 - D_J[J(J+1)]^2 - D_{JK}J(J+1)K^2 - D_KK^4$$

For the observed transitions, $\Delta J = +1$, $\Delta K = 0$, and so absorption lines are expected at the wavenumbers:

$$\tilde{\nu} = 2B(J+1) - 4D_J(J+1)^3 - 2D_{JK}K^2(J+1)$$

The first two terms are the same as for a linear molecule and correspond to its rotation about the axis with the largest moment of inertia and the centrifugal distortion
about this axis; the third term corresponds to the shift observed as the moment of
inertia about the molecule's symmetry axis changes with rotation. Note that at this
level of approximation, it is not possible to deduce the axial parameters, $C-B$ and
D_K.

Taking the $J = 17 \leftarrow 16$ transition, there are $2J + 1 = 33$ values of K, but transitions with $\pm K$ occur at the same wavenumber (because the K quantum number
appears squared in the above formula: The energy is the same whether the molecule
is spinning clockwise or anticlockwise about its symmetry axis). We, therefore,
expect 17 lines ($K = 0, \pm 1, \pm 2, \cdots \pm 16$) in the multiplet for this transition (Figure 28.7).

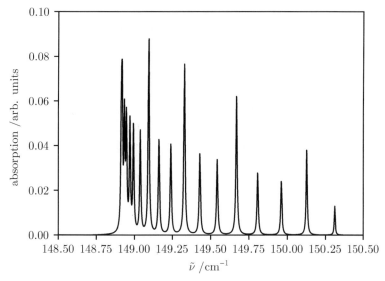

Figure 28.7   The $J = 17 \leftarrow 16$ transition in the rotational spectrum of PH_3.

```
plot_spec()
plt.xlim(148.5, 150.5)
plt.ylim(0, 0.1)
```

There are indeed 17 lines and their spacing increases with increasing wavenumber, so D_{JK} must be negative. The intensity alternation is due to nuclear spin effects and will not concern us here.

The above expression for $\tilde{\nu}$ can be used to fit the observed transition wavenumbers, which can be read in from the file PH3-lines.txt.

```
# Read in observed line positions and intensities, with lower-state
# J and K assignments
nu0, S, J, K = np.genfromtxt('PH3-lines.txt', unpack=True)
```

```
# Linear, least-squares fit to B, DJ and DJK.
x = np.stack((2*(J+1), -4*(J+1)**3, -2*(J+1)*K**2)).T
xx = np.linalg.lstsq(x, nu0, rcond=None)
B, DJ, DJK = xx[0]
# The fitted line positions.
nu_fit = 2*B*(J+1) - 4*DJ*(J+1)**3 - 2*DJK*(J+1)*K**2
rms = np.sqrt(np.sum((nu_fit-nu0)**2)/len(nu0))

print('Fitted parameters')
print(f'B   = {B:< 13.6f} cm-1')
print(f'DJ  = {DJ:13.6e} cm-1')
print(f'DJK = {DJK:13.6e} cm-1')
print(f'Root-mean-square error, rms = {rms:3f} cm-1')
```

```
Fitted parameters
B   =  4.451309      cm-1
DJ  =  1.236133e-04 cm-1
DJK = -1.562578e-04 cm-1
Root-mean-square error, rms = 0.014080 cm-1
```

This is not a particularly good fit and could be improved by considering other effects, including higher-order centrifugal distortion terms. The literature values[3] are typical of the high-precision that can be achieved in microwave spectroscopy from a fit to a total of about 20 parameters:

$$B = 4.5452417819 \text{ cm}^{-1}$$

$$D_J = 1.313039 \times 10^{-4} \text{ cm}^{-1}$$

$$D_{JK} = -1.724038 \times 10^{-4} \text{ cm}^{-1}$$

---

[3] H. S. P. Müller, *J. Quant. Spectrosc. Radiat. Transf.* **130**, 335 (2013).

29

Peak Finding

Many research activities in chemistry generate data in the form of a signal recorded as a function of time or some other independent variable, which must be analyzed for its properties, frequently in the presence of noise. Examples include obtaining the frequencies of absorption features in a spectrum, locating the positions of peaks in an NMR or mass spectrum, and obtaining the peak cathodic and anodic currents in a cyclic voltammogram for a reversible reaction.

Locating the position of signals in a data series is therefore often an exercise in peak-finding (locating and characterizing local maxima). The process is sometimes straightforward but more usually complicated by the presence of noise, baseline effects, broad and overlapping peaks and outliers.

29.1 Simple Peak-Finding

In the simplest (but rarest) case, a spectrum might consist only of signal peaks against a constant, noise-free background. For example, consider the mass spectrum of propane, shown in Figure 29.1.

In this case, one approach is simply to find the indexes of any elements in the intensity array that are nonzero. For example, consider the following text input file:

```
# (m/z) /u   Intensity /arb. units
        0      0.00
        1      0.00
        2      0.00
       ...
       23      0.00
       24      0.10
       25      0.50
       26      9.11
       27     41.94
       ...
       50      0.00
```

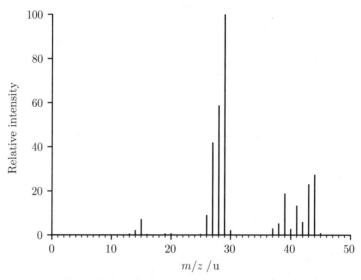

Figure 29.1   The noise-free mass spectrum of propane, C_3H_8.

This file is available as `propane-mass-spec.txt` at `https://scipython` `.com/chem/car/`. The two columns of data, the mass/charge ratio, M/z, and intensity, can be read into NumPy arrays with

```
In [x]: moz, intensity = np.genfromtxt('propane-mass-spec.txt',
                              skip_header=1, unpack=True)
```

The NumPy function `np.nonzeros` returns a tuple of arrays, one for each dimension of its input, containing the indexes of the nonzero elements. For example:

```
In [x]: arr = np.array(((1, 0, 2),(0, 1, 0)))
In [x]: arr
Out[x]:
array([[1, 0, 2],
       [0, 1, 0]])

In [x]: np.nonzero(arr)
Out[x]: (array([0, 0, 1]), array([0, 2, 1]))
```

indicates that the nonzero elements of `arr` are indexed at (0, 0), (0, 2) and (1, 1). Of course, for a one-dimensional array like `intensity`, the tuple has a single array of indexes in it, which can be retrieved with:

```
In [x]: idx = np.nonzero(intensity)[0]
```

Alternatively, the convenience function `np.flatnonzero` can be used to obtain this single array directly:

```
In [x]: idx = np.flatnonzero(intensity)
In [x]: idx
Out[x]:
array([12, 13, 14, 15, 16, 19, 20, 21, 24, 25, 26, 27, 28, 29, 30, 36, 37,
       38, 39, 40, 41, 42, 43, 44, 45])
```

The peaks are identified at the following positions:

```
In [x]: print('Peaks detected at:')
   ...: print('(m/z) /u   Relative Intensity')
   ...: for i in idx:
   ...:     print(f'{moz[i]:7.0f} {intensity[i]:9.2f}')
   ...:
   ...:
Peaks detected at:
(m/z) /u   Relative Intensity
     12       0.30
     13       0.60
     14       2.20
     15       7.21
     16       0.40
    ...
     43      23.12
     44      27.42
     45       0.90
```

In the presence of noise, just subtract some threshold value above the maximum noise value (reducing any negative values to zero) and proceed in the same way (of course, any signals with intensity below the threshold are invisible to this procedure). For example, with reference to Figure 29.2:

```
In [x]: moz, intensity = np.genfromtxt('propane-mass-spec-noisy.txt',
   ...:                                 skip_header=1, unpack=True)

In [x]: threshold = 6   # obtain the threshold by inspecting the data plot.

In [x]: intensity -= threshold

In [x]: intensity = intensity.clip(0)        # ensure negative values go to 0.

In [x]: idx = np.nonzero(intensity)[0]       # everything left is a peak.
In [x]: print('Peaks detected at:')
In [x]: print('(m/z) /u   Relative Intensity')
In [x]: for i in idx:
   ...:     # NB add back on the threshold we subtracted from the signal.
   ...:     print(f'{moz[i]:7.0f} {intensity[i]+threshold:9.2f}')
   ...:
Peaks detected at:
(m/z) /u   Relative Intensity
     15       7.13
     26       8.02
     27      40.34
     28      58.70
```

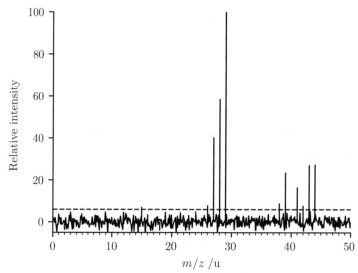

Figure 29.2   A noisy mass spectrum of propane, C_3H_8. The dashed line indicates the threshold intensity above which peaks are to be identified.

```
29     100.00
38       8.94
39      23.73
41      16.66
42       7.86
43      27.12
44      27.45
```

29.2 `scipy.signal.find_peaks`

Identifying broadened peaks, especially in the presence of noise, is a more complex task. The `scipy.signal` package contains a helpful function, `find_peaks`, that can be customized with various arguments, as described in Table 29.1. `find_peaks` returns an array of the indexes corresponding to the maxima and a dictionary of properties of the peaks where customization has been requested (e.g., the peak heights).

In the absence of noise or baseline effects, for the case when all peaks are needed, the function can be used with the array of data, `arr` as its only argument. For example, Figure 29.3 shows the intensity measured on a screen due to diffraction of monochromatic light by two narrow slits.

```
In [x]: from scipy.signal import find_peaks
In [x]: y, I = np.genfromtxt('double-slit-data.txt', unpack=True)
In [x]: idx, props = find_peaks(I)
In [x]: print('        y              Imax')
   ...: print(np.vstack((y[idx], I[idx])).T)
```

```
             y            Imax
[[-4.84746187e+00  2.80454179e-03]
 [-4.60240060e+00  9.24051034e-03]
 [-4.35233808e+00  1.52306505e-02]
  ...
 [ 4.35233808e+00  1.52306505e-02]
 [ 4.60240060e+00  9.24051034e-03]
 [ 4.84746187e+00  2.80454179e-03]]

In [x]: print(f'{len(idx)} peaks found.')
43 peaks found.
```

Table 29.1   Optional arguments to `scipy.signal.find_peaks`

`height`	Minimum height of peaks (if given as a single value), or the peak height range (if given as a sequence of two elements)
`threshold`	Minimum threshold of peaks (vertical distance to its neighbouring sample points), or the threshold range (if given as a sequence of two elements)
`distance`	Minimum horizontal distance between peaks, measured as a number of sample points
`prominence`	Minimum prominence between peaks (if given as a single value), or the prominence range (if given as a sequence of two elements)
`width`	Minimum width measured as a number of sample points, or the width range (if given as a sequence of two elements)

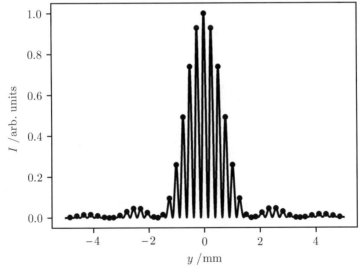

Figure 29.3   All peaks detected by `scipy.signal.find_peaks` in a double-slit diffraction experiment.

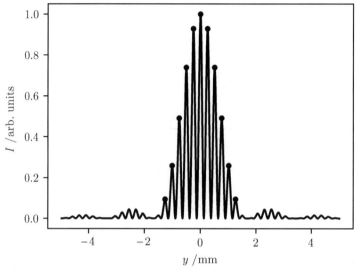

Figure 29.4 All peaks with intensities greater than 0.05 detected by `scipy.signal.find_peaks` in a double-slit diffraction experiment.

Of course, one may not require the position of every tiny ripple in the data: The `height` argument can be used to select all maxima above a certain height. This is particularly useful for eliminating a noisy baseline or selecting only the strongest signals (see Figure 29.4):

```
In [x]: # Only look for maximum values > 0.05. The data are normalized such that
In [x]: # the strongest peak has intensity 1, so this is a 5% threshold.
In [x]: idx, props = find_peaks(I, height=0.05)

In [x]: print('        y              Imax')
In [x]: print(np.vstack((y[idx], I[idx])).T)
In [x]: print(f'{len(idx)} peaks found.')

          y              Imax
[[-1.26406602e+00   9.40957272e-02]
 [-1.01400350e+00   2.58770527e-01]
 [-7.61440360e-01   4.91502176e-01]
 [-5.08877219e-01   7.38759743e-01]
 [-2.53813453e-01   9.28873405e-01]
 [-1.25031258e-03   1.00000000e+00]
 [ 2.53813453e-01   9.28873405e-01]
 [ 5.08877219e-01   7.38759743e-01]
 [ 7.61440360e-01   4.91502176e-01]
 [ 1.01400350e+00   2.58770527e-01]
 [ 1.26406602e+00   9.40957272e-02]]
11 peaks found.
```

If the baseline is sloping, this approach won't work very well. Instead, the `prominence` argument can be used to specify the minimum drop in height between a candidate peak and any higher one (Figure 29.5).

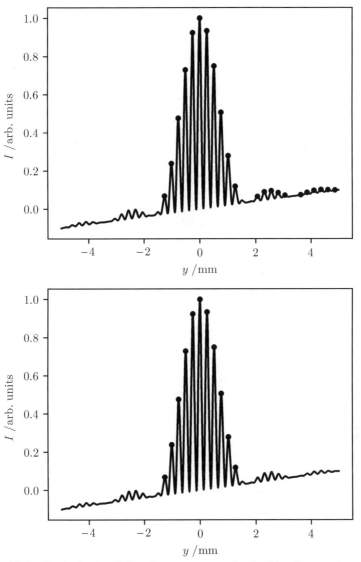

Figure 29.5 Peaks in the diffraction spectrum of a double-slit experiment, re-solved in the presence of a sloping baseline, using `scipy.signal.find_peaks` with the arguments `height=0.05` (upper panel) and `prominence=0.05` (lower panel).

```
In [x]: y, I = np.genfromtxt('double-slit-sloping.txt', unpack=True)
In [x]: idx, props = find_peaks(I, prominence=0.05)

In [x]: print('        y           Imax')
In [x]: print(np.vstack((y[idx], I[idx])).T)
In [x]: print(f'{len(idx)} peaks found.')
```

```
        y               Imax
[[-1.26406602e+00   6.88126861e-02]
 [-1.01400350e+00   2.38484494e-01]
 [-7.61440360e-01   4.76261460e-01]
 [-5.08877219e-01   7.28563980e-01]
 [-2.53813453e-01   9.23774036e-01]
 [ 1.25031258e-03   1.00000000e+00]
 [ 2.53813453e-01   9.33926320e-01]
 [ 5.08877219e-01   7.48918560e-01]
 [ 7.61440360e-01   5.06718312e-01]
 [ 1.01400350e+00   2.79043620e-01]
 [ 1.26406602e+00   1.19374062e-01]]
11 peaks found.
```

Finally, the `distance` argument can be very helpful when noise is present in the data: It sets a minimum number of data points between neighboring peaks: Noise presenting as smaller spikes between the signal peaks are therefore (hopefully) eliminated, as demonstrated by the following example.

29.3 Example

E29.1

In a particular double-slit experiment, a monochromatic light of wavelength $\lambda = 512\,\mathrm{nm}$ is directed at two slits, each of width a separated by a distance d. The resulting diffraction pattern is displayed on a screen a distance $D = 10\,\mathrm{cm}$ from the slits.

The file `double-slit-noisy.txt`, available at `https://scipython.com/chem/ard` contains the intensity of the light signal measured on the screen as a function of distance, y, perpendicular to the original light beam (Figure 29.6).

```python
import numpy as np
from scipy.signal import find_peaks
import matplotlib.pyplot as plt

y, I = np.genfromtxt('double-slit-data-noisy.txt', unpack=True)
idx, props = find_peaks(I, distance=40, prominence=0.05)
```

```python
plt.plot(y, I)
plt.scatter(y[idx], I[idx])
plt.xlabel(r'$y\;/\mathrm{mm}$')
plt.ylabel('Intensity /arb. units')
```

`idx` contains the indexes of the peaks. The positions (on the y-axis of the screen) are given by `y[idx]`, and the spacing between the peaks can be calculated using the function `numpy.diff`:

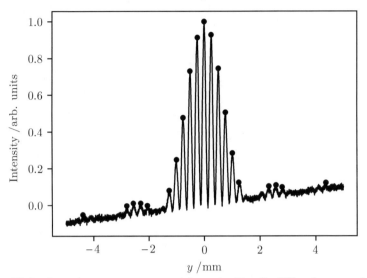

Figure 29.6   Intensity pattern measured for a double-slit diffraction experiment, showing noise and baseline artefacts.

```
peak_spacing = np.diff(y[idx])
peak_spacing
```

```
array([1.57539385, 0.24006002, 0.27006752, 0.24006002, 0.79019755,
       0.26756689, 0.24006002, 0.24756189, 0.27006752, 0.24256064,
       0.25006252, 0.26006502, 0.25756439, 0.24506127, 0.22755689,
       1.07776944, 0.25506377, 0.22005501, 1.56789197])
```

We are only interested in the spacing between peaks within each diffraction order, which seems to be around 0.25 mm:

```
peak_spacing = peak_spacing[peak_spacing < 0.3]
peak_spacing
```

```
array([0.24006002, 0.27006752, 0.24006002, 0.26756689, 0.24006002,
       0.24756189, 0.27006752, 0.24256064, 0.25006252, 0.26006502,
       0.25756439, 0.24506127, 0.22755689, 0.25506377, 0.22005501])
```

The diffraction formula gives the angular position of the maxima (interference fringes), measured from the midpoint between the slits:

$$n\lambda = d \sin \theta_n \approx \frac{dy_n}{D}, \quad n = 0, 1, 2, \ldots$$

Therefore, neighboring maxima are spaced by approximately

$$\Delta y \approx \frac{D\lambda}{d}.$$

Taking the average spacing value, we get an estimate of the distance between the slits, d:

```
sep = np.mean(peak_spacing)
# Light wavelength (mm)
lam = 512 * 1.e-6
# Distance of screen from the slits (mm)
D = 100
d = D * lam / sep
print(f'd = {d:.2f} mm')
```

```
d = 0.21 mm
```

30

Fitting the Vibrational Spectrum of CO

30.1 Analyzing a Rovibrational Spectrum

The high-resolution infrared absorption spectrum of a typical diatomic molecule will show signals due to rovibrational transitions between its energy levels. Real molecules are anharmonic oscillators, and bands corresponding to a change in vibrational quantum number, $\Delta v = 0, 1, 2, \ldots$ can be observed, given sufficient sensitivity, though the fundamental band $v = 0 \rightarrow v = 1$ is expected to be by far the strongest. Each band consists of individual transition lines involving a change in rotational quantum number, $\Delta J = \pm 1$ (for a closed-shell diatomic molecule). The collection of lines with $\Delta J = -1$ are referred to as the P-branch; those with $\Delta J = +1$ form the R-branch.

The files, `CO-v01.txt` and `CO-v02.txt`, which can be downloaded from `https://scipython.com/chem/cas/`, contain the infrared absorption spectrum of the principal isotopologue of carbon monoxide, $^{12}C^{16}O$, measured in the region of the fundamental and first overtone ($v = 0 \rightarrow v = 2$) bands, respectively.

Using these data, it is possible to obtain the spectroscopic constants corresponding to a model of the molecule which includes anharmonicity and centrifugal distortion. That is, the following equation for the rovibrational energy levels is taken:

$$E(v, J) = \omega_e(v + \tfrac{1}{2}) - \omega_e x_e(v + \tfrac{1}{2})^2$$
$$+ [B_e - \alpha_e(v + \tfrac{1}{2})]J(J + 1)$$
$$- D_e[J(J + 1)]^2.$$

The first two terms are the familiar Morse oscillator expression for the vibrational energy (see Chapter 24), whilst the third and fourth define the rotational energy, taking into account the vibrational-dependence of the rotational constant ($B_v = B_e - \alpha_e(v + \tfrac{1}{2})$) and centrifugal distortion (higher-order terms are omitted in this particular model).

For the general transition between energy levels $(v'', J'') \rightarrow (v', J')$, absorption lines are predicted to occur at the wavenumbers:

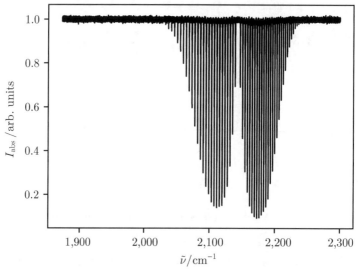

Figure 30.1   The fundamental band in the infrared spectrum of ^{12}C^{16}O. The individual rovibrational transitions in the P- and R-branches are well-resolved.

$$\tilde{\nu} = E(v', J') - E(v'', J'')$$
$$= \omega_e[(v' + \tfrac{1}{2}) - (v'' + \tfrac{1}{2})]$$
$$+ \omega_e x_e[(v'' + \tfrac{1}{2})^2 - (v' + \tfrac{1}{2})^2]$$
$$+ B_e[J'(J' + 1) - J''(J'' + 1)]$$
$$+ \alpha_e[(v'' + \tfrac{1}{2})J''(J'' + 1) - (v' + \tfrac{1}{2})J'(J' + 1)]$$
$$+ D_e\{[J''(J'' + 1)]^2 - [J'(J' + 1)]^2\}.$$

This model function is linear in the five spectroscopic parameters ω_e, $\omega_e x_e$, B_e, α_e and D_e, yet we might hope to observe far more than five individual absorption lines, so a linear least-squares fitting approach is justified.

After some imports, we will plot the fundamental infrared absorption band of ^{12}C^{16}O (Figure 30.1).

```
import numpy as np
from scipy.signal import find_peaks
import matplotlib.pyplot as plt
```

```
# Load the 0-1 spectrum as a function of wavenumber.
nu_v01, Iabs_v01 = np.genfromtxt('CO-v01.txt', unpack=True)
```

```
plt.plot(nu_v01, Iabs_v01, lw=1)
plt.xlabel(r'$\tilde{\nu}/\mathrm{cm^{-1}}$')
plt.ylabel('$I_\mathrm{abs}/\mathrm{arb.\;units}$')
```

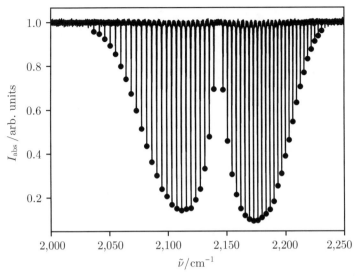

Figure 30.2   Rovibrational peaks identified in the infrared spectrum of $^{12}C^{16}O$.

We can use `scipy.signal.find_peaks` (see Chapter 29) to identify the transition wavenumbers as the centers of each peak in the spectrum. Since the spectrum is measured in absorption, the peaks present as local minima, so we need to find the maxima in the negative of the spectrum array, `-Iabs_v01`. There is some noise, but with a little trial-and-error, we can find suitable parameters prominence and width to pass to `scipy.signal.find_peaks` in order to capture most of the signals (Figure 30.2):

```
idx, _ = find_peaks(-Iabs_v01, prominence=0.04, width=5)
nu_peaks_v01, I_peaks_v01 = nu_v01[idx], Iabs_v01[idx]
```

```
plt.plot(nu_v01, Iabs_v01, lw=1)
plt.xlim(2000, 2250)
plt.scatter(nu_peaks_v01, I_peaks_v01)
```

These absorption lines are usually labeled by the lower-state rotational quantum number: $P(J'')$ and $R(J'')$. We could assign the peaks by hand, but the presence of a gap between the $P(1)$ and $R(0)$ lines (where the Q-branch, corresponding to $\Delta J = 0$ transitions is missing for this type of molecule) suggests an automated approach: find the maximum spacing between the peaks: The lines to lower wavenumber are $P(1), P(2), P(3) \ldots$; those at higher wavenumbers are $R(0), R(1), R(2) \ldots$

```
def assign_J(nu_peaks):
    """

    Return arrays Jpp and Jp: rotational quantum numbers for the lower and upper
    states respectively for the vibrational transitions located at the
    wavenumbers nu_peaks.
```

```
    """

    npeaks = len(nu_peaks)
    print(npeaks, 'peaks provided to assign')

    # The spacing between consecutive elements in nu_peaks:
    dnu = np.diff(nu_peaks)
    # Find the index of the maximum spacing, corresponding to the band center.
    idx = dnu.argmax()

    # We will store the rotational quantum numbers in these arrays.
    Jpp = np.empty(npeaks, dtype=int)
    Jp = np.empty(npeaks, dtype=int)

    # P-branch lines: everything up to (and including) idx,
    # decreasing with increasing index (and hence wavenumber).
    Jpp[:idx+1] = np.arange(idx+1, 0, -1, dtype=int)
    Jp[:idx+1] = Jpp[:idx+1] - 1       # P-branch: J' = J"-1
    # R-branch lines: everything from idx to the end of the array,
    # increasing with increasing index (and hence wavenumber).
    n = npeaks - idx - 1
    Jpp[idx+1:] = np.arange(0, n, 1)
    Jp[idx+1:] = Jpp[idx+1:] + 1       # R-branch: J' = J"+1


    return Jpp, Jp
```

```
Jpp_v01, Jp_v01 = assign_J(nu_peaks_v01)
print('Jpp:', Jpp_v01)
print('Jp :', Jp_v01)
```

```
51 peaks provided to assign
Jpp: [25 24 23 22 21 20 19 18 17 16 15 14 13 12 11 10  9  8  7  6  5  4  3  2
  1  0  1  2  3  4  5  6  7  8  9 10 11 12 13 14 15 16 17 18 19 20 21 22
 23 24 25]
Jp : [24 23 22 21 20 19 18 17 16 15 14 13 12 11 10  9  8  7  6  5  4  3  2  1
  0  1  2  3  4  5  6  7  8  9 10 11 12 13 14 15 16 17 18 19 20 21 22 23
 24 25 26]
```

To check we're on the right track, let's produce a labeled spectrum of the region
close to the band center (Figure 30.3).

```
plt.plot(nu_v01, Iabs_v01)
xmin, xmax = 2120, 2165
plt.xlim(xmin, xmax)
plt.scatter(nu_peaks_v01, I_peaks_v01)
# Get the indexes of the peaks within the wavenumber range of interest.
idx = np.where((xmin < nu_peaks_v01) & (nu_peaks_v01 < xmax))[0]
for i in idx:
    branch = 'P' if Jpp_v01[i] > Jp_v01[i] else 'R'
    # Label the peak with P(J") or R(J") with a line between the label
    # and the peak minimum. shrinkB backs off the line by 10 pts so it doesn't
    # actually connect with the peak: this looks better.
    plt.annotate(f'{branch}({Jpp_v01[i]})', xytext=(nu_peaks_v01[i], -0.1),
```

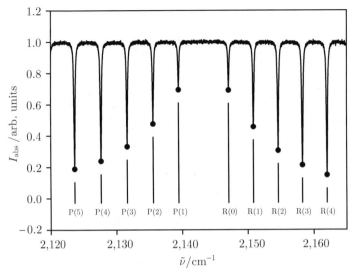

Figure 30.3   Assigned lines near the center of the fundamental band in the infrared spectrum of $^{12}C^{16}O$.

```
                    xy=(nu_peaks_v01[i], I_peaks_v01[i]), ha='center',
                    arrowprops={'arrowstyle': '-', 'shrinkB': 10})
plt.ylim(-0.2,1.2)
plt.xlabel(r'$\tilde{\nu}/\mathrm{cm^{-1}}$')
plt.ylabel('$I_\mathrm{abs}/\mathrm{arb.\;units}$')
```

We can follow the same process for the first overtone band, using the data in the file CO-v02.txt:

```
nu_v02, Iabs_v02 = np.genfromtxt('CO-v02.txt', unpack=True)
idx, _ = find_peaks(-Iabs_v02, prominence=0.003, width=5)
nu_peaks_v02, I_peaks_v02 = nu_v02[idx], Iabs_v02[idx]

Jpp_v02, Jp_v02 = assign_J(nu_peaks_v02)
print('Jpp:', Jpp_v02)
print('Jp :', Jp_v02)
```

```
40 peaks provided to assign
Jpp: [19 18 17 16 15 14 13 12 11 10  9  8  7  6  5  4  3  2  1  0  1  2  3  4
  5  6  7  8  9 10 11 12 13 14 15 16 17 18 19 20]
Jp : [18 17 16 15 14 13 12 11 10  9  8  7  6  5  4  3  2  1  0  1  2  3  4  5
  6  7  8  9 10 11 12 13 14 15 16 17 18 19 20 21]
```

30.2 Fitting the Spectrum

First, we need to combine the arrays of peak positions and quantum number assignments:

```
# All the transitions in a single array.
nu_peaks = np.hstack((nu_peaks_v01, nu_peaks_v02))
# All the lower and upper state rotational quantum numbers.
Jpp = np.hstack((Jpp_v01, Jpp_v02))
Jp = np.hstack((Jp_v01, Jp_v02))
# All the lower and upper state vibrational quantum numbers.
# The bands are 0-1 and 0-2 so the lower state is always v=0.
vpp = np.hstack((np.zeros(Jpp_v01.shape), np.zeros(Jpp_v02.shape)))
# The upper state is v=1 or v=2 for the two bands.
vp = np.hstack((np.ones(Jpp_v01.shape), 2 * np.ones(Jpp_v02.shape)))
```

Using these arrays, we can build the design matrix, **X**, based on the above expression for the transition wavenumbers (see Chapter 19):

```
def make_X(vpp, Jpp, vp, Jp):
    return np.vstack(((vp+0.5) - (vpp+0.5),
                      (vpp+0.5)**2 - (vp+0.5)**2,
                      Jp*(Jp+1) - Jpp*(Jpp+1),
                      (vpp+0.5)*Jpp*(Jpp+1) - (vp+0.5)*Jp*(Jp+1),
                      (Jpp*(Jpp+1))**2 - (Jp*(Jp+1))**2)).T

X = make_X(vpp, Jpp, vp, Jp)
```

The fit itself proceeds smoothly:

```
res = np.linalg.lstsq(X, nu_peaks, rcond=None)
prm_vals = res[0]
sq_resid = res[1]
rank = res[2]
sing_vals = res[3]
print(f'Squared residuals: {sq_resid} cm-1, rms error: {np.sqrt(sq_resid)} cm-1\n'
      f'Rank of X: {rank} (should be = {X.shape[1]})\n'
      f'Singular values of X:{sing_vals}')
```

```
Squared residuals: [0.00457467] cm-1, rms error: [0.06763629] cm-1
Rank of X: 5 (should be = 5)
Singular values of X:[2.10195973e+05 3.23559535e+03 1.16404896e+02 3.04587234e+01
 1.95244413e+00]
```

Note the relatively small residuals: The rms (root-mean-squared) error is a reasonable measure of the goodness of fit. The matrix rank of **X** is the same as the number of parameters to fit, meaning that the model does not contain linearly dependent terms.

```
prm_names = ('we', 'wexe', 'Be', 'alphae', 'De')
print(f'The fit parameters (cm-1):')
for prm_name, prm_val in zip(prm_names, prm_vals):
    print(f'{prm_name} = {prm_val:.6g}')
```

```
The fit parameters (cm-1):
we = 2169.75
```

```
wexe = 13.2396
Be = 1.93137
alphae = 0.0175024
De = 6.20704e-06
```

These compare quite well to the literature values, based on a more complete fit to a more extensive model function:

$$\omega_e = 2169.81358 \text{ cm}^{-1}$$

$$\omega_e x_e = 13.28831 \text{ cm}^{-1}$$

$$B_e = 1.93128087 \text{ cm}^{-1}$$

$$\alpha_e = 0.01750441 \text{ cm}^{-1}$$

$$D_e = 6.12147 \times 10^{-6} \text{ cm}^{-1}$$

The fitted parameters can be used to predict the locations of lines that are not observed in the measured spectrum. For example:

```python
def predict(vpp, Jpp, vp, Jp):
    """Predict the wavenumber of the transition(s) (vpp, Jpp) -> (vp, Jp)."""
    X = make_X(vpp, Jpp, vp, Jp)
    nu_pred = X @ prm_vals
    return nu_pred

# Predict the position of the (0,0) -> (3, 1) line (R(1) of the second overtone):
print(f'nu(0,0 -> 3,1) = {predict(0, 0, 3, 1)} cm-1')
```

```
nu(0,0 -> 3,1) = [6354.10656502] cm-1
```

Using this function, the residuals for all of the fitted lines can be plot on a scatter plot (Figure 30.4).

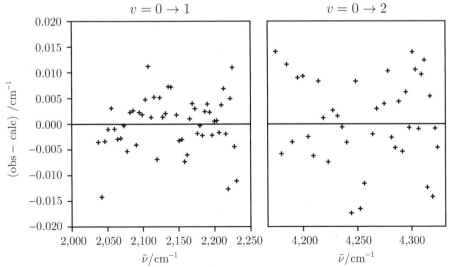

Figure 30.4   Fit residuals for the fundamental and first overtone band of $^{12}C^{16}O$.

```
resid = nu_peaks - predict(vpp, Jpp, vp, Jp)

fig, (ax_v01, ax_v02) = plt.subplots(nrows=1, ncols=2)
ax_v01.scatter(nu_peaks[vp==1], resid[vp==1], marker='+')
ax_v01.set_xlim(2000, 2250)
ax_v01.set_ylim(-0.02, 0.02)
ax_v01.set_xlabel(r'$\tilde{\nu}/\mathrm{cm^{-1}}$')
ax_v01.set_ylabel(r'$\mathrm{(obs - calc)\;/cm^{-1}}$')
ax_v01.axhline(0, c='k', lw=1)
ax_v01.set_title(r'$v = 0 \rightarrow 1$')
ax_v02.scatter(nu_peaks[vp==2], resid[vp==2], marker='+')
ax_v02.set_ylim(-0.02, 0.02)
ax_v02.set_xlabel(r'$\tilde{\nu}/\mathrm{cm^{-1}}$')
ax_v02.set_yticks([])
ax_v02.axhline(0, c='k', lw=1)
ax_v02.set_title(r'$v = 0 \rightarrow 2$')
```

31

pandas

pandas is a free, open-source library for manipulating and analyzing tabular data. Unlike NumPy's `array`, pandas' basic array-like data structure, the `DataFrame`, can contain heterogeneous data types (floats, integers, strings, dates, etc.); pandas data can exist in a hierarchy and is accessed through a large number of vectorized functions for transforming and aggregating it.

The pandas homepage, `https://pandas.pydata.org/`, contains details of the latest release and how to download and install pandas. In this chapter, we follow the common convention of importing the library as the alias pd:

```
import pandas as pd
```

The key pandas data structures are `Series` and `DataFrame`, representing a one-dimensional sequence of values and a data table, respectively. Their basic properties and use will be described in this chapter, with an emphasis on reading in, cleaning and plotting data.

31.1 Series

In its simplest form, a `Series` may be created in the same way as a one-dimensional NumPy array:

```
In [x]: DcH = pd.Series([-890, -1560, -2220, -2878, -3537])
In [x]: DcH
Out[x]:
0    -890
1   -1560
2   -2220
3   -2878
4   -3537
dtype: int64
```

The `Series` can be given a name and `dtype`:

```
In [x]: DcH = pd.Series(
   ...:     [-890, -1560, -2220, -2878, -3537],
   ...:     name="Enthalpy of Combustion /kJ.mol-1",
   ...:     dtype=float,
   ...: )

In [x]: DcH
Out[x]:
0     -890.0
1    -1560.0
2    -2220.0
3    -2878.0
4    -3537.0
Name: Enthalpy of Combustion /kJ.mol-1, dtype: float64
```

Unlike a NumPy array, however, each element in a pandas `Series` is associated with an *index*. Since we did not set the index explicitly here, a default integer sequence (starting at 0) is used for the index:

```
In [x]: DcH.index
Out[x]: RangeIndex(start=0, stop=5, step=1)
```

`RangeIndex` is a pandas object that works in a memory-efficient way like Python's `range` built-in to provide a monotonic integer sequence. However, it is often useful to refer to the rows of a `Series` with some other label than an integer index. Explicit indexing of the entries can be achieved by passing a sequence as the `index` argument or by creating the `Series` from a dictionary:

```
In [x]: DcH = pd.Series(
   ...:     [-890, -1560, -2220, -2878, -3537],
   ...:     name="Enthalpy of Combustion /kJ.mol-1",
   ...:     index=["CH4", "C2H6", "C3H8", "C4H10", "C5H12"],
   ...: )
```

or:

```
In [x]: DcH = pd.Series(
   ...:     {"CH4": -890,
   ...:      "C2H6": -1560,
   ...:      "C3H8": -2220,
   ...:      "C4H10": -2878,
   ...:      "C5H12": -3537
   ...:     },
   ...:     name="Enthalpy of Combustion /kJ.mol-1",
   ...: )

In [x]: DcH
Out[x]:
CH4      -890
C2H6    -1560
C3H8    -2220
```

```
C4H10    -2878
C5H12    -3537
Name: Enthalpy of Combustion /kJ.mol-1, dtype: int64
```

This allows a nicely expressive way of referring to `Series` entries using the index labels instead of integers, either individually:

```
In [x]: DcH["C2H6"]
Out[x]: -1560
```

instead of `DcH[1]` (indexing the second element of the `Series`). A `list` of index labels can also be used:

```
In [x]: DcH[["CH4", "C3H8", "C5H12"]]
Out[x]:
CH4      -890
C3H8    -2220
C5H12   -3537
Name: Enthalpy of Combustion /kJ.mol-1, dtype: int64
```

(though `DcH[[0, 2, 4]]` would also give the same result). Python-style slicing also works as expected:

```
In [x]: DcH[2::-1]
Out[x]:
C3H8    -2220
C2H6    -1560
CH4      -890
Name: Enthalpy of Combustion /kJ.mol-1, dtype: int64
```

It is even possible to use a slice-like notation for the index labels, but note that in this case, the endpoint is *inclusive*:

```
In [x]: DcH["C3H8":"CH4":-1]    # backwards, from propane to methane, inclusive
Out[x]:
C3H8    -2220
C2H6    -1560
CH4      -890
Name: Enthalpy of Combustion /kJ.mol-1, dtype: int64
```

Providing the index label is a valid Python identifier, one can also refer to a row as an *attribute* of the `Series`:

```
In [x]: DcH.C2H6
Out[x]: -1560
```

It is, of course, possible to do numerical operations on `Series` data, in a vectorized fashion, as for NumPy arrays:

```
In [x]: DcH /= [16, 30, 44, 58, 72]    # divide by the hydrocarbon molar masses
In [x]: DcH.name = "Enthalpy of Combustion /kJ.g-1"
```

```
In [x]: DcH
Out[x]:
CH4       -55.625000
C2H6      -52.000000
C3H8      -50.454545
C4H10     -49.620690
C5H12     -49.125000
Name: Enthalpy of Combustion /kJ.g-1, dtype: float64
```

In the above, we have also updated the `Series`' name attribute. Note that the `dtype` has also changed appropriately from `int64` to `float64` to accommodate the new values.

Comparison operations and filtering a `Series` with a boolean operation creates a new `Series` (which can be used to index into the original `Series`):

```
In [x]: abs(DcH) > 50
Out[x]:
CH4           True
C2H6          True
C3H8          True
C4H10        False
C5H12        False
Name: Enthalpy of Combustion /kJ.g-1, dtype: bool

In [x]: DcH[abs(DcH) > 50]
Out[x]:
CH4       -55.625000
C2H6      -52.000000
C3H8      -50.454545
Name: Enthalpy of Combustion /kJ.g-1, dtype: float64
```

Tests for *membership* of a `Series` examine the *index*, not the *values*:

```
In [x]: "C2H6" in DcH
Out[x]: True
```

`Series` can be *sorted*, either by their index or their values, using `Series.sort_index` and `Series.sort_values`, respectively. By default, these methods return a new `Series`, but they can also be used to update the original `Series` with the argument `inplace=True`. A further argument, `ascending`, can be `True` (the default) or `False` to set the ordering. For example:

```
In [x]: Tm = pd.Series(
   ...:        {"Sc": 1541, "Ti": 1670, "V": 1910, "Cr": 1907, "Mn": 1246},
   ...:        name="Melting point, degC",
   ...:        dtype=float,
   ...: )

In [x]: Tm
Out[x]:
Sc     1541.0
```

```
Ti      1670.0
V       1910.0
Cr      1907.0
Mn      1246.0
Name: Melting point, degC, dtype: float64

In [x]: Tm.sort_values(ascending=False)    # Returns a new Series
Out[x]:
V       1910.0
Cr      1907.0
Ti      1670.0
Sc      1541.0
Mn      1246.0
Name: Melting point, degC, dtype: float64

In [x]: Tm.sort_values(inplace=True)        # Sorts in-place, returns None

In [x]: Tm
Out[x]:
Mn      1246.0
Sc      1541.0
Ti      1670.0
Cr      1907.0
V       1910.0
Name: Melting point, degC, dtype: float64

In [x]: Tm.sort_index()  # sort rows by index (alphabetically) instead of value
Out[x]:
Cr      1907.0
Mn      1246.0
Sc      1541.0
Ti      1670.0
V       1910.0
Name: Melting point, degC, dtype: float64
```

When two series are combined, they are aligned by index label.

```
In [x]: DfusH = pd.Series(
   ...:        {"Sc": 14.1, "V": 21.5, "Mn": 12.9, "Fe": 13.8},
   ...:        name="Enthalpy of fusion /kJ.mol-1"
   ...: )

In [x]: DfusH
Out[x]:
Sc      14.1
V       21.5
Mn      12.9
Fe      13.8
Name: Enthalpy of fusion /kJ.mol-1, dtype: float64

In [x]: DfusS = DfusH / (Tm + 273.15) * 1000

In [x]: DfusS.name = "Entropy of fusion /J.K-1.mol-1"

In [x]: DfusS
```

```
Out[x]:
Cr          NaN
Fe          NaN
Mn     8.491591
Sc     7.772235
Ti          NaN
V      9.848155
Name: Entropy of fusion /J.K-1.mol-1, dtype: float64
```

Note that where no correspondence can be made within the indexes (an index label in one `Series` is missing from the other), the result is "Not a Number" (NaN). The methods `isnull` and `notnull` test for this:

```
In [x]: DfusS.notnull()
Out[x]:
Cr     False
Fe     False
Mn      True
Sc      True
Ti     False
V       True
dtype: bool
```

In this `Series`, we have valid data values for Mn, Sc and V.

The useful `dropna` method returns a `Series` with only the non-null values in it:

```
In [x]: DfusS.dropna()        # or index with DfusS[DfusS.notnull()]
Out[x]:
Mn     8.491591
Sc     7.772235
V      9.848155
dtype: float64
```

Finally, to return a *NumPy array* of the data in a `Series`, use the `values` property:

```
In [x]: DfusS.values
Out[x]:
array([       nan,        nan, 8.49159069, 7.77223493,        nan,
       9.84815519])
```

31.2 DataFrame

31.2.1 Defining a DataFrame

A `DataFrame` is a two-dimensional table of data that can be thought of as an ordered set of `Series` columns, which all have the same index. To create a simple `DataFrame` from a dictionary, assign key-value pairs with the column name as the key and a sequence, such as a `list`, as the value:

```
In [x]: lanthanides = pd.DataFrame(
   ...:     index=["La", "Ce", "Pr", "Nd", "Pm"],
   ...:     data={
   ...:         "atomic number": [57, 58, 59, 60, 61],
   ...:         "atomic radius /pm": [188, 182, 183, 182, 181],
   ...:         "melting point /degC": [920, 795, 935, 1024, None],
   ...:     },
   ...: )

In [x]: lanthanides
Out[x]:
    atomic number  atomic radius /pm  melting point /degC
La             57                188                920.0
Ce             58                182                795.0
Pr             59                183                935.0
Nd             60                182               1024.0
Pm             61                181                  NaN
```

As with a `Series`, the `index` labels the rows of a `DataFrame`, and in this case, we have three columns of data for each row. Values which were `None` in the data have been assigned to NaN in the `DataFrame`.

To rename a column or row, call `rename`, declaring which axis (`'index'` [the same as `'rows'`, and the default] or `'columns'`[1]) contains the label(s) to be renamed, and passing a dictionary mapping each original label to its replacement. Remember to set `inplace=True` if you want the original `DataFrame` modified rather than a new copy returned. For example:

```
In [x]: lanthanides.rename({"atomic number": "Z"}, axis="columns", inplace=True)

In [x]: lanthanides
Out[x]:
     Z  atomic radius /pm  melting point /degC
La  57                188                920.0
Ce  58                182                795.0
Pr  59                183                935.0
Nd  60                182               1024.0
Pm  61                181                  NaN
```

31.2.2 Accessing Rows, Columns, and Cells

Directly indexing a `DataFrame` with a name accesses all the rows in the corresponding *column*:

```
In [x]: lanthanides["Z"]            # or lanthanides.Z
Out[x]:
La    57
```

---

[1] One can also refer to the rows and columns of a `DataFrame` with `axis=0` and `axis=1`, respectively.

```
Ce    58
Pr    59
Nd    60
Pm    61
Name: Z, dtype: int64
```

Since this column is just a pandas `Series`, individual values can be retrieved by position or reference to the index label:

```
In [x]: lanthanides["Z"]["Nd"]      # or lanthanides.Z.Nd
Out[x]: 60
```

This is an example of "chained indexing" (`[..][..]`), and whilst it is unproblematic for *accessing* data, it should be avoided for *assigning* to a `DataFrame`: Depending on how the data are stored in memory, it is possible for the indexing expression to yield a *copy* of the data rather than a view. Assigning to this copy would leave the original `DataFrame` unchanged, which is presumably not what is wanted. For this reason, pandas raises a warning:

```
In [x]: # Set the unknown melting point of Pm to an estimated value
In [x]: lanthanides["melting point /degC"]["Pm"] = 1042
<ipython-input-95-019333525c8c>:1: SettingWithCopyWarning:
A value is trying to be set on a copy of a slice from a DataFrame

See the caveats in the documentation: https://pandas.pydata.org/...
  lanthanides["melting point /degC"]["Pm"] = 1042

In [x]: lanthanides
Out[x]:
        Z  atomic radius /pm  melting point /degC
La    57                188                920.0
Ce    58                182                795.0
Pr    59                183                935.0
Nd    60                182               1024.0
Pm    61                181               1042.0
```

In this case, despite the warning, the assignment has worked; however, to avoid uncertainty, the best way to reliably access and assign to columns, rows and cells is to use the two `DataFrame` methods, `loc` and `iloc`. `loc` selects by row and column *labels*. The first index identifies one or more rows by the index name or list of names:

```
In [x]: lanthanides.loc["La"]
Out[x]:
Z                      57.0
atomic radius /pm     188.0
melting point /degC   920.0
Name: La, dtype: float64

In [x]: lanthanides.loc[["La", "Ce", "Pr"]]
Out[x]:
        Z  atomic radius /pm  melting point /degC
```

```
La  57                   188                  920.0
Ce  58                   182                  795.0
Pr  59                   183                  935.0
```

The second index identifies the column(s):

```
In [x]: lanthanides.loc["Ce", "atomic radius /pm"]
Out[x]: 182

In [x]: lanthanides.loc["Ce", ["Z", "atomic radius /pm"]]
Out[x]:
Z                      58.0
atomic radius /pm     182.0
Name: Ce, dtype: float64
```

Slicing, "fancy" indexing and boolean indexing are all supported by `loc`:

```
In [x]: # all lanthanides up to Pr; columns from atomic radius to Z (backwards)
In [x]: lanthanides.loc[lanthanides.Z <= 59, "atomic radius /pm":"Z":-1]
Out[x]:
    atomic radius /pm   Z
La               188   57
Ce               182   58
Pr               183   59
```

The second method, `iloc`, retrieves data by numerical index position:

```
In [x]: lanthanides.iloc[1]     # data for the second lanthanide
Out[x]:
Z                      58.0
atomic radius /pm     182.0
melting point /degC   795.0
Name: Ce, dtype: float64

In [x]: lanthanides.iloc[:, [1, 2]] # all rows, second and third columns
Out[x]:
    atomic radius /pm   melting point /degC
La               188                 920.0
Ce               182                 795.0
Pr               183                 935.0
Nd               182                1024.0
Pm               181                1042.0

In [x]: lanthanides.iloc[-1, 0]     # last row, first column (Z for Pm)
Out[x]: 61
```

To delete a row, use the **drop** method (with **inplace=True** if you want the operation to be carried out on the DataFrame itself rather than for a new DataFrame to be returned):

```
In [x]: lanthanides.drop("Pm", inplace=True)

In [x]: lanthanides
Out[x]:
     Z   atomic radius /pm   melting point /degC
```

```
La   57              188              920.0
Ce   58              182              795.0
Pr   59              183              935.0
Nd   60              182             1024.0
```

To delete a column, use either the `del` keyword or the `drop` method with `axis=`
`'columns'` or `axis=1`:

```
In [x]: # either use del lanthanides["atomic radius /pm"] or:
In [x]: lanthanides.drop("atomic radius /pm", axis="columns", inplace=True)

In [x]: lanthanides
Out[x]:
        Z  melting point /degC
La   57                920.0
Ce   58                795.0
Pr   59                935.0
Nd   60               1024.0
```

Columns can be added to a `DataFrame` by providing a `list` (array, tuple, etc.) or
dictionary of the correct length:

```
In [x]: lanthanides["boiling point /degC"] = [3469, 3468, 3127, 3027]

In [x]: lanthanides
Out[x]:
        Z  melting point /degC  boiling point /degC
La   57                920.0                 3469
Ce   58                795.0                 3468
Pr   59                935.0                 3127
Nd   60               1024.0                 3027
```

Alternatively, a pandas `Series` can be assigned to the `DataFrame`:

```
In [x]: I = pd.Series({"La": 538, "Pr": 523, "Nd": 529})
In [x]: lanthanides["ionization energy /kJ.mol-1"] = I

In [x]: lanthanides
Out[x]:
        Z  melting point /degC  boiling point /degC  ionization energy /kJ.mol-1
La   57                920.0                 3469                        538.0
Ce   58                795.0                 3468                          NaN
Pr   59                935.0                 3127                        523.0
Nd   60               1024.0                 3027                        529.0
```

Note that the `Series` index does not have to match that of the `DataFrame`: Rows
in the `Series` that are not in the `DataFrame` are (silently) dropped; rows in the
`DataFrame` that are not in the `Series` are assigned to NaN:

```
In [x]: I = pd.Series({"La": 538, "Pr": 523, "Nd": 529, "Pm": 536})
In [x]: lanthanides["ionization energy /kJ.mol-1"] = I

In [x]: lanthanides
```

```
Out[x]:
       Z   melting point /degC   boiling point /degC   ionization energy /kJ.mol-1
La    57                 920.0                  3469                         538.0
Ce    58                 795.0                  3468                           NaN
Pr    59                 935.0                  3127                         523.0
Nd    60                1024.0                  3027                         529.0
```

Table 31.1   Important pandas data summary methods. Applied to a
DataFrame, the argument `axis` defaults to 0 (`'rows'`); to calculate the same
quantity evaluated over the columns, set `axis=1` (or `'columns'`)

min	Minimum
argmin	Index position of minimum value in `Series`
idxmin	Index label (or column name) of minimum value for each column or row in a `DataFrame`
max	Maximum
argmax	Index (or column) position of maximum value in `Series`
idxmax	Index label (or column name) of maximum value for each column or row in a `DataFrame`
sum	Sum of each column or row
prod	Product of each column or row
mean	Mean average along each column or row
median	Median average along each column or row
std	Standard deviation along each column or row
var	Variance along each column or row

More sophisticated methods for joining `DataFrames`, such as `join` and `merge`,
which mirror the behavior of relational databases are also available – see the documentation for details.[2]

31.2.3 Sorting and Statistics

pandas provides many methods for describing the data contained in a `DataFrame`;
the more important of these are listed in Table 31.1 and demonstrated by example
in this section.

We will work initially with data concerning the properties of the halogens in the
following `DataFrame`.

```
In [x]: df = pd.DataFrame()
In [x]: df.index = "F Cl Br I At".split()
In [x]: df["Z"] = [9, 17, 35, 53, 85]
In [x]: df["X-X bond strength /kJ.mol-1"] = [159, 243, 193, 151, None]
In [x]: df["melting point /K"] = [53.5, 171.6, 265.8, 386.9, 575]
In [x]: df
```

[2] https://pandas.pydata.org/docs/reference/frame.html.

```
Out[x]:
     Z  X-X bond strength /kJ.mol-1  melting point /K
F    9                        159.0              53.5
Cl  17                        243.0             171.6
Br  35                        193.0             265.8
I   53                        151.0             386.9
At  85                          NaN             575.0
```

By default, the `min` and `max` methods return the minimum and maximum values for each column, ignoring NaN values, which is appropriate for this `DataFrame`:

```
In [x]: df.min()
Out[x]:
Z                               9.0
X-X bond strength /kJ.mol-1   151.0
melting point /K               53.5
dtype: float64
```

Note that these data are a `Series` of the minimum values for each column and do not correspond to the data in a single row. The `dtype` of the `Series` has been set to the `float64` to accommodate the necessary values, which is why the atomic number appears as a real number, `9.0`.

To return the index labels corresponding to the minimum or maximum values in each column, use `idxmin` and `idxmax`. For example,

```
In [x]: df.idxmax()
Out[x]:
Z                             At
X-X bond strength /kJ.mol-1   Cl
melting point /K              At
dtype: object
```

That is, the row containing the maximum value of Z is the one with the index label `'At'`, that forming the strongest homoatomic bond is `'Cl'`, etc. The entire row of a `DataFrame` corresponding to the maximum in one of its columns can be retrieved by selecting this index label:

```
In [x]: idx = df["X-X bond strength /kJ.mol-1"].idxmax()

In [x]: idx
Out[x]: 'Cl'

In [x]: df.loc[idx]
Out[x]:
Z                             17.0
X-X bond strength /kJ.mol-1  243.0
melting point /K             171.6
Name: Cl, dtype: float64
```

`argmin` and `argmax` are similar, but only work with `Series` (i.e., single columns or rows of a `DataFrame`).

```
In [x]: df["X-X bond strength /kJ.mol-1"].argmin()
Out[x]: 3
```

(the row indexed at 3, corresponding to the label `'I'`, forms the weakest homoatomic bond).

Averaging and other statistical aggregation work according to a similar syntax.
Consider the following `DataFrame` of pH measurements on three water samples:

```
In [x]: pH = pd.DataFrame(
   ...:     {
   ...:         "Sample 1": [8.6, 8.5, 9.0, 8.8, None, 8.3, 9.1, 8.5],
   ...:         "Sample 2": [8.6, 9.0, 9.1, 9.0, 8.9, 8.4, None, 8.3],
   ...:         "Sample 3": [8.5, 9.1, 8.8, 9.2, 9.0, 8.6, 9.0, 8.5],
   ...:     },
   ...:     index="A B C D E F G H".split(),
   ...: )
In [x]: pH.index.name = "Measurement"
   ...:

In [x]: pH
Out[x]:
             Sample 1  Sample 2  Sample 3
Measurement
A                 8.6       8.6       8.5
B                 8.5       9.0       9.1
C                 9.0       9.1       8.8
D                 8.8       9.0       9.2
E                 NaN       8.9       9.0
F                 8.3       8.4       8.6
G                 9.1       NaN       9.0
H                 8.5       8.3       8.5
```

The `mean` function can be applied over rows (the default) or columns:

```
In [x]: pH.mean()
Out[x]:
Sample 1    8.685714
Sample 2    8.757143
Sample 3    8.837500
dtype: float64

In [x]: pH.mean(axis=1)
Out[x]:
Measurement
A    8.566667
B    8.866667
C    8.966667
D    9.000000
E    8.950000
F    8.433333
G    9.050000
H    8.433333
dtype: float64
```

The functions median, std (standard deviation) and var (variance) work in a similar way:

```
In [x]: pH.std()
Out[x]:
Sample 1     0.291139
Sample 2     0.320713
Sample 3     0.277424
dtype: float64
```

Finally, there is a handy method, describe, to summarize the contents of a Data Frame, including the above statistics as well as percentile information:

```
In [x]: pH.describe()
Out[x]:
         Sample 1   Sample 2   Sample 3
count    7.000000   7.000000   8.000000
mean     8.685714   8.757143   8.837500
std      0.291139   0.320713   0.277424
min      8.300000   8.300000   8.500000
25%      8.500000   8.500000   8.575000
50%      8.600000   8.900000   8.900000
75%      8.900000   9.000000   9.025000
max      9.100000   9.100000   9.200000
```

31.3 Reading and Writing Series and DataFrames

31.3.1 Reading Text Files

The core method for reading text files of data into a DataFrame is pd.read_csv. This works in much the same way as NumPy's genfromtxt method but with additional functionality for naming columns and setting the DataFrame index. It takes no fewer than 52 possible arguments, but the most important are as follows:[3]

- filepath_or_buffer (required): The path to the file to read: this can be a local file or a URL for fetching data from the internet.
- sep: The column delimiter; by default ',', but use '\s+' for whitespace-delimited columns, '\t' for tab-delimiters, or None to force pandas to try to infer the delimiter. See also delim_whitespace.
- delimiter: An alias for sep.
- header: The row numbers (indices) to use for the column names. The default is header=0: use the first row for the column names. Note: if the file does not have a header, specify header=None and set the column names with the names argument.

---

[3] See the documentation at https://pandas.pydata.org/pandas-docs/stable/reference/api/pandas.read_csv.html for a complete description.

- `names`: A sequence of unique column names to use. If the file contains no header, set `header=None` in addition to setting `names`.
- `index_col`: The column(s) to use as the row labels in the `DataFrame`.
- `usecols`: A sequence of column indices (as for NumPy's `genfromtxt` method) or column names identifying the columns to be read into the `DataFrame`.
- `squeeze`: If the data required consist of a single column, then `squeeze=True` will return a `Series` instead of the default, a `DataFrame`.
- `converters`: A dictionary of functions for converting the values in specified columns in the input file into data values for the `DataFrame`. The dictionary keys can be column indices or column names.
- `skiprows`: An integer giving the number of lines at the start of the file to skip over before reading the data or a sequence giving the indices of rows to skip.
- `skipfooter`: The number of rows at the bottom of the file to skip (by default, 0).
- `nrows`: The number of rows of the file to read: This is useful for reading a subset of lines from a very large file for testing or exploring its data.
- `na_values`: A string or sequence of strings to treat as NaN values, in addition to the default values which include `'NaN'`, `'NA'`, `'NULL'` and `'#N/A'` (see the documentation for a full list).
- `parse_dates`: Set to `True` to parse the index column(s) as a sequence of `date time` objects; other options are available for this argument (see the online documentation).
- `comment`: Specify a single character, such as `'#'`, which, when found at the start of a line, signals that the whole line is to be ignored.
- `skip_blank_lines`: The default, `True`, skips over blank lines in the input file; set to `False` to interpret these as a row of NaN values instead.
- `delim_whitespace`: Can be set to `True` instead of specifying `sep='\s+'` to indicate that the data columns are separated by whitespace.

31.3.2 Writing Text Files

The `DataFrame` method `to_csv` outputs its data to a text file, formatted according to the arguments summarized below.[4]

- `path_or_buf`: A file path or file object to output to; if `None`, the `DataFrame` is returned as string.
- `sep`: The single-character field-delimiter (defaults to `','`).
- `na_rep`: The string to use to represent missing data (defaults to the empty string, `''`).

---

[4] Full documentation is available at `https://pandas.pydata.org/pandas-docs/stable/reference/api/pandas.DataFrame.to_csv.html`.

- `float_format`: The C-style format specifier for floating-point numbers.
- `columns`: A sequence identifying the columns to output.
- `header`: By default, `True`, indicating that column names should be output; can also be set to `False` or a list of column names.
- `index`: By default, `True`, indicating that row names should be output.
- `compression`: One of `'infer'`, `'gzip'`, `'bz2'`, `'zip'`, `'xz'`, `None` to specify whether and how to compress the output file. The default is `'infer'`: pandas determines the intended compression method from the filename extension.

31.4 Examples

E31.1

The Melting and Boiling Points of the Elements

The file `element-data.csv` available at `https://scipython.com/chem/egia/`, contains comma-separated, tabular data concerning the properties of the elements. Missing data are indicated with a hyphen character, `'-'`.

```
import pandas as pd
df = pd.read_csv('element-data.csv', index_col=0, na_values='-')
# Show the first 5 rows of the DataFrame.
df.head()
```

```
          name  Z  ...  abundance  melting point /K  boiling point /K
Symbol                  ...
H       Hydrogen  1  ...  1.40e-03             13.99            20.271
He        Helium  2  ...  8.00e-09              1.80             4.222
Li       Lithium  3  ...  2.00e-05            453.70          1603.000
Be     Beryllium  4  ...  2.80e-06           1560.00          2742.000
B          Boron  5  ...  1.00e-05           2349.00          4200.000

[5 rows x 8 columns]
```

The abundance is that of the elements in the Earth's crust as a fractional value; we might prefer to express it in parts per million (ppm) for this example:

```
df['abundance'] *= 1.e6
# Also rename the column to include the "units".
df.rename(columns={'abundance': 'abundance /ppm'}, inplace=True)
```

The standard NumPy-like operations can also be carried out on a pandas DataFrame. For example, to determine which elements are liquids at room temperature (and, implicitly, standard pressure):

```
Troom = 298
# Liquids are those with a melting point below room temperature but
# a boiling point above room temperature.
df[(df['melting point /K'] < Troom) & (df['boiling point /K'] > Troom)]
```

```
            name   Z  ... melting point /K  boiling point /K
Symbol                 ...
Br       Bromine  35  ...           265.8            332.00
Hg       Mercury  80  ...           234.3            629.88
Fr      Francium  87  ...           281.0            890.00

[3 rows x 8 columns]
```

We can also determine the densest element and the least-dense solid element in a similar way:

```
symbol = df['density /kg.m-3'].idxmax()
rho_max = df.loc[symbol]['density /kg.m-3']
print(f'Element with the greatest density: {symbol} ({rho_max} kg.m-3)')

# First extract only the solid elements from our DataFrame.
solids_df = df[df['melting point /K'] > Troom]
symbol = solids_df['density /kg.m-3'].idxmin()
rho_min = df.loc[symbol]['density /kg.m-3']
print(f'Solid element with the lowest density: {symbol} ({rho_min} kg.m-3)')
```

```
Element with the greatest density: Os (22590.0 kg.m-3)
Solid element with the lowest density: Li (534.0 kg.m-3)
```

The abundance can be plotted as a line chart; in this case, it is most useful to use a logarithmic scale for the y-axis (Figure 31.1).

```
import numpy as np
import matplotlib.pyplot as plt
```

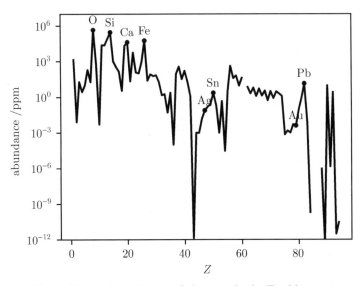

Figure 31.1   Abundances of elements in the Earth's crust.

```
fig, ax = plt.subplots()
# Since we're going to take the log of the abundance, replace the zeros with NaN.
ax.plot(df['Z'], df['abundance /ppm'].replace(0, np.nan))
ax.set_yscale('log')
ax.set_ylim(1e-12, 1e7)

# Add labels and markers for some significant elements.
symbols = ['O', 'Si', 'Ca', 'Fe', 'Ag', 'Sn', 'Au', 'Pb']
for symbol in symbols:
    x, y = df.loc[symbol, ['Z', 'abundance /ppm']]
    ax.text(x, y, symbol, ha='center', va='bottom')
ax.scatter(*df.loc[symbols, ['Z', 'abundance /ppm']].T.values)
ax.set_xlabel(r'$Z$')
ax.set_ylabel(r'abundance /ppm')
```

E31.2

Reaction Gibbs Free Energy and Equilibrium Constants

The file `thermo-data.csv`, available at `https://scipython.com/chem/gic`, contains gas-phase thermodynamic data on several compounds. The comma-separated columns are: chemical formula, InChIKey, standard molar enthalpy of formation ($\Delta_f H_m^{\ominus}$, in kJ mol^{-1} at 298 K), standard molar entropy ($S_m^{\ominus}$, in J K^{-1} mol^{-1} at 298 K), and the coefficients, $A - H$ for the temperature dependence of $\Delta_f H_m^{\ominus}$, $S_m^{\ominus}$ and the standard molar heat capacity, $C_{p,m}^{\ominus}$, according to the following *Shomate equations*:

$$C_{p,m}^{\ominus} = A + Bt + Ct^2 + Dt^3 + \frac{E}{t^2},$$

$$\Delta_f H_m^{\ominus} = \Delta_f H_m^{\ominus}(298 \text{ K}) + At + \frac{Bt^2}{2} + \frac{Ct^3}{3} + \frac{Dt^4}{4} - \frac{E}{t} + F - H,$$

$$S_m^{\ominus} = A \ln t + Bt + \frac{Ct^2}{2} + \frac{Dt^3}{3} - \frac{E}{2t^2} + G,$$

where $t = (T / \text{K})/1,000$.

We will read in the data file with pandas' `read_csv` method. Comments are denoted by the '#' character, and we would like to index the rows by the chemical formula (column 0):

```
import numpy as np
from scipy.constants import import R    # Gas constant, J.K-1.mol-1
import pandas as pd
```

```
df = pd.read_csv('thermo-data.csv', comment='#', index_col=0)
```

```
df
```

	InChIKey	DfHo_298	So_298	A	...	H
formula						
CH4	VNWKT...SA-N	-74.87	188.660	-0.703029	...	-74.8731
CO	UGFAI...SA-N	-110.53	197.660	25.567590	...	-110.5271
...	...	...	...	...	...	...
O2	MYMOF...SA-N	0.00	205.152	30.032350	...	0.0000

The column headers are taken from the first (non-comment) row of the file, but the spaces after the commas in this line have found their way into the column names:

```
df.columns
```

```
Index([' InChIKey', ' DfHo_298', ' So_298', ' A', ' B', ' C', ' D', ' E', ' F',
       ' G', ' H'],
      dtype='object')
```

There is an easy way to strip these stray spaces: call the **str**.strip method on them:

```
df.columns = df.columns.str.strip()
```

```
df.columns
```

```
Index(['InChIKey', 'DfHo_298', 'So_298', 'A', 'B', 'C', 'D', 'E', 'F', 'G',
       'H'],
      dtype='object')
```

The data are now accessible using regular pandas syntax:

```
df.loc['CO']        # Thermodynamic data relating to CO
```

```
InChIKey        UGFAIRIUMAVXCW-UHFFFAOYSA-N
DfHo_298                          -110.53
So_298                             197.66
A                                25.56759
B                                 6.09613
C                                4.054656
D                               -2.671301
E                                0.131021
F                               -118.0089
G                                227.3665
H                               -110.5271
Name: CO, dtype: object
```

It will be helpful to define a couple of functions to calculate $\Delta_f H_m^{\ominus}$ and $S_m^{\ominus}$ for a gas-phase compound specified by its formula, at some arbitrary temperature T, using the above Shomate equations:

```
def get_DfH(formula, T):
    """Return the standard molar enthalpy of formation of formula at T K."""
    s = df.loc[formula]
    t = T / 1000
    return (s.DfHo_298 + s.A * t + s.B * t**2 / 2 + s.C * t**3 / 3
            + s.D * t**4 / 4 - s.E / t + s.F - s.H)

def get_S(formula, T):
    """Return the standard molar entropy of formula at T K."""
    s = df.loc[formula]
    t = T / 1000
    return (s.A * np.log(t) + s.B * t + s.C * t**2 / 2 + s.D * t**3 / 3
            - s.E / 2 / t**2 + s.G)
```

These formulas can be used to calculate the standard reaction Gibbs free energy change, $\Delta_r G_m^\ominus$, and hence the equilibrium constant of reactions at different temperatures. For example, consider the *water-gas shift reaction*, which can be used to produce hydrogen gas:

$$CO(g) + H_2O(g) \rightleftharpoons H_2(g) + CO_2(g)$$

$$\Delta_r H_m^\ominus = \Delta_f H_m^\ominus(H_2(g)) + \Delta_f H_m^\ominus(CO_2(g)) - \Delta_f H_m^\ominus(CO(g)) - \Delta_f H_m^\ominus(H_2O(g)),$$
$$\Delta_r S_m^\ominus = S_m^\ominus(H_2(g)) + S_m^\ominus(CO_2(g)) - S_m^\ominus(CO(g)) - S_m^\ominus(H_2O(g)),$$
$$\Delta_r G_m^\ominus = \Delta_r H_m^\ominus - T\Delta_r S_m^\ominus,$$
$$K = \exp\left(-\frac{\Delta_r G_m^\ominus}{RT}\right),$$

where the relevant quantities are calculated at the temperature of interest.

Using the above functions, the molar enthalpy of formation and standard molar entropy of the reactants and products can be calculated at $T = 850$ K:

```
# Temperature, K
T = 850
# Get the enthalpy and entropy of formation
DfH, S = {}, {}
species = ('CO', 'H2O', 'H2', 'CO2')
for formula in species:
    DfH[formula] = get_DfH(formula, T)
    S[formula] = get_S(formula, T)
```

From this, the standard reaction enthalpy, entropy and Gibbs free energy changes can be calculated:

```
# Standard reaction enthalpy and entropy:
DrH = DfH['H2'] + DfH['CO2'] - DfH['CO'] - DfH['H2O']
DrS = S['H2'] + S['CO2'] - S['CO'] - S['H2O']
# Standard Gibbs free energy; NB convert DrS to kJ.K-1.mol-1:
DrG = DrH - T * DrS / 1000
```

```
print(f'DrH({T} K) = {DrH:.1f} kJ.mol-1')
print(f'DrS({T} K) = {DrS:.1f} J.K-1.mol-1')
print(f'DrG({T} K) = {DrG:.1f} kJ.mol-1')
```

```
DrH(850 K) = -36.3 kJ.mol-1
DrS(850 K) = -33.4 J.K-1.mol-1
DrG(850 K) = -7.9 kJ.mol-1
```

Finally, the equilibrium constant is;

```
K = np.exp(-DrG * 1000 / R / T)
print(f'K = {K:.2f}')
```

```
K = 3.05
```

The above code is easily generalized to any reaction:

```
def get_K(species, T):
    """
    Given a set of stoichiometric numbers (a, b, c, ...) and species formulas
    (A, B, C, ...) specifying a gas-phase reaction: aA + bB + ... <-> cC + dD + ...
    where reactant stoichiometric numbers are negative and products are positive,
    return the equilibrium constant for this reaction at temperature T (in K).
    e.g. for CO + H2O <-> H2 + CO2,
    species = [(-1, 'CO'), (-1, 'H2O'), (1, 'H2'), (1, 'CO2')]

    """

    DrH = sum(n * get_DfH(formula, T) for n, formula in species)
    DrS = sum(n * get_S(formula, T) for n, formula in species)
    DrG = DrH - T * DrS / 1000
    K = np.exp(-DrG * 1000 / R / T)
    return DrH, DrS, DrG, K
```

```
DrH, DrS, DrG, K = get_K([(-1, 'CO'), (-1, 'H2O'), (1, 'H2'), (1, 'CO2')], 850)
print(K)
```

```
3.0532234828745612
```

For a reaction in which the equilibrium lies far to the right (products), consider the combustion of methane:

$$CH_4(g) + 2\,O_2(g) \longrightarrow CO_2(g) + 2\,H_2O(g)$$

```
DrH, DrS, DrG, K = get_K([(-1, 'CH4'), (-2, 'O2'), (1, 'CO2'), (2, 'H2O')], 850)
print(f'DrH({T} K) = {DrH:.1f} kJ.mol-1')
print(f'DrS({T} K) = {DrS:.1f} J.K-1.mol-1')
print(f'DrG({T} K) = {DrG:.1f} kJ.mol-1')
print(f'K = {K:.2e}')
```

```
DrH(850 K)  =  -799.9 kJ.mol-1
DrS(850 K)  =  0.7 J.K-1.mol-1
DrG(850 K)  =  -800.5 kJ.mol-1
K = 1.56e+49
```

As expected for a highly exothermic reaction which also involves an increase in system entropy, the equilibrium constant, K, is huge, and this reaction proceeds to completion.

E31.3

Simulated Rovibrational Spectra of Diatomics

The file `diatomic-data.tab`, available at `https://scipython.com/chem/egib`, contains parameters, in cm^{-1}, describing the energy levels of a selection of diatomic molecules. These energies can be written in the form

$$S(v, J) = G(v) + F(v, J),$$

where the anharmonic vibrational energies are parameterized as:

$$G(v) = \omega_e \left(v + \tfrac{1}{2}\right) - \omega_e x_e \left(v + \tfrac{1}{2}\right)^2 + \omega_e y_e \left(v + \tfrac{1}{2}\right)^3$$

and the rotational energies include centrifugal distortion and vibration–rotation interaction terms:

$$F(v, J) = B_v J(J + 1) - D_v [J(J + 1)]^2$$
$$B_v = B_e - \alpha_e \left(v + \tfrac{1}{2}\right) + \gamma_e \left(v + \tfrac{1}{2}\right)^2$$
$$D_v = D_e - \beta_e \left(v + \tfrac{1}{2}\right).$$

In this example, we will use these data to plot the rovibrational (IR) spectrum of the fundamental band ($v = 1 \leftarrow 0$) of $^1H^{35}Cl$.

Some definitions, library imports and some useful physical constants from SciPy's `physical_constants` module:

```
import numpy as np
import matplotlib.pyplot as plt
import pandas as pd

from scipy.constants import physical_constants
# The second radiation constants, c2 = hc/kB, in cm.K
c2 = physical_constants['second radiation constant'][0] * 100
# Speed of light (m.s-1), atomic mass constant (kg) and Boltzmann constant (J.K-1)
c = physical_constants['speed of light in vacuum'][0]
u = physical_constants['atomic mass constant'][0]
kB = physical_constants['Boltzmann constant'][0]
```

The data are in fixed-width fields, separated by whitespace, with the asterisk character used as a placeholder for missing values.

```
df = pd.read_csv('diatomic-data.tab', sep='\s+', index_col=0, na_values='*')
df
```

parameter	(12C)(16O)	(1H)(35Cl)	...	(1H)(81Br)	(16O)2
			...		
we	2169.813600	2990.946300	...	2648.975000	1580.193000
wexe	13.288310	52.818600	...	45.217500	11.981000
weye	NaN	0.224370	...	-0.002900	0.047470
Be	1.931281	10.593416	...	8.464884	1.437677
alpha_e	0.017504	0.307181	...	0.233280	0.015930

```
[5 rows x 6 columns]
```

First, we are going to need a function to calculate the energy (in cm^{-1}) of a rovibrational state using the above formula for $S(v, J)$.

```
def get_E(v, J, molecule):
    """Return the energy, in cm-1, of the state (v, J) for a molecule."""

    # Get the parameters, replacing NaN values with 0
    prms = df[molecule].fillna(0)
    vfac, Jfac = v + 0.5, J * (J+1)
    Bv = prms['Be'] - prms['alpha_e'] * vfac + prms['gamma_e'] * vfac**2
    Dv = prms['De'] - prms['beta_e'] * vfac
    F = Bv * Jfac - Dv * Jfac**2
    G = prms['we'] * vfac - prms['wexe'] * vfac**2 + prms['weye'] * vfac**3
    return G + F
```

Next, identify the molecule by the column heading in the `prms` DataFrame, and decide how many energy levels to calculate; here up to $J = 50$ for each of the vibrational states, $v = 0, 1, 2$. These energy levels, and their degeneracies $(2J + 1)$ are stored in a new `DataFrame` called `states`, with a `MultiIndex`: Each row is identified by the tuple (`v`, `J`).

```
T = 296
molecule = '(1H)(35Cl)'
states_vmax, states_Jmax = 2, 50
index = pd.MultiIndex.from_tuples(
    ((v, J) for v in range(states_vmax+1) for J in range(states_Jmax+1)),
    names=['v', 'J'])
states = pd.DataFrame(index=index, columns=('E', 'g'), dtype=float)
for v, J in index:
    # Set the energy and the degeneracy columns.
    states.loc[(v, J), 'E'] = get_E(v, J, molecule)
    states.loc[(v, J), 'g'] = 2 * J + 1
# Convert the dtype of the degeneracy column to int.
states = states.convert_dtypes()
states
```

```
                    E      g
v  J
0  0       1482.296546       1
   1       1503.174069       3
   2       1544.916349       5
   3       1607.497853       7
   4        1690.88028       9
...               ...      ...
2  46      25906.990015      93
   47      26609.673157      95
   48      27317.604419      97
   49      28030.171006      99
   50      28746.747356     101

[153 rows x 2 columns]
```

Calculate the partition sum, q, from the calculated state energies and degeneracies:

$$q = \sum_i g_i \exp\left(-\frac{hc(E_i - E_0)}{k_B T}\right)$$

```
E, g = states['E'], states['g']
E0 = get_E(0, 0, molecule)
# In the partition sum, measure energies from the zero-point level.
q = np.sum(g * np.exp(-c2 * (E - E0) / T))
```

In calculating the transitions, it is not necessary to use all of the states above (since the higher-J energy levels will be barely populated at this temperature and the line strengths correspondingly small): Set a new rotational quantum number maximum for calculating the spectrum, and construct a new DataFrame to hold the transition data. For a closed-shell diatomic molecule there are two branches to the rovibrational spectrum, $P(J'')$ and $R(J'')$, corresponding to $\Delta J = J' - J'' = -1$ and $+1$, respectively. The index for this DataFrame will be a string corresponding to this designation.

```
trans_Jmax = 20
Jpp_P = np.arange(trans_Jmax, 0, -1)
Jpp_R = np.arange(0, trans_Jmax)
Pbranch = [f'P({Jpp})' for Jpp in Jpp_P]
Rbranch = [f'R({Jpp})' for Jpp in Jpp_R]
trans = pd.DataFrame(index=np.concatenate((Pbranch, Rbranch)))
trans['Jpp'] = np.concatenate((Jpp_P, Jpp_R))
trans['Jp'] = np.concatenate((Jpp_P-1, Jpp_R+1))
vp, vpp = 1, 0
```

Now, iterate over the values of J'' and J', calculating the transition wavenumber and relative line strengths from the energies in the states table. The line strengths are given by

$$S \propto \tilde{\nu}_0 \frac{H(J'')}{q} \exp\left(-\frac{c_2 E''}{T}\right)\left[1 - \exp\left(-\frac{c_2 \tilde{\nu}_0}{T}\right)\right],$$

where $H(J'')$ is a Hönl–London factor equal to $J'' + 1$ for the R-branch lines and J'' for the P-branch.

```
for br, (Jpp, Jp) in trans[['Jpp', 'Jp']].iterrows():
    # Store the transition's upper and lower state energies in the trans table,
    # measured from the zero-point level.
    Epp = states.loc[(vpp, Jpp), 'E'] - E0
    Ep = states.loc[(vp, Jp), 'E'] - E0
    nu = Ep - Epp
    # Also calculate the relative intensity, including a Honl-London factor,
    # P-branch: H(J") = J", R-branch: H(J") = J"+1
    HLfac = Jpp
    if Jp - Jpp == 1:
        HLfac = Jpp + 1
    rS = nu * HLfac / q * np.exp(-c2 * Epp / T) * (1 - np.exp(-c2 * nu / T))
    trans.loc[br, ['Epp', 'Ep', 'nu0', 'rS']] = Epp, Ep, nu, rS


trans.head()
```

	Jpp	Jp	Epp	Ep	nu0	rS
P(20)	20	19	4290.892494	6659.631076	2368.738583	0.000002
P(19)	19	18	3890.321554	6289.184891	2398.863337	0.000014
P(18)	18	17	3508.202491	5936.818786	2428.616295	0.000086
P(17)	17	16	3144.777869	5602.762558	2457.984688	0.000478
P(16)	16	15	2800.277487	5287.233239	2486.955751	0.002430

To calculate a spectrum, first establish a grid of wavenumber points covering the range of transition frequencies (with a bit of padding).

```
# Maximum and minimum transition wavenumbers.
numin, numax = trans['nu0'].min(), trans['nu0'].max()
# Pad the wavenumber grid by 5 cm-1 at each end
numin, numax = round(numin) - 5, round(numax) + 5
nu = np.arange(numin, numax, 0.001)
```

Also define some normalized lineshape functions: a Gaussian profile, f_G (for Doppler broadening), and a Lorentzian profile, f_L (for where pressure-broadening dominates), in terms of their respective half-widths at half-maximum (HWHM), α_D and γ_L:

$$f_G(\tilde{\nu}; \tilde{\nu}_0, \alpha_D) = \sqrt{\frac{\ln 2}{\alpha_D^2 \pi}} \exp\left[-\ln 2 \frac{(\tilde{\nu} - \tilde{\nu}_0)^2}{\alpha_D^2}\right],$$

$$f_L(\tilde{\nu}; \tilde{\nu}_0, \gamma_L) = \frac{\gamma_L/\pi}{(\tilde{\nu} - \tilde{\nu}_0)^2 + \gamma_L^2}.$$

```
def fG(nu, nu0, alphaD):
    """Normalized Gaussian function with HWHM alphaD, centered on nu0."""
```

```
    N = np.sqrt(np.log(2) / np.pi) / alphaD
    return N * np.exp(-np.log(2) * ((nu - nu0) / alphaD)**2)

def fL(nu, nu0, gammaL):
    """Normalized Lorentzian function with HWHM gammaL, centered on nu0."""
    N = 1 / np.pi
    return N * gammaL / ((nu - nu0)**2 + gammaL**2)
```

Next, we can determine the Gaussian and Lorentzian HWHM values, α_D and γ_L. At 296 K, the Doppler width for HCl,

$$\alpha_D = \frac{\tilde{\nu}}{c}\sqrt{\frac{2\ln 2 k_B T}{m}}$$

is small (about $0.003\ \mathrm{cm}^{-1}$):

```
# Molecular mass of (1H)(35Cl)
m = (1 + 35) * u
alphaD = nu.mean() / c * np.sqrt(2 * np.log(2) * kB * T / m)
print(alphaD)
```

```
0.0028377760138302835
```

Therefore, here we will set a largeish value for the pressure broadened width and use that with a Lorentzian line shape.

```
gammaL = 1
```

Finally, calculate the contribution to the spectrum from each line and normalize it to the maximum intensity.

```
spec = np.zeros_like(nu)
# Iterate over the trans DataFrame, one row at a time.
for br, row in trans.iterrows():
    spec += row['rS'] * fL(nu, row['nu0'], gammaL)
spec /= spec.max()
```

```
plt.plot(nu, spec)
plt.xlabel(r'$\tilde{\nu}\;/\mathrm{cm^{-1}}$')
plt.ylabel(r'$\sigma$ (arb. units)')
```

This plot is given in the Figure 31.2.

31.5 Exercise

P31.1   Plot the simulated spectrum of $^1H^{35}Cl$ from the previous example on the same Axes as that produced using the experimentally derived line parameters (provided in the file HITRAN_1H-35Cl_1-0.csv, available from

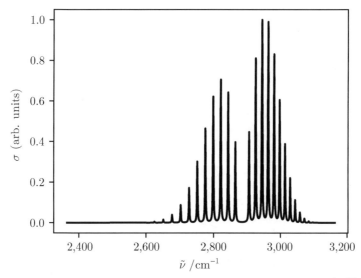

Figure 31.2 Simulated fundamental band in the IR spectrum of $^1H^{35}Cl$.

https://scipython.com/chem/exia/) for two temperatures: 100 K and 600 K. The relevant fields in this file are identified in the header line as nu (the transition wavenumber, $\tilde{\nu}_0$, in cm^{-1}), sw (the line intensity, $S(T_{ref})$, in cm molec^{-1}) and elower (the lower state energy, E'', in cm^{-1}). The line intensities are given for a reference temperature of $T_{ref} = 296$ K but can be scaled according to:

$$S(T) = S(T_{ref}) \frac{q(T)}{q(T_{ref})} \frac{\exp\left(-\frac{c_2 E''}{T}\right)\left[1 - \exp\left(-\frac{c_2 \tilde{\nu}_0}{T}\right)\right]}{\exp\left(-\frac{c_2 E''}{T_{ref}}\right)\left[1 - \exp\left(-\frac{c_2 \tilde{\nu}_0}{T_{ref}}\right)\right]},$$

where $c_2 = hc/k_B$ is the second radiation constant. Use a Lorentzian lineshape with $\gamma_L = 2\,cm^{-1}$.

32

Simulating a Powder Diffraction Spectrum

X-ray powder diffraction is a method of determining the crystal structure of a material by measuring the angles through which X-ray radiation is scattered by planes of atoms. Where a set of crystal planes, separated by a distance d is encountered at an incident angle, θ, the scattered X-rays interfere constructively where the path length $2d \sin \theta$ is an integer multiple of the wavelength λ. That is, bright spots appear in the diffraction pattern for angles, θ satisfying $2d \sin \theta = n\lambda$ (Bragg's law). X-rays are used because they have wavelengths comparable to the lattice spacing of the atoms.

Diffraction by a single crystal is anisotropic: A single set of crystal planes, labeled $\{hkl\}$, are accessible to the incoming radiation. Powder diffraction (also called polycrystalline diffraction) is isotropic: The sample is assumed to consist of every possible crystalline orientation, randomly arranged in the finely ground powder.

The relative integrated intensity of the hkl X-ray reflection by a powdered crystalline sample as a function of angle, θ, may be written:

$$I(\theta) \propto \frac{p}{V^2} \left[\frac{1 + \cos^2 2\theta}{\sin^2 \theta \cos \theta} \right] |F_{hkl}|^2 \exp\left(-\frac{2B \sin^2 \theta}{\lambda^2} \right),$$

where:

- θ is the Bragg angle for X-ray wavelength λ and hkl interplanar spacing, d;
- $|F_{hkl}|$ is the magnitude of the structure factor, as described below;
- p is the multiplicity of the hkl reflection, also described below;
- V is the unit cell volume;
- B is the isotropic temperature factor;
- The *Lorentz-polarization factor* $\left[\frac{1+\cos^2 2\theta}{\sin^2 \theta \cos \theta} \right]$ accounts for further angle-dependent effects on the scattered intensity.

The isotropic temperature factor, B, accounts for the effect of the atoms' thermal motion on the scattering intensity and is beyond the scope of this example: Here, we will take $B = 0$.

The volume of the crystal unit cell can be calculated from the lattice parameters a, b, c, α, β and γ:

$$V = abc\sqrt{1 - \cos^2 \alpha - \cos^2 \beta - \cos^2 \gamma + 2 \cos \alpha \cos \beta \cos \gamma}.$$

It is straightforward to write a function to calculate the cell volume:

```python
def calc_V(a, b, c, alpha, beta, gamma):
    """Calculate and return the unit cell volume.

    a, b, c are the cell edge dimensions
    alpha, beta, gamma are the cell angles in radians.

    """

    ca, cb, cg = np.cos(alpha), np.cos(beta), np.cos(gamma)
    return a*b*c * np.sqrt(1 - ca**2 - cb**2 - cg**2 + 2*ca*cb*cg)
```

The *structure factor*, F_{hkl}, relates the amplitude and phase of the diffracted beam to the (hkl) indexes of the crystal planes and is calculated as

$$F_{hkl} = \sum_{j=1}^{N} f_j \exp[-2\pi i(hx_j + ky_j + lz_j)]$$

where N is the number of atoms in the unit cell, (x_j, y_j, z_j) is the position of atom j, and f_j is the *atomic form factor* for atom j (which depends on the atom type and the magnitude of the scattering vector, q).

The atomic form factor is described by the empirical formula

$$f_j(q) = \sum_{k=1}^{4} a_k \exp\left[-b_k \left(\frac{q}{4\pi}\right)^2\right] + c,$$

where a_k, b_k and c are coefficients tabulated in the comma-separated file form-factors.csv, available at https://scipython.com/chem/cat/ for 211 atoms and ions and the magnitude of the scattering vector, q, is in nm^{-1}.

The *scattering vector*, $\mathbf{q}$, has magnitude

$$q = \frac{4\pi \sin \theta}{\lambda} = \frac{2\pi}{d}$$

where the interplanar spacing (the *d-spacing*) is given by

$$\frac{1}{d^2} = \frac{1}{V^2}(S_{11}h^2 + S_{22}k^2 + S_{33}l^2 + 2S_{12}hk + 2S_{23}kl + 2S_{13}hl).$$

In this equation,

$$S_{11} = b^2 c^2 \sin^2 \alpha, \tag{32.1}$$

$$S_{22} = a^2 c^2 \sin^2 \beta, \tag{32.2}$$

$$S_{33} = a^2 b^2 \sin^2 \gamma, \tag{32.3}$$

$$S_{12} = abc^2(\cos \alpha \cos \beta - \cos \gamma), \tag{32.4}$$

$$S_{23} = a^2 bc(\cos \beta \cos \gamma - \cos \alpha), \tag{32.5}$$

$$S_{13} = ab^2 c(\cos \gamma \cos \alpha - \cos \beta) \tag{32.6}$$

First, we need to calculate q for a given reflection:

```python
def calc_q(h, k, l):
    """Calculate and return the magnitude of the scattering vector, q.

    |q| is calculated for the reflection indexed at (h,k,l).
    The cell parameters: a, b, c, alpha, beta, gamma are to be
    resolved in the containing scope.

    """

    S11 = (b * c * np.sin(alpha))**2
    S22 = (a * c * np.sin(beta))**2
    S33 = (a * b * np.sin(gamma))**2
    ca, cb, cg = np.cos(alpha), np.cos(beta), np.cos(gamma)
    S12 = a * b * c**2 * (ca*cb - cg)
    S23 = a**2 * b * c * (cb*cg - ca)
    S13 = a * b**2 * c * (cg*ca - cb)
    V = calc_V(a, b, c, alpha, beta, gamma)
    d = V / np.sqrt(S11 * h**2 + S22 * k**2 + S33 * l**2 +
                    2 * (S12*h*k + S23*k*l + S13*h*l))
    q = 2 * np.pi / d
    return q
```

To complete the calculation of the structure factor, we need to read in the form factors from the provided CSV file. We can use pandas to do this and to extract the relevant rows, corresponding to the ions Na^+ and Cl^-.

```python
import numpy as np
import pandas as pd
import matplotlib.pyplot as plt

ff_coeffs = pd.read_csv('form-factors.csv', index_col=0)
# Strip stray spaces from the column names.
ff_coeffs.columns = ff_coeffs.columns.str.strip()

# For example the form factor coefficients for Na+ and Cl-:
ff_coeffs.loc[['Na+', 'Cl-']]
```

	a1	b1	a2	...	c
atom					
Na+	3.2565	2.6671	3.9362	...	0.404
Cl-	18.2915	0.0066	7.2084	...	-16.378

The form factor is readily calculated from the coefficients:

```python
def calc_f(atom, q):
    qfac = (q / 4 / np.pi)**2
    coeffs = ff_coeffs.loc[atom]
    f = 0
    for j in range(1,5):
        aj, bj = coeffs[f'a{j}'], coeffs[f'b{j}']
        f += aj * np.exp(-bj * qfac)
    f += coeffs['c']
    return f
```

To calculate the structure factor, we need the position of the atoms in the crystal unit cell. For this example, NaCl, we will use the following dictionary, corresponding to ion positions (given as coordinates in fractions of the unit cell lattice dimensions a, b, c):

$$Na^+ : (0, 0, 0), \left(\tfrac{1}{2}, \tfrac{1}{2}, 0\right), \left(\tfrac{1}{2}, 0, \tfrac{1}{2}\right), \left(0, \tfrac{1}{2}, \tfrac{1}{2}\right)$$
$$Cl^- : \left(0, \tfrac{1}{2}, 0\right), \left(\tfrac{1}{2}, 0, 0\right), \left(0, 0, \tfrac{1}{2}\right), \left(\tfrac{1}{2}, \tfrac{1}{2}, \tfrac{1}{2}\right)$$

```python
atoms = {'Na+': [(0, 0, 0), (0.5, 0.5, 0), (0.5, 0, 0.5), (0, 0.5, 0.5)],
         'Cl-': [(0, 0.5, 0), (0.5, 0, 0), (0, 0, 0.5), (0.5, 0.5, 0.5)]
        }
```

For a face-centered cubic crystal like this, the structure factor, F_{hkl}, has a simple form, but we will calculate it explicitly:

```python
def calc_absF(hkl, atoms, q=None):
    """Return the magnitude of the structure factor for the hkl reflection.

    hkl is a NumPy array of (h, k, l)
    atoms is a list of (atom_symbol, atom_position) tuples where atom_symbol
    identifies the atom or ion and atom_position is its fractional coordinates
    in the unit cell.
    q is the magnitude of the scattering vector for the hkl reflection: if it
    is not provided, it will be calculated.

    """

    if q is None:
        q = calc_q(*hkl)

    F = 0
    for atom, positions in atoms.items():
```

```
    f = calc_f(atom, q)
    F += f * np.sum(np.exp(-1j * 2*np.pi * np.sum(hkl * positions, axis=1)))
return np.abs(F)
```

There is one further complication, which is that in a powder diffraction pattern a signal is measured at a single angle due to all crystal planes related by symmetry-equivalent reflections (since all crystal orientations are equally-likely). It is convenient, therefore to work with families of planes, {*hkl*}, and consider the peak intensities to be proportional to a *multiplicity*, *p*, equal to the number of planes related in this way. For example, in a cubic crystal, the multiplicity of the {100} planes is $p = 6$, since the following are all equivalent:

$$(100), (010), (001), (\bar{1}00), (0\bar{1}0), (00\bar{1})$$

For a cubic crystal,

$$p = 3! \frac{2^{3-m}}{n!},$$

where *n* is the number of repeated indexes and *m* is the number of zeros (a zero-index reduces the multiplicity because $\bar{0} = 0$).

```
def calc_p(h, k, l):
    """Return the multiplicity of the hkl reflection in a cubic crystal"""
    # 0!, 1!, 2! and 3!
    nfactorial = {0: 1, 1: 1, 2: 2, 3: 6}
    miller = h, k, ell
    # The number if zeros out of h, k, l.
    m = sum((h==0, k==0, ell==0))
    # The number of repeated indexes (including zeros).
    n = len(miller) - len(set(miller)) + 1

    p = 6 * 2**(3-m) / nfactorial[n]
    return p
```

Since for a cubic crystal, there will only be a diffraction signal for reflections where *h*, *k* and *l* have the same parity, we only have to worry about this case. And because reflections arising from permutations of these Miller indexes all belong to the same family, we might as well label them {*hkl*} with $h \geq k \geq l$. The multiplicities for $h < 5$ can therefore be checked:

```
hmax = 4
reflections = []
# All values of h = 1, 2, ..., hmax
for h in range(1, hmax+1):
    # All values of k = 1, 3, ..., h or k = 0, 2, 4, ..., h
    for k in range(h%2, h+1, 2):
        # All values of l = 1, 3, ..., h or k = 0, 2, 4, ..., h
```

```
        for ell in range(h%2, k+1, 2):
            reflections.append((h, k, ell))
            print(h, k, ell, ':', calc_p(h, k, ell))
reflections = np.array(reflections)
```

```
1 1 1 : 8.0
2 0 0 : 6.0
2 2 0 : 12.0
2 2 2 : 8.0
3 1 1 : 24.0
3 3 1 : 24.0
3 3 3 : 8.0
4 0 0 : 6.0
4 2 0 : 24.0
4 2 2 : 24.0
4 4 0 : 12.0
4 4 2 : 24.0
4 4 4 : 8.0
```

Finally, we can put all this together in a single function that calculates the intensity of a given powder diffraction signal:

```
def calc_I(hkl, B=0, lam=0.154):
    """Return the Bragg angle and integrated intensity of reflection hkl.

    hkl is a NumPy array of (h, k, l)
    B is the Isotropic temperature factor in nm2.
    lam is the X-ray wavelength in nm.

    Returns:
    thetaB: the Bragg angle for the reflection hkl.
    I: the relative integrated intensity of reflection hkl.

    """

    # Magnitude of the scattering vector, nm-1.
    q = calc_q(*hkl)
    # Bragg angle, rad.
    thetaB = np.arcsin(lam * q / 4 / np.pi)
    # Unit cell volume, nm3.
    V = calc_V(a, b, c, alpha, beta, gamma)
    # Magnitude of the structure factor.
    F = calc_absF(hkl, atoms, q)
    # Multiplicity of the {hkl} family of reflections.
    p = calc_p(*hkl)
    # Lorentz-Polarization factor.
    LP = (1 + np.cos(2*thetaB)**2) / np.sin(thetaB)**2 / np.cos(thetaB)
    # Scattered intensity.
    I = p / V**2 * LP * np.exp(-2*B*(q/4/np.pi)**2) * F**2

    return thetaB, I
```

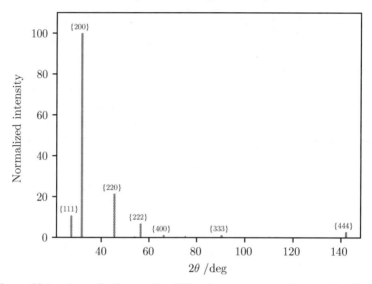

Figure 32.1   A synthetic powder diffraction spectrum calculated for NaCl.

Using NaCl as our example, the lattice parameters are:

```
# Unit cell dimensions (nm) and angles (rad).
a = b = c = 0.5639
alpha = beta = gamma = np.radians(90)
```

The scattered intensities for the {*hkl*} reflections can be calculated and normalized as follows:

```
thetaB, I = np.array([[(calc_I(hkl)) for hkl in reflections]]).T
I *= 100 / np.nanmax(I)
I
```

```
array([1.07093794e+01, 1.00000000e+02, 2.14035504e+01, 6.79879647e+00,
       4.54358969e-01, 1.49165332e-01, 1.15864871e+00, 1.11356744e+00,
       6.01633088e-01, 6.63315265e-02, 1.02695404e-01, 3.08640114e-01,
       2.77355324e+00])
```

Finally, the powder diffraction spectrum can be depicted as a bar chart (Figure 32.1).

```
angles = np.degrees(2*thetaB)
plt.bar(angles, I)
plt.xlabel(r'$2\theta\;/\mathrm{deg}$')
plt.ylabel(r'Normalized intensity')

# Label the strong reflections.
for x, y, hkl in zip(angles, I, reflections):
    if y > 1:
        label = '{' + '{}{}{}'.format(*hkl) + '}'
        plt.text(x, y, s=label)
```

33

The Hückel Approximation

33.1 Theory

The molecular orbital (MO) energies and wavefunctions for the conjugated π electronic systems of planar, unsaturated hydrocarbon molecules can be approximated using the Hückel method. In this approach, the π MO wavefunctions, ψ_i, are expressed as linear combinations of the 2p atomic orbitals (AOs), ϕ_j, corresponding to the atoms $j = 1, 2, \ldots, n$:

$$\psi = \sum_{j=1}^{n} c_j \phi_j$$

The coefficients, c_j, which best approximate the true MO wavefunction are those which minimize its expected energy in accordance with the variational principle. Applying this condition leads to the n so-called *secular equations*:

$$\sum_{j=1}^{n} \left(H_{jk} - E S_{jk} \right) c_j = 0,$$

where $H_{jk} = \langle \phi_j | \hat{H} | \phi_k \rangle$ and the overlap integral, $S_{jk} = \langle \phi_j | \phi_k \rangle$. The Hückel approximations further simplify these equations as follows:

- The overlap integrals are set equal to zero: $S_{jk} = 0$ if $j \neq k$. This is reasonable, since the 2p atomic orbitals in question are directed perpendicular to the plane of the molecule. Since the AO basis functions are normalized, $S_{jj} = 1$;
- All the Coulomb integrals are set equal to the same quantity, $H_{jj} = \alpha$;
- The resonance integrals, H_{jk} ($j \neq k$) are set to:

$$H_{jk} = \begin{cases} \beta & \text{if atom } j \text{ is bonded to atom } k, \\ 0 & \text{otherwise.} \end{cases}$$

In Hückel theory, therefore, the π orbital energies and wavefunctions are determined by the connectivity of the σ-bonded molecular framework, which is assumed to be known.

For example, the secular equations (in matrix form) for 1,3-butadiene, $CH_2 = CH - CH = CH_2$ simplify from:

$$\begin{pmatrix} H_{11} - ES_{11} & H_{12} - ES_{12} & H_{13} - ES_{13} & H_{14} - ES_{14} \\ H_{21} - ES_{21} & H_{22} - ES_{22} & H_{23} - ES_{23} & H_{24} - ES_{24} \\ H_{31} - ES_{31} & H_{32} - ES_{32} & H_{33} - ES_{33} & H_{34} - ES_{34} \\ H_{41} - ES_{41} & H_{42} - ES_{42} & H_{43} - ES_{43} & H_{44} - ES_{44} \end{pmatrix} \begin{pmatrix} c_1 \\ c_2 \\ c_3 \\ c_4 \end{pmatrix} = 0$$

to

$$\begin{pmatrix} \alpha - E & \beta & 0 & 0 \\ \beta & \alpha - E & \beta & 0 \\ 0 & \beta & \alpha - E & \beta \\ 0 & 0 & \beta & \alpha - E \end{pmatrix} \begin{pmatrix} c_1 \\ c_2 \\ c_3 \\ c_4 \end{pmatrix} = 0$$

To have a nontrivial solution, the *secular determinant* must be zero, in this case:

$$\begin{vmatrix} \alpha - E & \beta & 0 & 0 \\ \beta & \alpha - E & \beta & 0 \\ 0 & \beta & \alpha - E & \beta \\ 0 & 0 & \beta & \alpha - E \end{vmatrix} = 0.$$

The problem then becomes one of solving this equation for E (i.e., finding the roots of the characteristic polynomial equation it expands out to) and then, for each E, substituting it back into the secular equations to solve for the coefficients, c_j. This is a staple of undergraduate physical chemistry courses but can only be done by hand for relatively small systems (though it may be possible to reduce the complexity by adopting some symmetry-adapted linear combination of the atomic orbitals as the basis).

However, since the above equations are just $(A - EI)c = 0$, the energies E_i and coefficients, c_{ij}, are straightforward to compute as the eigenvalues and eigenvectors, respectively, of the matrix

$$A = \begin{pmatrix} \alpha & \beta & 0 & 0 \\ \beta & \alpha & \beta & 0 \\ 0 & \beta & \alpha & \beta \\ 0 & 0 & \beta & \alpha \end{pmatrix}.$$

Furthermore, if we are only interested in the molecular orbital energies *relative to* the C 2p atomic orbital energy, we can set $\alpha = 0$; the matrix then becomes

$$\beta \begin{pmatrix} 0 & 1 & 0 & 0 \\ 1 & 0 & 1 & 0 \\ 0 & 1 & 0 & 1 \\ 0 & 0 & 1 & 0 \end{pmatrix}.$$

The eigenvalues and eigenvectors of matrices like these can be readily calculated for any reasonably sized system using NumPy's `linalg` module, as demonstrated in the following example.

33.2 Examples

E33.1

1,3-Butadiene

We will label the atoms $j = 0, 1, 2, 3$ from one end of the molecule. The secular equations are

$$\begin{pmatrix} \alpha - E & \beta & 0 & 0 \\ \beta & \alpha - E & \beta & 0 \\ 0 & \beta & \alpha - E & \beta \\ 0 & 0 & \beta & \alpha - E \end{pmatrix} \begin{pmatrix} c_1 \\ c_2 \\ c_3 \\ c_4 \end{pmatrix} = 0.$$

As shown previously, finding the MO energies, E_j (in units of β, relative to the carbon 2p energy, α) and coefficients, c_{ij} then amounts to finding the eigenvalues and eigenvectors of the matrix

$$\begin{pmatrix} 0 & 1 & 0 & 0 \\ 1 & 0 & 1 & 0 \\ 0 & 1 & 0 & 1 \\ 0 & 0 & 1 & 0 \end{pmatrix}.$$

```
import numpy as np
import matplotlib.pyplot as plt
```

The matrix defining the problem is defined by the connectivity of the molecule. It is symmetric, so we only need to set half of it, for example, the upper triangular part:

```
M = np.zeros((4, 4))
# C0 is bonded to C1, C1 to C2, and C2 to C3.
M[0,1] = M[1,2] = M[2,3] = 1
```

```
# Calculate the eigenvalues (energies) and eigenvectors.
# Setting UPLO='U' uses the upper-triangular part of the matrix M.
E, eigvecs = np.linalg.eigh(M, UPLO='U')
E, eigvecs
```

```
(array([-1.61803399, -0.61803399,  0.61803399,  1.61803399]),
 array([[ 0.37174803, -0.60150096, -0.60150096,  0.37174803],
        [-0.60150096,  0.37174803, -0.37174803,  0.60150096],
        [ 0.60150096,  0.37174803,  0.37174803,  0.60150096],
        [-0.37174803, -0.60150096,  0.60150096,  0.37174803]]))
```

The molecular orbital energies are therefore $\alpha - 1.618\beta$, $\alpha - 0.618\beta$, $\alpha + 0.618\beta$ and $\alpha + 1.618\beta$. The coefficients defining the molecular orbitals are defined by the corresponding eigenvectors (the columns of `eigvecs`).

There are four 2p electrons, and in the ground state they occupy the lowest energy, bonding molecular orbitals with energies $\alpha + 1.618\beta$ and $\alpha + 0.618\beta$ (recall that the resonance integral, β, is negative). The total π electron energy is therefore:

```
Epi = 2*E[3] + 2*E[2]
print(Epi)
```

```
4.472135954999579
```

That is, $E_\pi = 4\alpha + 4.472\beta$. Two isolated double bonds would have an energy of $4\alpha + 4\beta$, so the resonance stabilization is therefore 0.472β.

The MOs can be visualized by plotting circles with areas in proportion to the coefficient magnitude (Figure 33.1).

```
# Atom positions as (x, y) coordinates in Angstroms.
# Huckel theory cannot tell us these: we need to known them from somewhere else.
pos = np.array([[-1.3707, 1.1829], [-0.6771, 0.0376],
                 [0.6771, -0.0376], [1.3707, -1.1829], ])
```

Matplotlib provides a set of functions for drawing shapes as *patches* and adding them to a plot. Here, we will use `plt.Circle` to draw circles representing the atomic orbitals.

```
# Scale the AO circles by this amount to keep them from overlapping.
scale = 0.8
# Negative coefficients are red, positive ones blue.
colours = {-1: 'tab:red', 1: 'tab:blue'}
fig, axes = plt.subplots(nrows=4, ncols=1, figsize=(2,8))

for i, coeffs in enumerate(eigvecs.T):
    # The molecular framework.
    axes[i].plot(*pos.T, c='k', lw=2)
    for j, coeff in enumerate(coeffs):
        radius = scale * np.sqrt(np.abs(coeffs[j]))
        colour = colours[np.sign(coeffs[j])]
        axes[i].add_patch(plt.Circle(pos[j], radius=radius, fc=colour))
        # Label the diagram with the MO energy
        axes[i].text(2, 0, r'$\alpha ' + f'{E[i]:+.3f}' + r'\beta$', fontsize=18)
    axes[i].axis('square')
    axes[i].axis('off')
```

The π-bond order can also be calculated from the orbital coefficients as

$$P_{ik} = \sum_j n_j^{occ} c_{ij} c_{kj},$$

where n_j^{occ} is the number of electrons occupying the MO labeled i. In this case:

```
nocc = np.array([0, 0, 2, 2])
P01 = np.abs(np.sum(nocc * eigvecs[:,0] * eigvecs[:,1]))
P12 = np.abs(np.sum(nocc * eigvecs[:,1] * eigvecs[:,2]))
P23 = np.abs(np.sum(nocc * eigvecs[:,2] * eigvecs[:,3]))
print(f'P01 = {P01:.3f}')
print(f'P12 = {P12:.3f}')
print(f'P23 = {P23:.3f}')
```

```
P01 = 0.894
P12 = 0.447
P23 = 0.894
```

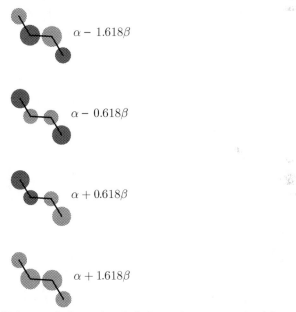

Figure 33.1   Orbital coefficients of the π-bonded 2p carbon atoms in 1,3-butadiene, estimated by Hückel theory.

There is greater π-bonding character to the terminal carbon-carbon bonds than in the center one.

Finally, the π electronic charge on each atom i is given by

$$Q_i = \sum_j n_j^{occ} |c_{ij}|^2$$

```
for i in range(4):
    Qi = np.sum(nocc * eigvecs[i,:]**2)
    print(f'Q{i} = {Qi:.1f}')
```

```
Q0 = 1.0
Q1 = 1.0
Q2 = 1.0
Q3 = 1.0
```

That is, the π electrons are evenly distributed across the four carbon atoms in 1,3-butadiene.

E33.2

Benzene

The six carbon atoms of benzene are labeled $j = 0, 1, \ldots, 5$ consecutively around the ring. The secular matrix describing this structure is then

$$\mathbf{A} = \begin{pmatrix} \alpha - E & \beta & 0 & 0 & 0 & \beta \\ \beta & \alpha - E & \beta & 0 & 0 & 0 \\ 0 & \beta & \alpha - E & \beta & 0 & 0 \\ 0 & 0 & \beta & \alpha - E & \beta & 0 \\ 0 & 0 & 0 & \beta & \alpha - E & \beta \\ \beta & 0 & 0 & 0 & \beta & \alpha - E \end{pmatrix} \Rightarrow \mathbf{M} = \begin{pmatrix} 0 & 1 & 0 & 0 & 0 & 1 \\ 1 & 0 & 1 & 0 & 0 & 0 \\ 0 & 1 & 0 & 1 & 0 & 0 \\ 0 & 0 & 1 & 0 & 1 & 0 \\ 0 & 0 & 0 & 1 & 0 & 1 \\ 1 & 0 & 0 & 0 & 1 & 0 \end{pmatrix}$$

```
M = np.zeros((6, 6))
# We only need to define the upper triangular part of the matrix M
# since it is symmetric.
M[0,1] = M[1,2] = M[2,3] = M[3,4] = M[4,5] = M[0,5] = 1
```

```
E, eigvecs = np.linalg.eigh(M, UPLO='U')
E, eigvecs
```

```
(array([-2., -1., -1.,  1.,  1.,  2.]),
 array([[ 4.08248290e-01,  4.89264905e-01, -3.06517840e-01,
         -5.00000000e-01,  2.88675135e-01,  4.08248290e-01],
        [-4.08248290e-01, -5.10084688e-01, -2.70456917e-01,
         -5.00000000e-01, -2.88675135e-01,  4.08248290e-01],
        [ 4.08248290e-01,  2.08197832e-02,  5.76974757e-01,
          4.93432455e-17, -5.77350269e-01,  4.08248290e-01],
        [-4.08248290e-01,  4.89264905e-01, -3.06517840e-01,
          5.00000000e-01, -2.88675135e-01,  4.08248290e-01],
        [ 4.08248290e-01, -5.10084688e-01, -2.70456917e-01,
          5.00000000e-01,  2.88675135e-01,  4.08248290e-01],
        [-4.08248290e-01,  2.08197832e-02,  5.76974757e-01,
         -6.16790569e-18,  5.77350269e-01,  4.08248290e-01]]))
```

```
colours = {-1: 'tab:red', 1: 'tab:blue'}
# Arrange the atoms equally around a circle to make a hexagon.
# The first point is added again at the end as well to close the ring in the plot.
theta = np.radians(np.arange(0, 420, 60))
pos = np.array((np.sin(theta), np.cos(theta)))

def plot_MOs(i, ax, pos, eigvecs):
    # Plot the molecular framework.
    ax.plot(*pos, c='k', lw=2)
    for j, coeff in enumerate(eigvecs[:,i]):
        ax.add_patch(plt.Circle(pos.T[j], radius=coeff,
                                fc=colours[np.sign(coeff)]))
    ax.axis('square')
    ax.axis('off')

# Plot the MO wavefunctions on a 2x3 grid of Axes.
fig, axes = plt.subplots(nrows=2, ncols=3)
for i in range(6):
    plot_MOs(i, axes[i // 3, i % 3], pos, eigvecs)
```

E33.3

Buckminsterfullerene

Buckminsterfullerene, C_{60}, is a famous fused-ring, cage-like molecule in the shape
of a soccer ball. Its π electron molecular orbitals and energies can be modeled with
Hückel theory, though only approximately since the simple theory takes no account
of the three-dimensional geometry of the atoms, only their connectivity.

To construct the secular matrix, we need to know which atoms are connected to
which others. Each carbon is bonded to three others, and numbering them from 1
– 60, their connectivity is summarized in the dictionary below. For example, entry
1: [60, 59, 58] means that carbon atom number 1 is bonded directly to atoms
labeled 60, 59 and 58.

To avoid any typographical and typing errors, the adjacency dictionary is checked
to ensure that if atom *i* is specified as bonded to atom *j*, then atom *j* is also bonded
to atom *i*.

```
# Adjacency matrix for C60, corrected from Trinajstic et al., Croatica Chemica
# Acta 68 (1), 241 (1995).
adjacency = {1: [60, 59, 58],    2: [60, 57, 56],    3: [59, 57, 55],    4: [60, 54, 53],
             5: [58, 54, 52],    6: [59, 51, 50],    7: [55, 51, 49],    8: [57, 48, 47],
             9: [56, 48, 46],   10: [54, 45, 44],   11: [53, 45, 43],   12: [52, 42, 41],
            13: [51, 40, 39],   14: [42, 40, 38],   15: [39, 38, 37],   16: [49, 36, 35],
            17: [48, 34, 33],   18: [36, 34, 32],   19: [33, 32, 31],   20: [43, 31, 30],
            21: [44, 29, 28],   22: [38, 27, 26],   23: [32, 27, 25],   24: [29, 26, 25],
            25: [31, 24, 23],   26: [28, 24, 22],   27: [37, 23, 22],   28: [42, 26, 21],
            29: [30, 24, 21],   30: [45, 29, 20],   31: [25, 20, 19],   32: [23, 19, 18],
            33: [46, 19, 17],   34: [35, 18, 17],   35: [47, 34, 16],   36: [37, 18, 16],
            37: [36, 27, 15],   38: [22, 15, 14],   39: [49, 15, 13],   40: [41, 14, 13],
            41: [50, 40, 12],   42: [28, 14, 12],   43: [46, 20, 11],   44: [52, 21, 10],
            45: [30, 11, 10],   46: [43, 33,  9],   47: [55, 35,  8],   48: [17,  9,  8],
            49: [39, 16,  7],   50: [58, 41,  6],   51: [13,  7,  6],   52: [44, 12,  5],
            53: [56, 11,  4],   54: [10,  5,  4],   55: [47,  7,  3],   56: [53,  9,  2],
            57: [ 8,  3,  2],   58: [50,  5,  1],   59: [ 6,  3,  1],   60: [ 4,  2,  1]}

# Check the integrity of the adjacency matrix.
for i, js in adjacency.items():
    for j in js:
        if i not in adjacency[j]:
            print(f'Error at atom {i}')
```

The matrix itself is easily constructed from this adjacency dictionary, with the small complication that we must subtract one from the atom labels because NumPy arrays are indexed from zero:

```
M = np.zeros((60, 60))
for i, js in adjacency.items():
    for j in js:
        M[i-1, j-1] = 1
```

As before, the energies and MO coefficients are given by the eigenvalues and eigenvectors of this matrix.

```
E, eigvecs = np.linalg.eigh(M)
```

There is a fair amount of degeneracy in the energies, so collect the eigenvalues together in a dictionary, `level_degeneracies`, and count the number of occurrences of each value:

```
this_E = 0
level_degeneracies = {}
# Iterate backwards because the lowest energy is last.
for Elevel in E[::-1]:
    if not np.isclose(Elevel, this_E):
        # A new energy level => a new entry in the dictionary.
        this_E = Elevel
```

```
            level_degeneracies[this_E] = 1
        else:
            # Another state with the same energy as the previous one.
            level_degeneracies[this_E] += 1

# Summarize the energy levels and their degeneracies.
print('E / |beta|  |   g')
print('-'*18)
for Elevel, g in level_degeneracies.items():
    print(f'{-Elevel:>9.3f}    | {g:3d}')
```

```
E / |beta|  |   g
------------------
   -3.000  |   1
   -2.757  |   3
   -2.303  |   5
   -1.820  |   3
   -1.562  |   4
   -1.000  |   9
   -0.618  |   5
    0.139  |   3
    0.382  |   3
    1.303  |   5
    1.438  |   3
    1.618  |   5
    2.000  |   4
    2.562  |   4
    2.618  |   3
```

We can plot a simple energy level diagram using Matplotlib as follows. Each state will be a horizontal line with y−coordinate equal to its energy. There are g states per energy level; These states will be plotted as lines with their centers at x−coordinate $-(g-1), -(g-3), \ldots, g-3, g-1$, and the lowest 30 energy levels annotated with paired red lines indicating their electron occupancy.

```
fig, ax = plt.subplots(figsize=(6,8))
nelec = 60
for Elevel, g in level_degeneracies.items():
    for x in range(-(g-1), g, 2):
        # Plot a line for this state.
        ax.plot([x-0.75, x+0.75], [-Elevel, -Elevel], c='k', lw=2)
        if nelec:
            # If we still have electrons to add to the plot, do it here.
            ax.plot([x-0.15, x-0.15], [-Elevel-0.08, -Elevel+0.08], c='r', lw=1)
            ax.plot([x+0.15, x+0.15], [-Elevel-0.08, -Elevel+0.08], c='r', lw=1)
            nelec -= 2

ax.set_xticks([])
ax.set_ylabel(r'Energy /$|\beta|$')
plt.show()
```

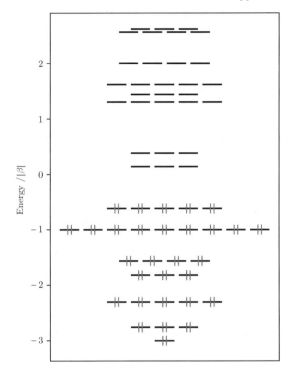

33.3 Exercises

P33.1 Calculate the π electron molecular orbital wavefunctions and energies for naphthalene within the Hückel approximation, and determine the bond orders and atom charges. The structure and suggested carbon atom labelings are given in Figure 33.2.

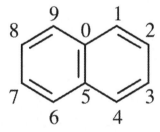

Figure 33.2   The structure and carbon atom labeling for naphthalene, $C_{10}H_8$.

P33.2 Heteroconjugated systems can also be studied with the Hückel approximation, with modified values for α and β. Assuming values of $\alpha = 0$ for C and $\alpha = 0.5$

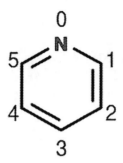

Figure 33.3   The structure and carbon atom labeling for pyridine, C_5H_5N.

for N, and $|\beta| = 1$ for the C − C and $|\beta| = 0.8$ for the C − N bonds, predict the MO energies and wavefunctions for the pyridine molecule (Figure 33.3).

34

Nonlinear Fitting and Constrained Optimization

Chapters 19 and 20 have described linear least squares fitting, used when the model function to fit is a linear function of its parameters (e.g., in a fit of the coefficients to a quadratic function, $f(x; a, b, c) = ax^2 + bx + c$). In general, this problem is well-defined and a unique solution can be obtained using linear algebra by solving the so-called normal equations which find the minimum in the sum of the squared values of the residuals (fit – model values).

If the model function is *nonlinear* in its parameters, however, a different approach must be taken and, generally, the fitting process is more complicated: An initial guess for the parameters is usually needed and there is typically more than one minimum in the residual function (which may not be a concave function of the parameters), so there is no guarantee that an iterative optimization algorithm will converge on the global minimum (if it converges at all).

There are two important routines in the `scipy.optimize` package to carry out nonlinear least squares fitting, `least_squares` and `curve_fit`.

34.1 `scipy.optimize.least_squares`

In its most basic usage, `least_squares` takes a function, `fun`, an initial guess for its best-fit parameters, `x0`, and any additional arguments the function requires, `args`:

```
scipy.optimize.least_squares(fun, x0, args=())
```

`fun` *should return the residuals* between the data to be modeled and the fit function's output; the algorithm will then attempt to minimize the residuals with respect to the parameters. For example, if the function to be fit is:

$$f(x; N, c, \Gamma) = \frac{N}{(x - c)^2 + \Gamma^2}$$

the definition of the function fun might be:

```
def fun(X, x, y):
    N, c, Gamma = X
    return N / ((x-c)**2 + Gamma**2) - y
```

The parameters to be fitted are held in the object X and unpacked to separate variable names for convenience. fun then returns the residuals between our function, evaluated at the points x and the measured values, y. These can be passed to least_squares as follows:

```
X0 = (9, 5, 1.5)                    # Initial guess for the parameters
ret = least_squares(fun, X0, args=(x, y))
```

The returned object, ret, contains information about how the optimization algorithm went, including a boolean flag, ret.success, indicating whether it thinks it succeeded, and ret.x, the optimized parameters:

```
if ret.success:
    N, c, Gamma = ret.x
else:
    print(f'Failure in least_squares: ret = {ret}')
```

An additional parameter, bounds, can be passed to least_squares to constrain the parameters to within a range of values. For example, if a parameter is required to be non-negative, its bounds can be set to the tuple (0, np.inf). bounds is passed a sequence of such tuples, one for each parameter: see Example E34.2.

34.2 scipy.optimize.curve_fit

The curve_fit routine provides a simpler interface for directly fitting a function to some data. Its basic call signature is:

```
curve_fit(f, xdata, ydata, p0=None)
```

where f is the model function to be fit, the observed values to be fit are (xdata, ydata), and p0 is an initial guess for the fit parameters (each is set to 1 if p0 is not provided). The function f takes the independent variable (x) as its first argument, and the parameters to be fit as the remaining arguments.

When called, curve_fit returns the best-fit parameters, popt, and estimated covariance matrix, pcov. The diagonal values of this matrix are the squares of the parameter fit standard deviations, as shown in Example E34.3.

34.3 scipy.optimize.minimize with Constraints

The optimization algorithms used in Chapter 22 allowed the parameters with respect to which the target function is to be minimized to take any real values. In application to scientific problems, however, there are frequently constraints on the parameters imposed by the physical situation that the function models. It can also be helpful to constrain the parameters to avoid unhelpful or trivial solutions (e.g., all parameters equal to zero). Constraints are allowed with the minimization algorithms `'COBYLA'` and `'SLSQP'`; for these methods, each constraint is provided as a dictionary in a list; each dictionary should define at least two keys: `'type'` defining the type of constraint (`'eq'` or `'ineq'` for equality and inequality constraints respectively) and `'fun'`, a function taking the parameters and returning a value which is zero in the case of an equality constraint being met and non-negative for inequality constraints.

For example, to determine the minimum surface area, S, of a rectangular box with side lengths a, b and c such that its total volume is $V_0 = 300 \, \text{cm}^3$:

```python
from scipy.optimize import minimize

def get_S(P):
    """Return the total surface area of the box."""
    a, b, c = P
    return 2 * (a*b + b*c + a*c)

def get_V(P):
    """Return the total volume of the box."""
    a, b, c = P
    return a * b * c

# The fixed, target volume.
V0 = 300
# Initial guesses for the parameters
a0 = b0 = c0 = 1
constraints = [
    {'type': 'eq', 'fun': (lambda P: get_V(P) - V0)}
]
minimize(get_S, (a0, b0, c0), method="SLSQP", constraints=constraints)
```

Here, the constraint on the volume is expressed as an anonymous function which returns 0 when the volume calculated for a given set of parameters, P = (a, b, c) is equal to the target volume, V_0:

```python
lambda P: get_V(P) - V0
```

The output is:

```
   fun: 268.88428473172553
   jac: array([26.77735138, 26.77737427, 26.77722549])
message: 'Optimization terminated successfully'
  nfev: 41
   nit: 10
  njev: 10
```

```
 status: 0
success: True
      x: array([6.69431263, 6.69430066, 6.69437521])
```

That is, a cube with sides $a = b = c = 6.69$ cm minimizes S.

To add an inequality constraint, suppose we want to ensure that $ab \geq 50$ cm^2. This can be expressed as another anonymous function that returns the difference between P[0]*P[1] and 50: The fit will be constrained to keep this quantity non-negative.

```
constraints = [
    {'type': 'eq', 'fun': (lambda P: get_V(P) - V0)},
    {'type': 'ineq', 'fun': (lambda P: P[0]*P[1] - 50)}
]
minimize(get_S, (x0, y0, z0), constraints=constraints)
```

produces:

```
    fun: 269.7056274847193
    jac: array([26.14213562, 26.14213562, 28.28427124])
message: 'Optimization terminated successfully'
   nfev: 21
    nit: 5
   njev: 5
 status: 0
success: True
      x: array([7.07106781, 7.07106781, 6.        ])
```

The sides a and b have increased in length to 7.07 cm to allow this additional condition to be met, whilst $c = 6$ cm is smaller to meet the required volume constraint.

Examples E34.1, below, and Section 37.4 use constrained minimization to solve problems relevant to chemistry.

34.4 Examples

E34.1

Example E31.2 demonstrated how to calculate the equilibrium constant of a reaction from thermodynamic data of its reactants and products. For a single process, it is sufficient to calculate the standard Gibbs free energy change for the reaction, $\Delta_r G^\ominus$, from which the equilibrium constant, K, can be obtained from the expression $\Delta_r G^\ominus = -RT \ln K$.

The derivation of this equation is usually introduced in undergraduate chemistry courses (with much notational sleight of hand) as a consequence of the reaction mixture reaching a minimum in its Gibbs free energy at equilibrium.

Consider the general reaction:

$$|\nu_A| [A] + |\nu_B| [B] + \cdots \rightarrow \nu_P[P] + \nu_Q[Q] + \cdots,$$

where the stoichiometric numbers, ν_A, ν_B, etc. are agreed to be negative for reactants and positive for products. If the reaction proceeds by an infinitesimal amount, $d\xi$, (at constant temperature and pressure), such that the amounts of the various species change by $dn_J = \nu_J\, d\xi$, then the total Gibbs free energy changes by

$$dG = \sum_J \mu_J dn_J = \sum_J \mu_J \nu_J\, d\xi \quad \Rightarrow \quad \left(\frac{\partial G}{\partial \xi}\right)_{p,T} = \sum_J \nu_J \mu_J,$$

where μ_J is the chemical potential of species J which varies with composition according to its activity, a_J:

$$\mu_J = \mu_J^{\ominus} + RT \ln a_J.$$

At the minimum of G, therefore,

$$\left(\frac{\partial G}{\partial \xi}\right)_{p,T} = \sum_J \nu_J \mu_J = 0 \quad \Rightarrow \quad \sum_J \nu_J \mu_J^{\ominus} + RT \sum_J \ln \nu_J a_J = 0.$$

From the definition of chemical potential and the properties of logarithms, it then follows that:

$$\Delta_r G^{\ominus} + RT \ln \prod_J a_J^{\nu_J} = \Delta_r G^{\ominus} + RT \ln K = 0$$

and hence $\Delta_r G^{\ominus} = -RT \ln K$.

Another perspective on this derivation can be obtained by directly minimizing the Gibbs free energy of the reaction, which can be calculated for an arbitrary composition. We return to the water–gas shift reaction,

$$CO(g) + H_2O(g) \longrightarrow H_2(g) + CO_2(g)$$

for which the equilibrium constant $K = 3.05$ at 850 K was calculated in Example E31.2.

The equilibrium condition can be thought of as the composition (molar amounts of reactants and products: n_{CO}, n_{H_2O}, n_{H_2}, n_{CO_2}) that gives a minimum in the total Gibbs free energy of the system, nG:

$$nG = \sum_J n_J G_J^{\ominus} + RT \sum_J n_J \ln a_J$$

$$= \sum_J n_J \Delta_f G_J^{\ominus} + RT \sum_J n_J \ln \left(x_J \frac{p_{tot}}{p^{\ominus}}\right),$$

where x_J is the mole fraction of component J and in the last line we have assumed the reactants to be ideal gases and are measuring the standard Gibbs free energies of the component compounds relative to their elements in their standard state so that $G_J^{\ominus} = \Delta_f G_J^{\ominus}$.

A numerical approach to determining the equilibrium composition therefore involves minimizing this expression for nG with respect to the quantities n_{CO}, n_{H_2O}, n_{H_2}, n_{CO_2}, *subject to the constraint* that the total number of carbon, oxygen and hydrogen atoms is conserved (and that none of the species amounts is negative).

The code below achieves this, with the aim of reproducing the result of Example E31.2. The Gibbs free energies of formation are taken from this example.

```python
import numpy as np
import pandas as pd
from scipy.constants import R

# Standard pressure, Pa
pstd = 1.e5

# Standard Gibbs free energy of formation of the reaction species at 850 K.
DfGo = {
    'CO': -288.59,
    'H2O': -414.14,
    'H2': -120.96,
    'CO2': -589.65
}
```

Imagine an initial composition of $n_{init} = n_{CO} = n_{H_2O} = 1$, $n_{H_2} = n_{CO_2} = 0$ at a total pressure of 1 atm = 101,325 Pa and temperature, $T = 850$ K.

```python
ninit = 1
# Temperature (K) and total pressure (Pa)
T, ptot = 850, 101325
species = ('CO', 'H2O', 'H2', 'CO2')
def nG(narr):
    ntot = sum(narr)
    ret = 0
    for i, n in enumerate(narr):
        formula = species[i]
        x = n / ntot
        ret += n * (DfGo[formula] * 1000 + R * T * np.log(x * ptot/pstd))
    return ret
```

```python
from scipy.optimize import minimize

# n[0]    n[1]    n[2]    n[3]
# CO    + H2O -> H2    + CO2
# We start with this number of moles of carbon, oxygen and hydrogen atoms
# and must keep them constant:
nC = ninit
nO = 2*ninit
nH = 2*ninit


# The constraints for the fit: conserve the number of moles of each
# atom and ensure that the amount of each compound is positive.
cons = [{'type': 'eq',   'fun': lambda n: n[0] + n[1] + 2*n[3] - nO},   # O atoms
        {'type': 'eq',   'fun': lambda n: n[0] + n[3] - nC},            # C atoms
```

```
        {'type': 'eq',   'fun': lambda n: 2*n[1] + 2*n[2] - nH},     # H atoms
        {'type': 'ineq', 'fun': lambda n: n[0]},          # nCO  > 0
        {'type': 'ineq', 'fun': lambda n: n[1]},          # nH2O > 0
        {'type': 'ineq', 'fun': lambda n: n[2]},          # nH2  > 0
        {'type': 'ineq', 'fun': lambda n: n[3]},          # nCO2 > 0
    ]


# Use the Sequential Least Squares Programming (SLSQP) algorithm, which allows
# constraints to be specified. Take the initial guess for the equilibrium
# composition to be 0.5 moles of each component species.
ret = minimize(nG, (0.5, 0.5, 0.5 ,0.5), method="SLSQP",
                constraints=cons)
print(ret)
```

```
     fun: -563890.3716842766
     jac: array([-219170.1875, -344720.1875,  -47600.1875, -516290.1875])
 message: 'Optimization terminated successfully'
    nfev: 26
     nit: 5
    njev: 5
  status: 0
 success: True
       x: array([0.36412663, 0.36412663, 0.63587337, 0.63587337])
```

```
print(f'T = {T} K, ptot = {ptot/1.e5} bar')
print('Equilibrium composition (mol)')
for i, formula in enumerate(species):
    print(f'{formula:>3s}: {ret.x[i]:.3f}')
```

```
T = 850 K, ptot = 1.01325 bar
Equilibrium composition (mol)
 CO: 0.364
H2O: 0.364
 H2: 0.636
CO2: 0.636
```

Finally, calculate the equilibrium constant from the equilibrium molar amounts of each component species:

```
x = ret.x
K = x[2] * x[3] / (x[0] * x[1])
print(f'K = {K:.2f}')
```

```
K = 3.05
```

The real benefit of this approach is in finding the equilibrium composition where more than one reaction may be determining (see Exercise P34.2).

E34.2

Fitting the STO-NG Basis Set

In computational chemistry, *Slater-type orbitals* (STOs) are functions which are used to represent atomic orbitals in the linear combination of atomic orbitals (LCAO) molecular orbital method of approximating the wavefunction of a molecule. The normalized family of STOs are defined as

$$\phi_n^S(r, \zeta) = (2\zeta)^n \sqrt{\frac{2\zeta}{(2n)!}} r^{n-1} e^{-\zeta r}, \quad n = 1, 2, 3, \ldots,$$

where ζ is a constant parameter related to the effective nuclear charge experienced by the electron. Quantities here are expressed in Hartree atomic units, so distances are in units of the Bohr radius, $a_0 = 5.292 \times 10^{-11}$ m.

STOs capture some of the features of the hydrogenic atomic orbital wavefunctions (e.g., exponential decay as $r \to \infty$, though they do not possess radial nodes.)

Unfortunately, they are not very convenient from a computational point of view: The integrals used in the LCAO approach are expensive to calculate and so early on in the history of computational chemistry, it became common to use Gaussian-type orbitals (GTOs) instead. The simplest, s-type GTO is

$$\phi_1^G(r, \alpha) = \left(\frac{2\alpha}{\pi}\right)^{\frac{3}{4}} e^{-\alpha r^2}.$$

GTOs have the nice property that the product of two GTOs centered on different positions is equal to the sum of GTOs centered on the line between them. However, as we shall see, simply replacing an STO with the single "closest" GTO does not give a very good basis function to use in constructing molecular orbitals, and so it is necessary to use a Gaussian-type basis in which each basis function is formed from a linear combination of "primitive" GTOs of the above form. This is the STO-NG Basis Set, where the N represents the number of primitive GTOs in the linear combination:

$$\psi^{NG} = \sum_{k=1}^{N} c_k \phi_1^G(r, \alpha_k)$$

ψ^{NG} is called a *contracted* Gaussian function. In this example, we will determine the optimum parameters c_k and α_k for the STO-1G, STO-2G and STO-3G basis set representation of a 1s orbital.

```
import numpy as np
from scipy.optimize import least_squares, minimize
```

```
from scipy.integrate import quad
import matplotlib.pyplot as plt
```

```
def STO1(r, zeta=1):
    """The normalized 1s Slater-type orbital."""
    return np.sqrt(zeta**3 / np.pi) * np.exp(-zeta * r)

def pGTO(r, alpha):
    """
    The normalized primitive Gaussian-type orbital with exponent parameter alpha.
    """
    return (2 * alpha / np.pi)**0.75 * np.exp(-alpha * r**2)
```

To start with the simplest case, for the STO-1G function, we are simply trying to find the parameter α that yields the function $\phi_1^G(r, \alpha)$ closest to $\phi_1^S(r, 1)$. There are different ways of defining "closest" in this context, but a popular choice is to select the GTO that maximizes the overlap with the target STO function:

$$I_1 = \int_0^\infty \phi_1^G(r, \alpha)\phi_1^S(r, 1)\, 4\pi r^2 dr.$$

The volume element, $4\pi r^2 dr$, is determined by the spherical symmetry of the orbital.

```
def overlap1(prms):
    """Return the negative of the overlap integral between pGTO and STO1."""

    # In this case, there is only one parameter.
    alpha = prms[0]

    # The function to integrate: the product of pGTO and STO1. The volume
    # element in spherical polars is 4.pi.r^2, but there's no need to include
    # the 4.pi factor if we just want to find the maximum overlap.
    f = lambda r: pGTO(r, alpha) * STO1(r) * r**2
    res = quad(f, 0, np.inf)
    return -res[0]
```

In order to maximize I_1, we can minimize its negative value. Using SciPy's minimize function with an initial guess of $\alpha = 0.1$ should converge fairly rapidly:

```
minimize(overlap1, [0.1])
```

```
      fun: -0.07785894757934787
 hess_inv: array([[3.84982329]])
      jac: array([7.63777643e-06])
  message: 'Optimization terminated successfully.'
     nfev: 8
      nit: 3
     njev: 4
   status: 0
```

```
success: True
       x: array([0.27097914])
```

That is, the optimum value is determined to be $\alpha = 0.2710\, a_0^{-2}$. A comparison of these approximate wavefunctions can be made in a plot as follows:

```
alpha_11 = 0.27097914
def plot_STO1_GTO1G():
    r = np.linspace(0, 5, 1000)
    plt.plot(r, STO1(r), label=r'STO: $\phi_1^\mathrm{S}(r, 1)$')
    plt.plot(r, pGTO(r, 0.27097914),
             label=r'STO-1G: $\phi^\mathrm{G}_n(r, \alpha)$')
    plt.xlabel(r'$r\,/a_0$')
    plt.ylabel(r'$\phi_1(r)\,/a_0^{-3/2}$')
    plt.legend()

plot_STO1_GTO1G()
```

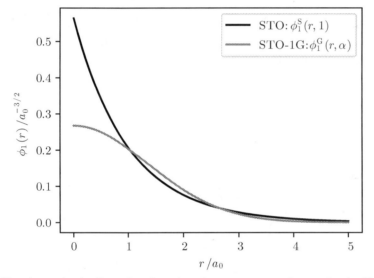

Clearly, a single Gaussian function does not approximate the 1s Slater-type orbital well and a linear expansion of primitive GTOs (pGTOs) is necessary:

$$\psi^{NG} = \sum_{k=1}^{N} c_k \phi_1^G(r, \alpha_k).$$

We will seek the expansion coefficients c_k and the exponent coefficients α_k that maximize the overlap of ψ^{NG} with $\phi_1^S(r, 1)$.

```
def GTO(r, prms):
    """

    Return the Gaussian-type orbital psi_NG as a linear expansion
    of primitive Gaussians. The coefficients are packed into prms as
    [c1, alpha1, c2, alpha2, ..., cN, alphaN].
```

```
    """

    N = len(prms) // 2
    psi = 0
    for k in range(N):
        c_k, alpha_k = prms[2*k:2*k+2]
        psi += c_k * pGTO(r, alpha_k)
    return psi
```

There is a minor complication in that, since ψ^{NG} must be normalized, we can only
fit $N - 1$ of the c_k coefficients, the remaining one being fixed by the normalization
requirement:

$$\langle \psi^{NG} | \psi^{NG} \rangle = \int_0^\infty |\psi^{NG}|^2 \, 4\pi r^2 dr = 1.$$

We will fit the parameters $c_2, c_3, \ldots, c_N$ and leave c_1 as fixed by the above condition. It will therefore be helpful to have a method that determines the value of c_1
that leaves ψ^{NG} normalized. This is a nonlinear fitting problem:

```
def get_c1(red_prms):
    """
    Determine the value of c1 that normalizes the STO-NG function described
    by the coefficients red_prms = [alpha1, c2, alpha2, ..., cN, alphaN].

    """

    def resid(c1):
        """Return the difference between <Psi_NG|Psi_NG> and unity."""
        prms = np.concatenate((c1, red_prms))
        I = quad(lambda r: GTO(r, prms)**2 * r**2, 0, np.inf)[0] * 4 * np.pi
        return I - 1

    c1guess = 0.1
    fit = least_squares(resid, [c1guess])
    return fit.x[0]
```

With this methodology we must modify the function that calculates the overlap
integral between our STO-NG wavefunction and $\phi_1^S(r, 1)$. This function is passed
the parameters to be fit, deduces the remaining coefficient c_1, and then returns the
negative of the overlap integral.

```
def overlap(fit_prms):
    c1 = get_c1(fit_prms)
    prms = np.concatenate(([c1], fit_prms))
    f = lambda r: GTO(r, prms) * STO1(r) * r**2
    res = quad(f, 0, np.inf)
    return -res[0]
```

```
N = 2
# Random initial guesses for alpha1, c2, alpha2
x0 = np.random.random(2*N - 1)
fit2 = minimize(overlap, x0)
fit2
```

```
      fun: -0.07945171539378691
 hess_inv: array([[   7.05300716,  -23.61129341,   50.76648821],
       [ -23.61129341,  103.31418071,  -215.78158982],
       [  50.76648821,  -215.78158982,  691.23060092]])
      jac: array([-3.83984298e-06,  8.38190317e-09, -6.51925802e-08])
  message: 'Optimization terminated successfully.'
     nfev: 100
      nit: 22
     njev: 25
   status: 0
  success: True
        x: array([0.15158048, 0.43027844, 0.85146645])
```

```
c1 = get_c1(fit2.x)
prms2 = np.concatenate(([c1], fit2.x))
prms2
```

```
array([0.43010152, 0.85203937, 0.67895534, 0.15163544])
```

Therefore,

$$\psi^{2G} = 0.6790\phi_1^G(r, 0.1516) + 0.4301\phi_1^G(r, 0.8520)$$

Similarly, for the STO-3NG function:

```
N = 3
# Initial guesses for alpha1, c2, alpha2, c3, alpha3.
x0 = [0.2, 0.5, 0.4, 0.5, 0.5]
# Set bounds on the fit parameters in this case, to constrain the
# fit from wandering off into unrealistic values of the coefficients.
fit3 = minimize(overlap, x0, bounds=[(0, 5)] * (2*N-1))
fit3
```

```
      fun: -0.07956431865440691
 hess_inv: <5x5 LbfgsInvHessProduct with dtype=float64>
      jac: array([ 1.33698608e-05,  7.69245774e-06,  1.03431153e-05,  1.38528078e-05,
       -7.35522758e-08])
  message: 'CONVERGENCE: REL_REDUCTION_OF_F_<=_FACTR*EPSMCH'
     nfev: 270
      nit: 39
     njev: 45
   status: 0
  success: True
        x: array([0.1099509 , 0.53494108, 0.40671275, 0.15398082, 2.23376247])
```

```
c1 = get_c1(fit3.x)
prms3 = np.concatenate(([c1], fit3.x))
prms3
```

```
array([0.44542232, 0.1099509 , 0.53494108, 0.40671275, 0.15398082,
       2.23376247])
```

Therefore,

$$\psi^{3G} = 0.4454\phi_1^G(r, 0.1100) + 0.5349\phi_1^G(r, 0.4067) + 0.1540\phi_1^G(r, 2.2338)$$

The improvement in the fit can be seen by plotting a comparison of these STO-1G, STO-2G and STO-3G orbitals with the target $\phi_1^S(r, 1)$ wavefunction:

```
def plot_STO_comparison():
    r = np.linspace(0, 5, 1000)
    plt.plot(r, STO1(r), label=r'$\phi_1^\mathrm{S}(r, 1)$')

    plt.plot(r, pGTO(r, 0.27097914), label='STO-1G')
    plt.plot(r, GTO(r, prms2), label='STO-2G')
    plt.plot(r, GTO(r, prms3), label='STO-3G')

    plt.xlabel(r'$r\,/a_0$')
    plt.ylabel(r'$\psi(r)\,/a_0^{-3/2}$')
    plt.legend()

plot_STO_comparison()
```

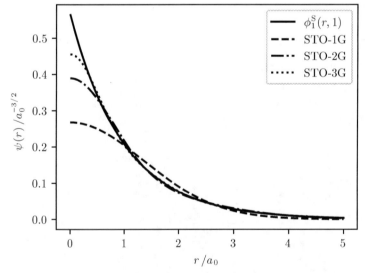

Another perspective on these basis functions can be obtained by comparing how well they represent the hydrogen atom, since the STO they approximate is the exact solution to the Schrödinger equation for the one-electron atom,

$$\hat{H}\phi_1^S(r, 1) = \left(\frac{1}{2}\nabla^2 - \frac{1}{r}\right)\phi_1^S(r, 1) = E\phi_1^S(r, 1),$$

with the eigenvalue $E = -\frac{1}{2}$.

We will calculate $\langle E \rangle = \langle \psi^{NG} | \hat{H} | \psi^{NG} \rangle$, the expectation value of the energy eigenvalue of the functions ψ^{NG}:

$$\langle \psi^{NG} | \hat{H} | \psi^{NG} \rangle = \int_0^\infty \psi^{NG} \sum_{k=1}^N c_k \left(\hat{H}\phi_1^G(r, \alpha_k)\right) 4\pi r^2 dr.$$

For the 1s orbital, we only need to consider the radial part of the Hamiltonian and it is straightforward to show that, for the primitive Gaussian function $\phi_1^G = \phi_1^G(r; \alpha_k)$:

$$\hat{H}\phi_1^G = \left(-\frac{1}{2}\nabla^2 - \frac{1}{r}\right)\phi_1^G = 3\alpha_k\phi_1^G - 2\alpha_k^2 r^2\phi_1^G - \frac{\phi_1^G}{r}.$$

```
def func(r, prms):
    """

    Return the integrand for the expectation value of the energy of the STO-NG
    function defined by prms = [c1, alpha1, c2, alpha2, ..., cN, alphaN],
    <phi_NG | H | phi_NG> as a function of the electron-nucleus distance, r.
    """

    phi_NG = GTO(r, prms)
    N = len(prms) // 2
    rhs = 0
    for k in range(N):
        ck, alphak = prms[2*k:2*k+2]
        rhs += ck * (3 * alphak - 2 * (alphak * r)**2 - 1/r) * pGTO(r, alphak)
    return phi_NG * rhs * 4 * np.pi * r**2

def get_E(prms):
    E = quad(func, 0, np.inf, args=(prms,))[0]
    return E

Eexact = -0.5
energies = {1: get_E([1, alpha_11]), 2: get_E(prms2), 3: get_E(prms3)}
for N, E in energies.items():
    error = abs((E - Eexact) / Eexact) * 100
    print(f'STO-{N}G: {E:.4f} Eh, error = {error:.2f}%')
```

```
STO-1G: -0.4242 Eh, error = 15.16%
STO-2G: -0.4812 Eh, error = 3.77%
STO-3G: -0.4949 Eh, error = 1.02%
```

In this analysis, the STO-3G function reproduces the 1s orbital with an error in the energy of approximately 1%, and in practice, this is considered to be the simplest useful basis function.

E34.3

Fitting a Gaussian Function

The Doppler broadening of spectral lines is caused by the distribution of their velocities relative to the absorbed or emitted photon and, in the absence of other broadening mechanisms (pressure broadening, lifetime broadening, instrument resolution effects, etc.) results in a Gaussian line profile:

$$f_G(\nu; \nu_0, \alpha_D) = \sqrt{\frac{\ln 2}{\alpha_D^2 \pi}} \exp\left[-(\ln 2)\frac{(\nu - \nu_0)^2}{\alpha_D^2}\right],$$

where ν_0 is the transition frequency (central frequency of the line), and the half-width at half-maximum,

$$\alpha_D = \sqrt{\frac{2k_B T \ln 2}{mc^2}} \nu_0$$

for an absorbing species of mass m and at temperature, T.

The high-resolution spectrum of a low-density gas can therefore be used as a thermometer if isolated spectral lines can be fit to a Gaussian function. Conversely, before its value was fixed by definition, such a fit to a spectrum taken at a known temperature was one way of measuring the Boltzmann constant, k_B.

The file `NH3-line.csv`, available at `https://scipython.com/chem/eghc/`, contains the high-resolution spectrum of the ν_2 saQ(6, 3) rovibrational line of the principal isotopologue of ammonia ($^{14}N^1H_3$), which is well-separated from other absorptions. We can analyze this spectrum to determine the temperature of the gas.

```
import numpy as np
from scipy.optimize import curve_fit
import matplotlib.pyplot as plt
```

```
def fG(nu, nu0, alphaD):
    """Normalized Gaussian function with HWHM alphaD, centered on nu0."""
    N = np.sqrt(np.log(2) / np.pi) / alphaD
    return N * np.exp(-np.log(2) * ((nu - nu0) / alphaD)**2)
```

```
# Read in the data file
nu, spec = np.genfromtxt('NH3-line.csv', delimiter=',', unpack=True, skip_header=1)
```

At this stage, it makes sense to plot the experimental data to see if we can make an educated guess for the values of the parameters we want to fit (Figure 34.1):

```
plt.plot(nu, spec, 'x')
plt.xlabel(r'$\nu\;/\mathrm{MHz}$')
plt.ylabel(r'Optical Depth')
```

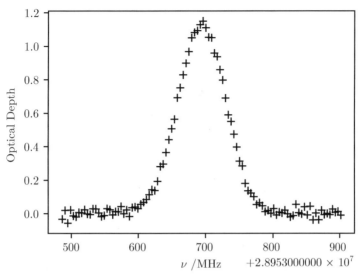

Figure 34.1 The experimental spectrum of the v_2 saQ(6, 3) rovibrational line of ammonia.

The function to fit is actually $f(v) = Af_G(v)$, since the optical depth depends on the product of the path length, absorber density, line strength and the lineshape function. We will absorb the other factors into the parameter, A:

```
def f(nu, A, nu0, alphaD):
    return A * fG(nu, nu0, alphaD)
```

There is some noise, but we can get good estimates of the parameter values from inspecting the spectrum:

```
# Spectrum maximum signal.
A_guess = 1.2
# Frequency of the spectrum maximum (MHz).
nu0_guess = nu[np.argmax(spec)]
# Guess for HWHM of line (MHz).
alphaD_guess = 40
popt, pcov = curve_fit(f, nu, spec, p0=(A_guess, nu0_guess, alphaD_guess))
A, nu0, alphaD = popt
A, nu0, alphaD
```

```
(100.33655618951202, 28953694.49912839, 41.66608989340652)
```

The 1σ standard deviations in these parameters are given by the square root of the diagonal elements of the covariance matrix:

```
pstd = np.sqrt(np.diag(pcov))
sigma_A, sigma_nu0, sigma_alphaD = pstd
sigma_A, sigma_nu0, sigma_alphaD
```

```
(0.5656240531671611, 0.23034846010693605, 0.27120953083941735)
```

The fitted parameters are therefore:

$$A = 100.34(57)$$

$$\nu_0 = 28953694.50(23) \text{ MHz}$$

$$\alpha_D = 41.67(27) \text{ MHz}$$

The temperature can now be retrieved from the definition of α_D.

```
from scipy.constants import c, k as kB, u

# Isotope masses, mass of (14N)(1H)3, in kg.
mH, mN = 1.007825031898 * u, 14.003074004 * u
m = mN + 3*mH

b = m * c**2 / 2 / kB / np.log(2)
T = b * (alphaD / nu0)**2
T
```

```
274.93872647125903
```

If we ignore correlations between the fitted parameters, the uncertainty in the re-trieved temperature can be calculated by propagating the fit uncertainties:

```
sigma_T = 2 * T * np.sqrt(
    (sigma_alphaD / alphaD)**2 + (sigma_nu0 / nu0)**2)
sigma_T
```

```
3.5792176902931563
```

We might report the temperature as $T = 274.9 \pm 3.6$ K therefore.

The fitted line profile can be plot along with the data as follows (Figure 34.2):

```
plt.plot(nu, spec, '+', label='Experiment')
plt.plot(nu, f(nu, *popt), label='Fitted line')
plt.xlabel(r'$\nu\;/\mathrm{MHz}$')
plt.ylabel(r'Optical Depth')
plt.legend()
plt.plot()
```

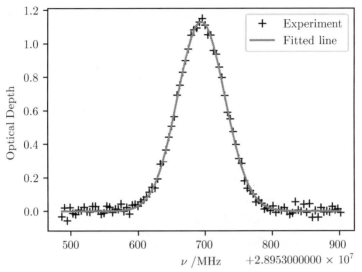

Figure 34.2 Measured and fitted spectral lineshape for the v_2 saQ(6, 3) transition of ammonia.

34.5 Exercises

P34.1 In the Example E34.2, the STO-3G contracted Gaussian function,

$$\psi^{3G} = \sum_{k=1}^{N=3} c_k \phi_k^G(r, \alpha_k)$$

was fit to the 1s Slater-type orbital, $\phi_1^S(r, 1)$, by maximizing the overlap between them.

(a) Instead of targeting this overlap function, perform a direct curve fit of ψ^{3G} to a discretized set of STO function values, $\phi_1^S(r_i, 1)$ for 500 evenly spaced values of r_i between 0 and $5a_0$:

```python
import numpy as np

def STO1(r, zeta=1):
    """The normalized 1s Slater-type orbital."""
    return np.sqrt(zeta**3 / np.pi) * np.exp(-zeta * r)

r = np.linspace(0, 5, 500)
phi_S = STO1(r)
```

Determine the percentage error in the expectation value of the energy for the fitted function, $\langle E \rangle = \langle \psi^{3G} | \hat{H} | \psi^{3G} \rangle$, as compared to the exact value, $E = \frac{1}{2} E_h$.

(b) Now, instead, find the fit coefficients, c_k and α_k that directly minimize $\langle E \rangle$. Again, determine the percentage error in $\langle E \rangle$ compared with $E = \frac{1}{2} E_h$.

P34.2 Example E34.1 considered the water–gas shift reaction as an equilibrium between CO(g), H_2O(g), H_2(g) and CO_2(g):

$$CO(g) + H_2O(g) \rightleftharpoons H_2(g) + CO_2(g)$$

In the presence of methane, an additional process occurs:

$$CH_4(g) + H_2O(g) \rightleftharpoons CO(g) + 3\,H_2(g)$$

Use the Shomate equations and coefficients in the file `thermo-data.csv` from Example E31.2 to determine the equilibrium composition at 1,000 K and a total pressure of 4 bar of a mixture initially composed of 1.5 mol H_2O and 2 mol CH_4.

35

SymPy

SymPy is a free, open source library for performing computer algebra (also known as symbolic computation): That is, it is able to manipulate, simplify and transform mathematical formulas through a representation of the relationship between their symbols rather than by direct numerical evaluation. Mathematical expressions are left in symbolic form from which they can be subsequently evaluated or further manipulated.

For example, the NumPy library and `math` module give a numerical approximation in evaluation $\sin\frac{\pi}{3}$:

```
In [x]: import numpy as np
In [x]: np.sin(np.pi / 3)
Out[x]: 0.8660254037844386
```

which is $\sqrt{3}/2$ to about 16 decimal places. In contrast, SymPy is able to report the exact representation as a surd:

```
In [x]: import sympy
In [x]: sympy.sin(sympy.pi / 3)
Out[x]: sqrt(3)/2
```

Some of the functionality included in SymPy includes:

- Basic arithmetic, including noncommutative operations; simplification, substitution, expansion of expressions;
- Logarithmic, exponential, trigonometric, hyperbolic and a variety of "special" functions (the Dirac delta, gamma, beta and Bessel functions, spherical harmonics and so on);
- Polynomial manipulations;
- Matrix algebra: arithmetic, eigenvalues, determinants and inversion;
- Calculus: differentiation, integration, limits and Taylor series;
- Solutions to systems of linear equations;
- Classical Mechanics;

- Quantum Mechanics, particularly commutators, anticommutators and angular momentum coupling constants;
- Function plotting.

A short introduction to the main features of SymPy is given here; for more detailed information, there is good documentation and a comprehensive tutorial on the SymPy website, `www.sympy.org/en/index.html`. SymPy can also be used through its online interface, SymPy Live: see `https://live.sympy.org/`.

This chapter will assume that SymPy is imported with the alias `sp`:

```
In [x]: import sympy as sp
```

35.1 Algebra and Mathematical Functions

35.1.1 Symbols and Expressions

SymPy builds representations of mathematical expressions out of objects it calls `Symbols`. A `Symbol` is assigned to a variable name but is also given a display name, which may be rendered using LaTeX in some interfaces. A single `Symbol` may be created by passing its display name to `sp.Symbol`; one or more `Symbols` can be defined by providing a name string of names, separated by spaces or commas to the `sp.symbols` function:

```
In [x]: a = sp.Symbol('a')
In [x]: a, b, c, x, y, z = sp.symbols('a, b, c, x, y, z')
In [x]: theta, phi = sp.symbols(r'theta phi')
```

It is possible to define a `Symbol` with certain assumptions about its nature with various boolean arguments, including `integer`, `real`, `complex`, `nonzero`, `odd`, `even`, `finite`, `negative`, `nonnegative`, `positive`, `nonpositive`. By default, `Symbols` are considered to be complex and to commute. For example:

```
In [x]: x * y == y * x          # By default, Symbols commute
Out[x]: True

In [x]: A, B = sp.symbols('A B', commutative=False)
In [x]: A * B == B * A
Out[x]: False
```

SymPy `Symbols` are combined with each other and with numbers into algebraic *expressions* using the usual Python arithmetical operators, +, -, * and /:

```
In [x]: expr1 = a * x**2 + b * x + c
In [x]: expr1
Out[x]: a*x**2 + b*x + c
```

Exponential, logarithmic, trigonometric and hyperbolic functions are all supported in expressions, but their SymPy implementations must be used:

```
In [x]: expr2 = sp.cos(sp.pi * x / 3)
In [x]: expr3 = sp.exp(-a * x**2)
```

Expressions can be further manipulated to create new expressions:

```
In [x]: ((x+1) / expr2)**2
Out[x]: (x + 1)**2/cos(pi*x/3)**2
```

Substitution (of a number or an expression into a `Symbol`) is achieved using the `subs` method:

```
In [x]: expr3.subs(x, z)    # substitute z for x in expr3
Out[x]: exp(-a*z**2)

In [x]: expr2.subs(x, 1)    # set x=1 in expr2: cos(pi / 3)
Out[x]: 1/2
```

The original expression is left unchanged (expressions are immutable objects): To store the resulting expression, it must be assigned to a new variable. This also means that expressions don't change if the variables originally assigned to their `Symbol`s are re-assigned to something else:

```
In [x]: x = sp.symbols("x")
In [x]: expr = x + 5
In [x]: x = 2             # reassign x to an integer
In [x]: expr
Out[x]: x + 5
```

Here, `expr` still refers to the original `Symbol` with the display name x, even though the variable that originally referred to that `Symbol` has been reassigned to the integer 2. That is, x the variable name and x the `Symbol` within the expression `expr` no longer have anything to do with one another.

It is possible to perform more than one substitution at once by passing a sequence of (`old, new`) pairs to `subs`:

```
In [x]: expr1
Out[x]: a*x**2 + b*x + c

In [x]: expr4 =  expr1.subs([(a, 1), (b, 0), (c, -1)])
In [x]: expr4
Out[x]: x**2 - 1
```

Alternatively, a dictionary can be used:

```
In [x]: expr1.subs({'a': 0, 'b': 2})
Out[x]: c + 2*x
```

Within an IPython session, or on a terminal, SymPy will, by default, display expressions using a Python-like syntax:

```
In [x]: expr = x**2 + 1 / x
In [x]: expr
Out[x]: x**2 + 1/x
```

Calling SymPy's `init_printing()` function will configure the output display to the best available for the current environment ("pretty printing"). For example:

```
In [x]: sp.init_printing()
In [x]: expr
Out[x]:
 2    1
x  + -
      x
```

In a Jupyter Notebook, SymPy output usually looks much better since it will use LaTeX and MathJax to render publication-quality equations.

To *evaluate* a Sympy expression to give a numerical result, call `evalf` as follows, optionally passing the number of significant digits as the argument n:

```
In [x]: expr =  sp.pi * x

In [x]: expr
Out[x]: pi*x

In [x]: expr.evalf(subs={x:2})          # return 2.pi to the default 15 s.f.
Out[x]: 6.28318530717959

In [x]: expr.evalf(subs={x:2}, n=50)    # return 2.pi to 50 s.f.
Out[x]: 6.2831853071795864769252867665590057683943387987502
```

35.1.2 Simplifying Expressions

SymPy has many methods for simplifying mathematical expressions. The most general-purpose one, `sp.simplify`, generally does a good job, though it can be slow. For example, to confirm the following identities:

$$\sin^2 x + \cos^2 x = 1$$
$$\sec^2 x - 1 = \tan^2 x$$
$$\frac{x^3 - 3x^2 + x - 3}{x^2 + 1} = x - 3$$

pass the corresponding SciPy expressions to `sp.simplify`:

```
In [x]: sp.simplify(sp.cos(x)**2 + sp.sin(x)**2)
Out[x]: 1

In [x]: sp.simplify(sp.sec(x)**2 - 1)
Out[x]: tan(x)**2

In [x]: sp.simplify((x**3 - 3*x**2 + x - 3) / (x**2 + 1))
Out[x]: x - 3
```

Calling the `simplify` method on an expression is also possible:

```
In [x]: expr = sp.sec(x) ** 2 - 1
In [x]: expr.simplify()
Out[x]: tan(x)**2
```

If you specifically want to expand or factor an expression, use the corresponding methods:

```
In [x]: sp.expand((x+1) * (x-2) * (3*x + 1))
Out[x]: 3*x**3 - 2*x**2 - 7*x - 2

In [x]: sp.factor(x**2 + 2*x + 1)
Out[x]: (x + 1)**2

In [x]: sp.factor(6*x**5 - 5*x**4 - 6*x**3 + 3*x**2 + 2*x)
Out[x]: x*(x - 1)**2*(2*x + 1)*(3*x + 2)
```

Other relevant SymPy methods are `cancel` and `collect` which simplify an expression by specifically cancelling common factors out of a rational function and collecting terms with common powers in an expression:

```
In [x]: sp.cancel((9*x**2 + 6*x) / (3 * x**3))
Out[x]: (3*x + 2)/x**2

In [x]: expr = 4*x**2 + x + 1 + 3*x + 5
In [x]: sp.collect(expr, x)
Out[x]: 4*x**2 + 4*x + 6
```

There is a further, nice function, `apart`, that performs a partial fraction decomposition on a rational function, for example:

$$\frac{x^2 - x + 2}{x(x+2)^2} = \frac{1}{2(x+2)} - \frac{4}{(x+2)^2} + \frac{1}{2x}.$$

```
In [x]: expr = (x**2 - x + 2) / (x * (x+2)**2)
In [x]: sp.apart(expr)
Out[x]: 1/(2*(x + 2)) - 4/(x + 2)**2 + 1/(2*x)
```

Table 35.1   Simplification methods in SymPy

`powsimp`	$x^a x^b \to x^{a+b}$
`expand_power_exp`	$x^{a+b} \to x^a x^b$
`expand_power_base`	$(xy)^a \to x^a y^a$
`powdenest`	$(x^a)^b \to x^{ab}$
`expand_log`	$\log(xy) \to \log x + \log y;\ \log(x^2) \to 2\log x$, etc.
`logcombine`	$\log x + \log y \to \log(xy)$, etc.

There are a couple of methods that can be faster for simplifying and expanding trigonometric and hyperbolic functions, `trigsimp` and `expand_trig`:

$$\cos\left(\theta - \frac{\pi}{2}\right) = \sin\theta$$

$$\frac{\sin\theta}{1 - \sin^2\theta} = \frac{\sin\theta}{\cos^2\theta}$$

$$\cos(\theta - \phi) = \sin\theta\sin\phi + \cos\theta\cos\phi$$

```
In [x]: sp.trigsimp(sp.cos(theta - sp.pi/2))
Out[x]: sin(theta)

In [x]: sp.trigsimp(sp.sin(theta) / (1 - sp.sin(theta)**2))
Out[x]: sin(theta)/cos(theta)**2

In [x]: sp.expand_trig(sp.cos(theta - phi))
Out[x]: sin(phi)*sin(theta) + cos(phi)*cos(theta)
```

A more complete list of SymPy's simplification methods is provided in Table 35.1.

Some of these methods will refuse to simplify expressions where the simplified expression is not equal, in general, to the original. For example:

```
In [x]: sp.powsimp(x**a * y**a)
Out[x]: x**a*y**a
```

Since it is not true, in general, that $x^a y^a = (xy)^a$ (consider, e.g., $x = y = -1$ and $a = \frac{1}{2}$), no simplification is performed. To coerce SymPy to simplify this expression in the way probably intended, one can pass the argument `force=True` or define the `Symbol` assumptions more closely:

```
In [x]: sp.powsimp(x**a * y**a, force=True)
Out[x]: (x*y)**a

In [x]: x, y = sp.symbols('x y', nonnegative=True)
In [x]: a = sp.Symbol('a', real=True)
In [x]: sp.powsimp(x**a * y**a)
Out[x]: (x*y)**a
```

35.1.3 Expression Equality

SymPy expressions are only considered equal if they have the same structure – in other words, in addition to being mathematically equal, they must be expanded to the same terms (in any order) before cancellation or simplification. For example:

```
In [x]: x + 1 == 1 + x
Out[x]: True

In [x]: x**2 - 1 == x**2 - 1
Out[x]: True

In [x]: x**2 - 1 == -1 + x**2
Out[x]: True

In [x]: x**2 - 1 == (x + 1) * (x - 1)
Out[x]: False
```

To test for expression equality, an approach that usually works is to simplify the difference between the two expressions and see if the result is zero:

```
In [x]: expr1 = x**2 - 1
In [x]: expr2 = (x + 1) * (x - 1)
In [x]: sp.simplify(expr1 - expr2) == 0
Out[x]: True
```

Alternatively, there is a method, `equals`, that tests two expressions for equality by evaluating them for random values of their arguments:

```
In [x]: expr1.equals(expr2)
Out[x]: True
```

35.2 Equation Solving

35.2.1 Single Equations

Mathematical equations in SymPy are represented using `sp.Eq`, which takes the equation's left- and right-hand side expressions as its arguments. For example, the equation $2x + 6 = 1$ could be represented by the `Equality` object created by the following:

```
In [x]: eqn = sp.Eq(2*x + 6, 1)
In [x]: eqn
Out[x]: Eq(2*x + 6, 1)
```

It is important to keep in mind that SymPy does not change any of Python's syntax: If x is a `Symbol`, writing `2x + 6 = 1` does not produce a symbolic equation: It is an (attempted) *assignment* of the value 1 to an invalid left-hand side target object `2x + 6`: This can't be done and so a `SyntaxError` would be raised.

Eq objects have a lefthand side (`lhs`) and a righthand side (`rhs`):

```
In [x]: eqn.lhs
Out[x]: 2*x + 6

In [x]: eqn.rhs
Out[x]: 1
```

However, as we shall see, in many cases, it is possible to avoid explicitly creating `Equality` objects: If the equation can be manipulated so that the right-hand side is zero, the left-hand side *expression* can be used in place of an `Equality` object. For example, in solving the above equation for *x* with SymPy's `solve` method:

```
In [x]: sp.solve(sp.Eq(2*x + 6, 1), x)
Out[x]: [-5/2]
```

can be shortened by rearranging the equation to $2x + 5 = 0$:

```
In [x]: sp.solve(2*x + 5, x)
Out[x]: [-5/2]
```

The recommended function for solving algebraic equations, however, is `sp.solveset`. This is quite a sophisticated solver, but its basic use is with the syntax:

> `sp.solveset(equation, variable, domain)`

The *domain* defaults to `sp.Complexes`; if you want to restrict your solutions (if any) to the real numbers, set `domain=sp.Reals`. For example, to solve the equation $x^3 = -x$, first rewrite it as an expression equal to zero:

$$x^3 + x = 0$$

There are three solutions within the complex domain, returned as a `FiniteSet` object, one of which is real:

```
In [x]: sp.solveset(x**3 + x, x)
Out[x]: FiniteSet(0, I, -I)              # all solutions: x = 0, i, -i

In [x]: sp.solveset(x**3 + x, x, sp.Reals)
Out[x]: FiniteSet(0)                      # one real solution: x = 0
```

If there is no solution, an `EmptySet` is returned:

```
In [x]: sp.solveset(x**2 + 1, x, sp.Reals)
Out[224]: EmptySet
```

(there is no real number, *x* for which $x^2 + 1 = 0$).

SymPy also does a good job of finding the set of solutions when more than one exists, as with the equation $\sin x = 1$:

```
In [x]: sp.solveset(sp.sin(x) - 1, x)
Out[x]: ImageSet(Lambda(_n, 2*_n*pi + pi/2), Integers)
```

Without going into the details, this is SymPy's way of returning the solution set

$$\left\{2n\pi + \tfrac{\pi}{2} \big| n \in \mathbb{Z}\right\}.$$

That is, $x = 2n\pi + \tfrac{\pi}{2}$ for $n = 0, \pm 1, \pm 2, \ldots$, the same as $x = \ldots, \tfrac{-7\pi}{2}, \tfrac{-3\pi}{2}, \tfrac{\pi}{2}, \tfrac{5\pi}{2}, \tfrac{9\pi}{2}, \ldots$

35.2.2 Linear and Nonlinear Systems of Equations

Systems of linear equations (a set of relations between variables which hold simultaneously and in which each of the terms involves a single variable or a constant) can be solved (if a solution exists) with the sp.linsolve method. For example, the following simple pair of linear equations in variables x and y:

$$4x - 3y = -7$$
$$6x - y = 0$$

can be solved by passing their representations (expressed using Eq or as simple expressions equal to zero) to sp.linsolve along with a sequence of the Symbols to be solved for:

```
In [x]: sp.linsolve((4*x - 3*y + 7, 6*x - y), (x, y))
Out[x]: FiniteSet((1/2, 3))
```

The order of the results matches the order of the Symbols passed to sp.linsolve (here, (x, y); that is, $x = \tfrac{1}{2}$, $y = 3$). The solution here is not a numerical one but, rather, deduced algebraically in symbolic form. For example, if other Symbols appear in the system of equations, the solution is returned in terms of those Symbols. For example, if a and b are constants, the pair of simultaneous equations:

$$ax - 3y = -7$$
$$6x - y = b$$

has the solution

$$x = \frac{3b + 7}{18 - a}, \quad y = \frac{ab + 42}{18 - a}.$$

```
In [x]: sp.linsolve((a*x - 3*y + 7, 6*x - y - b), (x, y))
Out[x]: FiniteSet((-(3*b + 7)/(a - 18), -(a*b + 42)/(a - 18)))
```

Systems of equations that are underdetermined result in a parametric solution in terms of the specified symbols. For example, consider the following pair of equations

$$7x + y - z = 0$$
$$2x + z = 4.$$

There are three unknown variables in two equations so there is no unique solution, but x and y can be written in terms of the third variable, z:

$$x = 2 - \frac{z}{2}, \quad y = \frac{9z}{2} - 14.$$

```
In [x]: sp.linsolve((7*x + y - z, 2*x + z - 4), (x, y, z))
Out[x]: FiniteSet((2 - z/2, 9*z/2 - 14, z))
```

The set of equations can also be provided in matrix form $Ax = b$:

$$\begin{pmatrix} 7 & 1 & -1 \\ 2 & 0 & 1 \end{pmatrix} \begin{pmatrix} x \\ y \\ z \end{pmatrix} = \begin{pmatrix} 0 \\ 4 \end{pmatrix}$$

```
In [x]: A = sp.Matrix(((7, 1, -1), (2, 0, 1)))
In [x]: b = sp.Matrix(((0,), (4,)))              # NB a row vector also works.
In [x]: sp.linsolve((A, b), (x, y, z))
Out[x]: FiniteSet((2 - z/2, 9*z/2 - 14, z))
```

Systems of equations that are nonlinear in one or more of their variables can, in some cases, be solved with `sp.nonlinsolve`. For example, to determine the co-ordinates where the unit circle centred at the origin intersects another centered at $(a, 0)$, the following quadratic equations must be solved for x and y:

$$x^2 + y^2 = 1$$
$$(x - a)^2 + y^2 = 1$$

giving

$$x = \frac{a}{2}, \quad y = \pm \frac{4 - a^2}{2}.$$

```
In [x]: x, y, a = sp.symbols('x y a', real=True)
In [x]: sp.nonlinsolve((x**2 + y**2 - 1, (x-a)**2 + y**2 - 1), (x,y))
In [x]: FiniteSet((a/2, -sqrt(4 - a**2)/2), (a/2, sqrt(4 - a**2)/2))
```

There is one further function, `sp.roots` specifically for determining the roots of polynomials:

```
In [x]: quintic = x**5 - 9*x**3 + 4*x**2 + 12*x
In [x]: sp.roots(quintic, x)
Out[x]: {0: 1, 2: 2, -1: 1, -3: 1}
```

The dictionary returned identifies the roots in its keys and their multiplicities as the corresponding values: in this case, the quintic polynomial factorizes as

```
In [x]: sp.factor(quintic)
Out[x]: x * (x - 2)**2 * (x + 1) * (x + 3)
```

35.3 Calculus

35.3.1 Differentiation

To take the derivative of an expression, either call the `sp.diff` function directly on an expression:

```
In [x]: sp.diff(sp.sin(a * x), x)      # differentiate sin(ax) with respect to x
Out[x]: a*cos(a*x)
```

or create a `Derivative` object and call its `doit` method:

```
In [x]: deriv = sp.Derivative(sp.sin(a * x), x)
In [x]: deriv.doit()
Out[x]: a*cos(a*x)
```

Multiple derivatives can be calculated by repeating the variable after the expression as many times as the order of derivative required or by passing a number after the variable:

```
In [x]: sp.diff(sp.exp(-a*x), x, x, x)      # third derivative of exp(-ax)
Out[x]: -a**3*exp(-a*x)

In [x]: sp.diff(sp.exp(-a*x), x, 3)         # third derivative of exp(-ax)
Out[x]: -a**3*exp(-a*x)
```

a similar syntax applies to partial differentiation:

```
In [x]: f = x*y**2 + y * x**2      # f(x,y) = xy^2 + yx^2

In [x]: sp.diff(f, x, x)           # d2f/dx2
Out[x]: 2*y

In [x]: sp.diff(f, x, y)           # d2f/dxdy
Out[x]: 2*(x + y)

In [x]: sp.diff(f, y, x)           # d2f/dydx
Out[x]: 2*(x + y)

In [x]: sp.diff(f, y, y)           # d2f/dy2
Out[x]: 2*x
```

35.3.2 Integration

The basic Sympy function for integration is `sp.integrate`. Both definite and indefinite integrals are supported. In the case of definite integrals, the variable to be integrated over and its limits are specified in a `tuple`:

```
In [x]: sp.integrate(x**2, x)      # indefinite integral of x^2 with respect to x
Out[x]: x**3/3
In [x]: sp.integrate(x**2, (x, 1, 3))    # definite integral between 1 and 3
Out[x]: 26/3
```

Note that no constant of integration is added to the indefinite integral solution.
Infinity is represented in SymPy by the quantity `sp.oo`:

```
In [x]: a = sp.Symbol('a', real=True, positive=True)
In [x]: sp.integrate(sp.exp(-x * a**2), (x, -sp.oo, sp.oo))
Out[x]: sqrt(pi)/sqrt(a)
```

When (as chemists, at least) we write

$$\int_{-\infty}^{\infty} \exp(-ax^2)\,dx = \sqrt{\frac{\pi}{a}},$$

we implicitly assume a to be a positive real number. SymPy does not assume this,
so we have to constrain the `Symbol` accordingly. Otherwise, it patiently divides
the domain of the complex number a into a region where the above formula holds
$(\mathrm{Re}(a) \geq 0)$ and everywhere else (where it gives up and returns the integral expression unchanged):

```
In [x]: a = sp.Symbol('a')        # NB a is a complex number
In [x]: sp.integrate(sp.exp(-(a * x**2)), (x, -sp.oo, sp.oo))
Out[x]: Piecewise((sqrt(pi)/sqrt(a), Abs(arg(a)) <= pi/2),     # ie Re(a) >= 0
                  (Integral(exp(-a*x**2), (x, -oo, oo)), True))
```

Alternatively, the argument `conds='none'` can be set to evaluate the integral as
(probably) intended:

```
In [x]: sp.integrate(sp.exp(-(a * x**2)), (x, -sp.oo, sp.oo), conds='none')
Out[x]: sqrt(pi)/sqrt(a)
```

Multiple integration is also supported:

$$\int_0^b \int_0^a \sin^2\left(\frac{\pi x}{a}\right) \sin^2\left(\frac{\pi y}{b}\right)\,dx\,dy = \frac{ab}{4}$$

$$\int_0^{2\pi} \int_0^{\pi} \int_0^a r^2 \sin\theta\,dr\,d\theta\,d\phi = \frac{4}{3}\pi a^3.$$

```
In [x]: expr = (sp.sin(sp.pi*x/a) * sp.sin(sp.pi*y/b))**2
In [x]: sp.integrate(expr, (x, 0, a), (y, 0, b))
Out[x]: a*b/4

In [x]: r, theta, phi = sp.symbols('r theta phi', real=True)
In [x]: sp.integrate(r**2 * sp.sin(theta),
                     (r, 0, a), (theta, 0, sp.pi), (phi, 0, 2*sp.pi))
Out[x]: 4*pi*a**3/3
```

As with derivatives, `Integral` objects can be created and evaluated later with `doit`:

```
In [x]: n = sp.Symbol('n', integer=True, positive=True)
In [x]: L = sp.Symbol('L', real=True)
In [x]: norm_integral = sp.Integral(sp.sin(sp.pi * n * x / L)**2, (x, 0, L))
In [x]: N = 1/sp.sqrt(norm_integral.doit())
Out[x]: sqrt(2)/sqrt(L)
```

35.3.3 Summation

To evaluate a finite or infinite sum of terms expressed using some variable, either use `sympy.summation` or create a `sympy.Sum` object and call `doit()`. The symbol to be summed over and the limits of the summation are expressed as a `tuple` in the same way as for integration:

$$\sum_{i=1}^{n} i = \frac{n(n+1)}{2}, \quad \sum_{i=1}^{n} i^2 = \frac{n(2n^2 + 3n + 1)}{6}$$

```
In [x]: i, n = sp.symbols("i n", integer=True)

In [x]: sp.summation(i, (i, 1, n)).simplify()
Out[x]: n*(n + 1)/2

In [x]: sp.Sum(i**2, (i, 1, n)).doit().simplify()
Out[x]: n*(2*n**2 + 3*n + 1)/6
```

35.3.4 Series Expansion

To calculate the Taylor series expansion of a function, `f`, around the point x_0 up to terms of order x^n, use `f.series(x, x0, n)`. `x0` and `n` take the default values `x0=0` and `n=6` if they are not specified; for example:

$$\ln(1+x) = x - \frac{x^2}{2} + \frac{x^3}{3} - \frac{x^4}{4} + \frac{x^5}{5} + O(x)$$

```
In [x]: sp.log(1 + x).series(x)
Out[x]: x - x**2/2 + x**3/3 - x**4/4 + x**5/5 + O(x**6)

In [x]: sp.cos(x).series(x, sp.pi, 4)          # expansion of cos(x) about pi
Out[x]: -1 + (x - pi)**2/2 + O((x - pi)**4, (x, pi))
```

The Landau order notation, $O(x^6)$, means that all terms with powers greater than x^6 are omitted from the expansion. It can be removed with `remove0()`:

```
In [x]: sp.log(1 + x).series(x).removeO()
Out[x]: x**5/5 - x**4/4 + x**3/3 - x**2/2 + x
```

35.3.5 Differential Equations

SymPy can also solve many types of differential equation. First, create a `Symbol` to represent a function (it isn't necessary to define what the function is):

```
In [x]: f = sp.Function('f')        # or f = sp.symbols('f', cls=sp.Function)
```

A differential equation is just a regular `Eq` object involving a function and its derivative(s). For example, to represent the ordinary differential equation:

$$\frac{df}{dx} = -kx$$

write it as follows:

```
In [x]: k = sp.Symbol('k', real=True)
In [x]: diffeq = sp.Eq(f(x).diff(x), -k*f(x))
```

(alternatively, the expression `f(x).diff(x) + k * f(x)` could be used as `diffeq`). The SymPy method `dsolve` returns the solution using the provided function and variable:

```
In [x]: sp.dsolve(diffeq, f(x))
Out[x]: Eq(f(x), C1*exp(-k*x))
```

That is, $f(x) = C_1 e^{-kx}$, where C_1 is some undetermined constant.

Care is sometimes needed with the parameters to more complicated differential equations to ensure that they are kept within the physically relevant domain (real numbers, non-negative numbers, etc). For example, the ordinary differential equation governing the motion of a damped harmonic oscillator can be written as

$$\frac{d^2 z}{dt^2} + 2\gamma\omega_0 \frac{dz}{dt} + \omega_0^2 z = 0,$$

where $z(t)$ is the position of the oscillator at time t, ω_0 is the undamped (angular) frequency, and γ is the *damping ratio*: For $\gamma > 1$, the oscillator is described as *overdamped* and returns to a steady state by exponential decay; for $\gamma < 1$, the system oscillates but with an exponential decaying amplitude.

The time coordinate, t, and the parameters that determine the oscillator's behavior, ω_0 and γ, are non-negative real numbers. SymPy can solve this differential equation for the displacement function, $z(t)$:

```
In [x]: z = sp.Function('z')
In [x]: t = sp.symbols('t', positive=True)
In [x]: w0 = sp.symbols('w0', positive=True)
In [x]: gamma = sp.symbols('gamma', positive=True)

In [x]: diffeq = z(t).diff(t,t) + 2*gamma*w0 * z(t).diff(t) + w0**2 * z(t) # =0
In [x]: soln = sp.dsolve(diffeq, z(t))
```

```
In [x]: soln
Out[x]: Eq(z(t), C1*exp(t*w0*(-gamma - sqrt(gamma - 1)*sqrt(gamma + 1)))
              + C2*exp(t*w0*(-gamma + sqrt(gamma - 1)*sqrt(gamma + 1))))
```

This general solution is equivalent to

$$z(t) = C_1 \exp\left(\omega_0 t(-\gamma - \sqrt{\gamma^2 - 1})\right) + C_2 \exp\left(\omega_0 t(-\gamma + \sqrt{\gamma^2 - 1})\right).$$

The oscillatory behavior occurs when $\gamma < 1$, and the arguments to the exponential functions are complex numbers. As well as depending on the parameters ω_0 and γ, the behavior of the system depends on the initial conditions: the displacement and velocity of the oscillator at time $t = 0$. These determine the constants C_1 and C_2 in the general solution above. Here, we will take $z(0) = 1$ and $dz/dt(0) = 0$ and obtain the constants by solving the pair of linear equations that result from substituting these two conditions into the general solution equation:

```
In [x]: eqn1 = sp.Eq(soln.rhs.subs(t, 0), 1)        # z(0) = 1
In [x]: eqn2 = sp.Eq(soln.rhs.diff(t).subs(t, 0), 0)  # dz/dt(0) = 0
In [x]: consts, = sp.linsolve((eqn1, eqn2), ('C1', 'C2'))
In [x]: print(consts)
(-gamma/(2*sqrt(gamma**2 - 1)) + 1/2, gamma/(2*sqrt(gamma**2 - 1)) + 1/2)
```

The constants are returned in a tuple containing C_1 and C_2 in that order. To substitute these values, corresponding to our initial conditions, into our general solution, we can pass a dictionary to subs:

```
In [x]: C1, C2 = consts
In [x]: z_t_soln = soln.subs({'C1': C1, 'C2': C2})
In [x]: z_t_soln
Out[x]: Eq(z(t), (-gamma/(2*sqrt(gamma**2 - 1)) + 1/2)
                  *exp(t*w0*(-gamma - sqrt(gamma - 1)*sqrt(gamma + 1)))
              + (gamma/(2*sqrt(gamma**2 - 1)) + 1/2)
                  *exp(t*w0*(-gamma + sqrt(gamma - 1)*sqrt(gamma + 1))))
```

That is,

$$z(t) = \frac{1}{2}\left[1 - \frac{\gamma}{\sqrt{\gamma^2 - 1}}\right]\exp\left[-\omega_0 t\left(\gamma - \sqrt{\gamma^2 - 1}\right)\right]$$
$$+ \frac{1}{2}\left[1 + \frac{\gamma}{\sqrt{\gamma^2 - 1}}\right]\exp\left[-\omega_0 t\left(\gamma + \sqrt{\gamma^2 - 1}\right)\right].$$

It is easy to show that this expression reduces to the undamped harmonic oscillator solution, $z(t) = \cos(\omega_0 t)$, for $\gamma = 0$.

We can now substitute different values of w0 and gamma into the right-hand side of this Equation to give the full solution for a particular configuration of the oscillator. For example, $\omega_0 = 2\pi$, $\gamma = 0.1$:

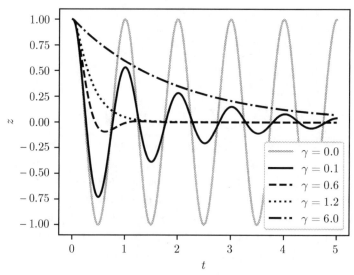

Figure 35.1   The motion of a damped oscillator with different damping ratios, γ.

```
In [x]: z_t_soln.rhs.subs({w0:2*np.pi, gamma:0.1})
Out[x]: (1/2 + 0.05025189*I) * exp(6.283185*t*(-0.1 - 0.994987*I))
      + (1/2 - 0.05025189*I) * exp(6.283185*t*(-0.1 + 0.994987*I))
```

(as expected, the function is complex for $\gamma < 1$: the system will oscillate).

If we want to plot this solution for a grid of time points, we need to turn it from a SymPy symbolic `Equation` into a function that can be called to return numerical values. This is what the `sp.lambdify` method does. In this case, we want the callable function to use NumPy methods, so it will be fast and vectorized when called on a NumPy array:

```
tt = np.linspace(0, 5, 1000)
z_t = sp.lambdify(t, z_t_sol.rhs.subs({w0:2*np.pi, gamma:0.1}), modules='numpy')
```

This function is plotted, with some other values for γ in Figure 35.1, which is produced by the following code

```
tt = np.linspace(0, 5, 1000)
for g in (0, 0.1, 0.6, 1.2, 6):
    z_t = sp.lambdify(t, z_t_sol.rhs.subs({w0:2*np.pi, gamma:g}), 'numpy')
    plt.plot(tt, z_t(tt).real, label=f'$\gamma = {g:.1f}$')
plt.xlabel('$t$')
plt.ylabel('$z$')
plt.show()
```

When $\gamma = 1$, the oscillator is said to be critically-damped: It returns to steady state at the fastest rate, without oscillating. Since the solution equation is singular for $\gamma = 1$, we cannot substitute this value directly into the solution:

```
In [x]: z_t_soln.rhs.subs(gamma, 1)
Out[x]: nan
```

However, SymPy does allow us to take the *limit* as $\gamma \to 1$:

```
In [x]: z_t_soln.rhs.limit(gamma, 1)
Out[x]: (t*w0 + 1)*exp(-t*w0)
```

That is, for critical damping, $z(t) = e^{-t\omega_0}(1 + \omega_0 t)$.

35.4 Example

E35.1

An Approximate Harmonic Oscillator Wavefunction

As described in Section 23.1, the one-dimensional harmonic oscillator Schrödinger equation:

$$\hat{H}\psi = -\frac{1}{2}\frac{d^2\psi}{dq^2} - \frac{1}{2}q\psi = E\psi$$

has an exact, analytical solution. In atomic units, the ground state wavefunction and energy are

$$\psi_0(q) = \pi^{-1/4}e^{-q^2/2}, \quad E_0 = \tfrac{1}{2}.$$

Suppose, however, we were to adopt the following variational wavefunction with a similar shape:

$$\psi = \frac{1}{a^2 + q^2},$$

and adjust its parameter, a, so as to minimize the expectation value of the energy,

$$\langle E \rangle = \frac{\langle \psi | \hat{H} | \psi \rangle}{\langle \psi | \psi \rangle}$$

What is the optimum value of a?

This problem can be solved by hand, but the calculus is a bit tedious. SymPy can handle the details for us:

```
import sympy as sp

# Define the principal symbols to use and the trial wavefunction.
q = sp.symbols('q', real=True)
a = sp.symbols('a', real=True, positive=True)
psi = 1 / (q**2 + a**2)
```

```
# Define the denominator in the energy expression, the normalization integral.
den = sp.integrate(psi**2, (q, -sp.oo, sp.oo))
den
```

$$\frac{\pi}{2a^3}$$

```
# Second derivative of the wavefunction.
d2psi_dq2 = sp.diff(psi, q, 2)
d2psi_dq2
```

$$\frac{2\left(\frac{4q^2}{a^2+q^2}-1\right)}{\left(a^2+q^2\right)^2}$$

```
# The function obtained by acting on psi with the Hamiltonian.
Hpsi = -d2psi_dq2 / 2 + q**2 * psi / 2
Hpsi
```

$$\frac{q^2}{2\left(a^2+q^2\right)}-\frac{\frac{4q^2}{a^2+q^2}-1}{\left(a^2+q^2\right)^2}$$

```
# The numerator, <psi | H | psi>.
num = sp.integrate(psi * Hpsi, (q, -sp.oo, sp.oo))
```

```
# The energy expectation.
E = num / den
E
```

$$\frac{2a^4+1}{4a^2}$$

To determine the minimum of E, solve $d\langle E\rangle/da = 0$.

```
# First find the derivative with respect to a.
dEda = sp.diff(E, a)
dEda
```

$$2a-\frac{2a^4+1}{2a^3}$$

```
# Next, solve for dE/da = 0
soln = sp.solve(dEda, a)
soln
```

```
[2**(3/4) / 2]
```

The optimum value of the variational parameter (suitably simplified) is therefore $a = 2^{-1/4}$.

```
E.subs(a, 2**(-1/4))
```

```
0.707106781186547
```

The variationally optimum energy with this wavefunction is therefore $1/\sqrt{2} \approx 0.707$, as compared with the true ground state energy, $E_0 = 0.5$.

E35.2

The Laplacian Operator in Spherical Polar Coordinates

The Laplacian operator,

$$\nabla^2 = \frac{\partial^2}{\partial x^2} + \frac{\partial^2}{\partial y^2} + \frac{\partial^2}{\partial z^2}$$

occurs frequently in quantum mechanics. It takes the above, simple form in Cartesian coordinates, (x, y, z), but is more usefully expressed in spherical polar coordinates, (r, θ, ϕ), for problems with spherical symmetry (e.g., determining the atomic orbitals).

Coordinate transformation is a fairly standard but lengthy exercise in differential calculus. The two coordinate systems are related to one another through the relations:

$$x = r \sin \theta \cos \phi, \tag{35.1}$$
$$y = r \sin \theta \sin \phi, \tag{35.2}$$
$$z = r \cos \theta, \tag{35.3}$$

where r is the radial distance from the origin, θ is the polar angle (from the z-axis), and ϕ is the azimuthal angle (the angle from the x-axis of the orthogonal projection of the position vector onto the xy plane).

We start with the differential operators $\partial/\partial x$, $\partial/\partial y$ and $\partial/\partial z$. By the chain rule, the relationship between these single partial derivatives in the two coordinate systems is given by:

$$\frac{\partial f}{\partial x} = \frac{\partial r}{\partial x}\frac{\partial f}{\partial r} + \frac{\partial \theta}{\partial x}\frac{\partial f}{\partial \theta} + \frac{\partial \phi}{\partial x}\frac{\partial f}{\partial \phi},$$

$$\frac{\partial f}{\partial y} = \frac{\partial r}{\partial y}\frac{\partial f}{\partial r} + \frac{\partial \theta}{\partial y}\frac{\partial f}{\partial \theta} + \frac{\partial \phi}{\partial y}\frac{\partial f}{\partial \phi},$$

$$\frac{\partial f}{\partial z} = \frac{\partial r}{\partial z}\frac{\partial f}{\partial r} + \frac{\partial \theta}{\partial z}\frac{\partial f}{\partial \theta} + \frac{\partial \phi}{\partial z}\frac{\partial f}{\partial \phi},$$

We therefore need the nine derivatives $\partial r/\partial x$, $\partial \theta/\partial x$, ... $\partial \phi/\partial z$.

The matrix of derivatives effecting a coordinate change like this is called the *Jacobian*, **J**. It is relatively easy to see from the above relations (Equations 35.1–35.3) that:

$$\mathbf{J} = \begin{pmatrix} \frac{\partial x}{\partial r} & \frac{\partial x}{\partial \theta} & \frac{\partial x}{\partial \phi} \\ \frac{\partial y}{\partial r} & \frac{\partial y}{\partial \theta} & \frac{\partial y}{\partial \phi} \\ \frac{\partial z}{\partial r} & \frac{\partial z}{\partial \theta} & \frac{\partial z}{\partial \phi} \end{pmatrix} = \begin{pmatrix} \sin\theta\cos\phi & r\cos\theta\cos\phi & -r\sin\theta\sin\phi \\ \sin\phi\sin\theta & r\cos\theta\sin\phi & r\sin\theta\cos\phi \\ \cos\theta & -r\sin\theta & 0 \end{pmatrix}.$$

SymPy can calculate the Jacobian from a collection of functions:

```
import sympy as sp

r, theta, phi = sp.symbols(r'r theta phi', negative=False, real=True)
x = r * sp.sin(theta) * sp.cos(phi)
y = r * sp.sin(theta) * sp.sin(phi)
z = r * sp.cos(theta)
```

```
J = sp.Matrix([x, y, z]).jacobian([r, theta, phi])
J
```

$$\begin{bmatrix} \sin(\theta)\cos(\phi) & r\cos(\phi)\cos(\theta) & -r\sin(\phi)\sin(\theta) \\ \sin(\phi)\sin(\theta) & r\sin(\phi)\cos(\theta) & r\sin(\theta)\cos(\phi) \\ \cos(\theta) & -r\sin(\theta) & 0 \end{bmatrix}$$

Incidentally, the determinant of this Jacobian matrix appears in the volume element in spherical polar coordinates, $dV = r^2 \sin\theta\, dr\, d\theta\, d\phi$:

```
sp.simplify(J.det())
```

$$r^2 \sin(\theta)$$

What we want, however, is the inverse relationship (derivatives of the spherical polar coordinates with respect to the Cartesian coordinates), which is given by the inverse of the matrix:

```
J_inv = sp.simplify(J.inv())
J_inv
```

$$
\begin{bmatrix}
\sin(\theta)\cos(\phi) & \sin(\phi)\sin(\theta) & \cos(\theta) \\
\dfrac{\cos(\phi)\cos(\theta)}{r} & \dfrac{\sin(\phi)\cos(\theta)}{r} & -\dfrac{\sin(\theta)}{r} \\
-\dfrac{\sin(\phi)}{r\sin(\theta)} & \dfrac{\cos(\phi)}{r\sin(\theta)} & 0
\end{bmatrix}
$$

That is,

$$
\boldsymbol{J}^{-1} = \begin{pmatrix}
\frac{\partial r}{\partial x} & \frac{\partial r}{\partial y} & \frac{\partial y}{\partial z} \\
\frac{\partial \theta}{\partial x} & \frac{\partial \theta}{\partial y} & \frac{\partial \phi}{\partial z} \\
\frac{\partial \phi}{\partial x} & \frac{\partial \phi}{\partial y} & \frac{\partial \phi}{\partial z}
\end{pmatrix} = \begin{pmatrix}
\sin\theta\cos\phi & \sin\theta\sin\phi & \cos\theta \\
\frac{\cos\theta\cos\phi}{r} & \frac{\cos\theta\sin\phi}{r} & -\frac{\sin\theta}{r} \\
-\frac{\sin\phi}{r\sin\theta} & \frac{\cos\phi}{r\sin\theta} & 0
\end{pmatrix}
$$

This enables us to write the chain rule expressions above in matrix form:

$$
\begin{pmatrix}
\frac{\partial f}{\partial x} \\
\frac{\partial f}{\partial y} \\
\frac{\partial f}{\partial z}
\end{pmatrix} = \left(\boldsymbol{J}^{-1}\right)^{T}
\begin{pmatrix}
\frac{\partial f}{\partial r} \\
\frac{\partial f}{\partial \theta} \\
\frac{\partial f}{\partial \phi}
\end{pmatrix}
$$

```
f = sp.Function('f')(r, theta, phi)
df_sphpol = sp.Matrix((f.diff(r), f.diff(theta), f.diff(phi)))
df_car = J_inv.T @ df_sphpol
df_car
```

$$
\begin{bmatrix}
\sin(\theta)\cos(\phi)\frac{\partial}{\partial r}f(r,\theta,\phi) - \dfrac{\sin(\phi)\frac{\partial}{\partial \phi}f(r,\theta,\phi)}{r\sin(\theta)} + \dfrac{\cos(\phi)\cos(\theta)\frac{\partial}{\partial \theta}f(r,\theta,\phi)}{r} \\
\sin(\phi)\sin(\theta)\frac{\partial}{\partial r}f(r,\theta,\phi) + \dfrac{\sin(\phi)\cos(\theta)\frac{\partial}{\partial \theta}f(r,\theta,\phi)}{r} + \dfrac{\cos(\phi)\frac{\partial}{\partial \phi}f(r,\theta,\phi)}{r\sin(\theta)} \\
\cos(\theta)\frac{\partial}{\partial r}f(r,\theta,\phi) - \dfrac{\sin(\theta)\frac{\partial}{\partial \theta}f(r,\theta,\phi)}{r}
\end{bmatrix}
$$

The function generated by the Laplacian acting on f is the sum of the second derivatives of f with respect to the Cartesian coordinates. We can therefore just repeat the above procedure, using df_car instead of f:

$$
\frac{\partial^2 f}{\partial x^2} = \frac{\partial r}{\partial x}\left[\frac{\partial}{\partial r}\frac{\partial f}{\partial x}\right] + \frac{\partial \theta}{\partial x}\left[\frac{\partial}{\partial \theta}\frac{\partial f}{\partial x}\right] + \frac{\partial \phi}{\partial x}\left[\frac{\partial}{\partial \phi}\frac{\partial f}{\partial x}\right],
$$

$$
\frac{\partial^2 f}{\partial y^2} = \frac{\partial r}{\partial y}\left[\frac{\partial}{\partial r}\frac{\partial f}{\partial y}\right] + \frac{\partial \theta}{\partial y}\left[\frac{\partial}{\partial \theta}\frac{\partial f}{\partial y}\right] + \frac{\partial \phi}{\partial y}\left[\frac{\partial}{\partial \phi}\frac{\partial f}{\partial y}\right],
$$

$$
\frac{\partial^2 f}{\partial z^2} = \frac{\partial r}{\partial z}\left[\frac{\partial}{\partial r}\frac{\partial f}{\partial z}\right] + \frac{\partial \theta}{\partial z}\left[\frac{\partial}{\partial \theta}\frac{\partial f}{\partial z}\right] + \frac{\partial \phi}{\partial z}\left[\frac{\partial}{\partial \phi}\frac{\partial f}{\partial z}\right].
$$

```
# 3 x 3 matrix of the derivatives in square brackets above.
d2f = sp.Matrix((df_car.diff(r).T, df_car.diff(theta).T, df_car.diff(phi).T,))
del2f = 0
for i in range(3):
    for j in range(3):
        del2f += J_inv[i, j] * d2f_car[i, j]
del2 = sp.simplify(del2)
del2
```

$$\frac{r^2 \frac{\partial^2}{\partial r^2} f(r, \theta, \phi) + 2r \frac{\partial}{\partial r} f(r, \theta, \phi) + \frac{\partial^2}{\partial \theta^2} f(r, \theta, \phi) + \frac{\frac{\partial}{\partial \theta} f(r,\theta,\phi)}{\tan(\theta)} + \frac{\frac{\partial^2}{\partial \phi^2} f(r,\theta,\phi)}{\sin^2(\theta)}}{r^2}$$

The Laplacian operator in the spherical polar coordinate system is therefore

$$\nabla^2 = \frac{\partial^2}{\partial r^2} + \frac{2}{r} \frac{\partial}{\partial r} + \frac{1}{r^2} \frac{\partial^2}{\partial \theta^2} + \frac{1}{r^2 \tan \theta} \frac{\partial}{\partial \theta} + \frac{1}{r^2 \sin^2 \theta} \frac{\partial^2}{\partial \phi^2}.$$

35.5 Exercises

P35.1

(a) Find the partial fraction decomposition of the expression

$$\frac{x^3 - 1}{x^2(x - 2)^3}$$

(b) Find the partial fraction decomposition, (as far as is possible without using complex numbers) of the expression

$$\frac{x^3 + 14}{x^3 - 3x^2 + 8x}$$

(c) Evaluate the definite integral

$$\int_{-\infty}^{2} \frac{1}{(4 - x)^2} \, dx$$

(d) Evaluate the integral (and simplify the result)

$$\int_{-\infty}^{\infty} e^{-ax^2} e^{-b(x-c)^2} \, dx$$

(e) Determine the value of the sum

$$\sum_{n=1}^{\infty} \frac{1}{n^2}$$

P35.2   Determine the normalization constant, N, of the particle-in-a-box wave-functions for the one-dimensional potential which is 0 for $0 < x < L$ and ∞ elsewhere,

$$\psi(x) = N \sin\left(\frac{n\pi x}{L}\right),$$

where the quantum number, $n = 1, 2, 3, \ldots$.

The expectation values of the position and momentum of the particle, $\langle x \rangle = \frac{L}{2}$ and $\langle p_x \rangle = 0$ by symmetry. Determine expressions for $\langle x^2 \rangle$ and $\langle p_x^2 \rangle$ and confirm that the wavefunctions satisfy the Heisenberg Uncertainty Principle, $\Delta x \Delta p_x \geq \frac{\hbar}{2}$.

Hints: The position and linear momentum operators are $\hat{x} = x \cdot$ and $\hat{p}_x = -i\hbar \frac{d}{dx}$, respectively; the standard deviation in an observable A is $\Delta A = \sqrt{\langle A^2 \rangle - \langle A \rangle^2}$.

36

Molecular Orbital Theory for H_2^+

The molecular ion, H_2^+, is one of the simplest molecular systems but, as a three-body problem (two nuclei, one electron), it is still a challenge to model analytically. One standard approximate approach is to write the molecular wavefunction as a linear combination of the hydrogen atoms' 1s atomic orbitals:

$$\Psi = c_a \psi_a + c_b \psi_b.$$

The expected energy of this wavefunction, $\langle E \rangle$, can be optimized variationally (the coefficients c_a and c_b chosen to minimize $\langle E \rangle$) for a *fixed bond length*, R (the Born-Oppenheimer approximation). Once $\langle E \rangle_{min}$ is known as a function of R, the equilibrium bond length can be estimated as that which minimizes $\langle E \rangle_{min}(R)$.

The hydrogen atoms are identical, so we will look for solutions satisfying $|c_a| = |c_b|$, that is:

$$\Psi_\pm = c_\pm (\psi_a \pm \psi_b)$$

where the normalized hydrogen 1s orbital wavefunctions are

$$\psi_a = \frac{1}{\sqrt{\pi}} e^{-r_a}, \qquad \psi_b = \frac{1}{\sqrt{\pi}} e^{-r_b}$$

(we will work in Hartree atomic units). If the molecular wavefunctions, $\Psi_\pm$ are to be normalized themselves, we must have:

$$\langle \Psi_\pm | \Psi_\pm \rangle = |c_\pm|^2 [\langle \psi_a | \psi_a \rangle + \langle \psi_b | \psi_b \rangle \pm \langle \psi_a | \psi_b \rangle \pm \langle \psi_b | \psi_a \rangle] = 1.$$

Now, $\langle \psi_a | \psi_a \rangle = \langle \psi_b | \psi_b \rangle = 1$ because the 1s orbitals are normalized, and $S = \langle \psi_a | \psi_b \rangle = \langle \psi_b | \psi_a \rangle$ because they are real. S is called the overlap integral and depends on the separation of the atomic centres (i.e., the bond length). Therefore,

$$|c_\pm|^2 [2 \pm 2S] = 1 \quad \Rightarrow \quad c_\pm = \frac{1}{\sqrt{2(1 \pm S)}}.$$

In simpler treatments, S is approximated to zero and the two molecular orbitals written as $\frac{1}{\sqrt{2}}(\psi_a \pm \psi_b)$. In this chapter, however, we will calculate this integral (and others like it) with the help of SymPy.

The expectation value of the energy is, for the normalized wavefunctions,

$$E_\pm = \langle \Psi_\pm | \hat{H} | \Psi_\pm \rangle,$$

where, in atomic units, the Hamiltonian operator for the molecular ion is

$$\hat{H} = -\frac{1}{2}\nabla^2 - \frac{1}{r_a} - \frac{1}{r_b} + \frac{1}{R}.$$

The first term here represents the electron's kinetic energy, the second and third its potential energy due to its attractive interactions with nuclei a and b and the final term is due to the repulsion of the nuclei (separated by the bond length, R). Using this operator in the above expression yields

$$\frac{1}{|c_\pm|^2}E_\pm = \langle \psi_a | -\tfrac{1}{2}\nabla^2 - \tfrac{1}{r_a} - \tfrac{1}{r_b} + \tfrac{1}{R} | \psi_a \rangle$$

$$+ \langle \psi_b | -\tfrac{1}{2}\nabla^2 - \tfrac{1}{r_b} - \tfrac{1}{r_b} + \tfrac{1}{R} | \psi_b \rangle$$

$$\pm \langle \psi_a | -\tfrac{1}{2}\nabla^2 - \tfrac{1}{r_b} - \tfrac{1}{r_b} + \tfrac{1}{R} | \psi_b \rangle$$

$$\pm \langle \psi_b | -\tfrac{1}{2}\nabla^2 - \tfrac{1}{r_b} - \tfrac{1}{r_b} + \tfrac{1}{R} | \psi_a \rangle$$

Now,

$$\left(-\frac{1}{2}\nabla^2 - \frac{1}{r_a} \right)\psi_a = E_{1s}\psi_a,$$

(and similarly for ψ_b) where E_{1s} is the energy of an isolated hydrogen atom. Also,

$$\langle \psi_a | \frac{1}{R} | \psi_a \rangle = \langle \psi_b | \frac{1}{R} | \psi_b \rangle = \frac{1}{R}$$

and

$$\langle \psi_a | \frac{1}{R} | \psi_b \rangle = \langle \psi_b | \frac{1}{R} | \psi_a \rangle = \frac{S}{R}.$$

There are two other types of integral,

$$\langle \psi_a | \frac{1}{r_b} | \psi_a \rangle = \langle \psi_b | \frac{1}{r_a} | \psi_b \rangle$$

and

$$\langle \psi_a | \frac{1}{r_b} | \psi_b \rangle = \langle \psi_b | \frac{1}{r_a} | \psi_a \rangle.$$

We will define two quantities: the *Coulomb Integral*,

$$J = -\langle \psi_a | \frac{1}{r_b} | \psi_a \rangle + \frac{1}{R}$$

and the *Exchange Integral*,

$$K = -\langle \psi_a | \frac{1}{r_b} | \psi_b \rangle + \frac{S}{R}$$

With these definitions in place, the energy expression is

$$\frac{1}{|c_\pm|^2} E_\pm = 2E_{1s} \pm 2SE_{1s} + 2J \pm 2K$$

and therefore (using the above definition of $c_\pm$)

$$E_\pm = E_{1s} + \frac{J \pm K}{1 \pm S}.$$

We will now examine the integrals S, J and K. They cannot be evaluated in carte-sian or spherical polar coordinates (the distances r_a and r_b are measured from dif-ferent centers). Instead, the problem can be transformed in to two-center *elliptical coordinates*:

The two-centre elliptical coordinates are

$$\lambda = \frac{r_a + r_b}{R}, \quad \mu = \frac{r_a - r_b}{R}, \quad \text{and } \phi$$

where $1 \leq \lambda \leq \infty$, $-1 \leq \mu \leq 1$ and ϕ is the azimuthal angle around the internuclear axis, $0 \leq \phi \leq 2\pi$. The H_2^+ ion is cylindrically symmetrical, so ϕ does not appear explicitly in the wavefunctions. The inverse transformation is

$$r_a = \frac{R}{2}(\lambda + \mu), \quad r_b = \frac{R}{2}(\lambda - \mu)$$

In this coordinate system, the volume element is

$$d\tau = \frac{R^3}{8} \left(\lambda^2 - \mu^2 \right) d\lambda \, d\mu \, d\phi$$

and the atomic orbital wavefunctions are

$$\psi_a = \frac{1}{\sqrt{\pi}} \exp\left[-\frac{R}{2}(\lambda + \mu) \right] \quad \text{and} \quad \psi_b = \frac{1}{\sqrt{\pi}} \exp\left[-\frac{R}{2}(\lambda - \mu) \right]$$

The overlap integral in these coordinates is

$$S = \langle \psi_a | \psi_b \rangle = \int_\tau \psi_a \psi_b \, d\tau$$

$$= \frac{1}{\pi} \int_0^{2\pi} \int_{-1}^1 \int_1^\infty \exp\left[-\frac{R}{2}(\lambda + \mu)\right] \exp\left[-\frac{R}{2}(\lambda - \mu)\right] \frac{R^3}{8} (\lambda^2 - \mu^2) \, d\lambda \, d\mu \, d\phi$$

$$= \frac{R^3}{4} \int_{-1}^1 \int_1^\infty e^{-R\lambda} (\lambda^2 - \mu^2) \, d\lambda \, d\mu.$$

This can be evaluated using standard techniques (including integration by parts) but is a tedious exercise with little physical insight to show for it. We will therefore use SymPy.

```
import sympy as sp

# Internuclear separation (atomic units).
R = sp.symbols('R', real=True, positive=True, nonzero=True)
# Two-centre elliptical coordinates lambda and mu
lam = sp.symbols('lambda', real=True, negative=False)
mu = sp.symbols('mu', real=True)

S = R**3 / 4 * sp.integrate(sp.exp(-R*lam) * (lam**2 - mu**2),
                            (lam, 1, sp.oo), (mu, -1, 1))
S = sp.simplify(S)
S
```

$$\left(\frac{R^2}{3} + R + 1\right) e^{-R}$$

Similarly, the Coulomb Integral,

$$J = -\frac{1}{\pi} \int_0^{2\pi} \int_{-1}^1 \int_1^\infty \frac{\exp\left[-R(\lambda + \mu)\right] R^3}{\frac{R}{2}(\lambda - \mu)} \frac{R^3}{8} (\lambda^2 - \mu^2) \, d\lambda \, d\mu \, d\phi + \frac{1}{R}$$

$$= -\frac{R^2}{2} \int_{-1}^1 \int_1^\infty \exp\left[-R(\lambda + \mu)\right] (\lambda + \mu) \, d\lambda \, d\mu + \frac{1}{R}.$$

```
J = -R**2 / 2 * sp.integrate(sp.exp(-R*(lam+mu)) / (lam - mu) * (lam**2 - mu**2),
        (lam, 1, sp.oo), (mu, -1, 1)) + 1/R
J = sp.simplify(J)
J
```

$$\frac{(R + 1) e^{-2R}}{R}$$

and the Exchange Integral,

$$
K = -\frac{1}{\pi} \int_0^{2\pi} \int_{-1}^{1} \int_{1}^{\infty} \frac{\exp\left[-\frac{R}{2}(\lambda+\mu)\right]\exp\left[-\frac{R}{2}(\lambda-\mu)\right]}{\frac{R}{2}(\lambda-\mu)}
$$
$$
\times \frac{R^3}{8}\left(\lambda^2-\mu^2\right)\, \mathrm{d}\lambda\, \mathrm{d}\mu\, \mathrm{d}\phi + \frac{S}{R}
$$
$$
= -\frac{R^2}{2} \int_{-1}^{1} \int_{1}^{\infty} e^{-R\lambda}(\lambda+\mu)\, \mathrm{d}\lambda\, \mathrm{d}\mu + \frac{S}{R}
$$

```
Kp = -R**2 / 2 * sp.integrate(sp.exp(-R*lam) * (lam + mu),
       (lam, 1, sp.oo), (mu, -1, 1))
Kp = sp.simplify(Kp)
K = Kp + S/R
K
```

$$
-(R+1)e^{-R} + \frac{\left(\frac{R^2}{3}+R+1\right)e^{-R}}{R}
$$

The three integrals are therefore

$$
S = \left(1+R+\frac{R^2}{3}\right)e^{-R},
$$
$$
J = \left(1+\frac{1}{R}\right)e^{-2R},
$$
$$
K = \frac{S}{R} - (1+R)e^{-R}.
$$

These expressions can then be used to plot a potential energy curve (Figure 36.1) for the bonding and antibonding states of H_2^+, taking the energy of the isolated hydrogen atoms to be $E_{1s} = 0$.

```
Ep, Em = (J + K) / (1 + S), (J - K) / (1 - S)
```

```
import numpy as np
import matplotlib.pyplot as plt
Rgrid = np.linspace(1, 8, 1000)
Ep_lam = sp.lambdify(R, Ep, modules='numpy')
Em_lam = sp.lambdify(R, Em, modules='numpy')
Epgrid = Ep_lam(Rgrid)
```

```
plt.plot(Rgrid, Epgrid)
plt.plot(Rgrid, Em_lam(Rgrid))
plt.ylim(-0.075, 0.1)
plt.xlabel(r'$R / a_0$')
plt.ylabel(r'$E / E_\mathrm{h}$')
```

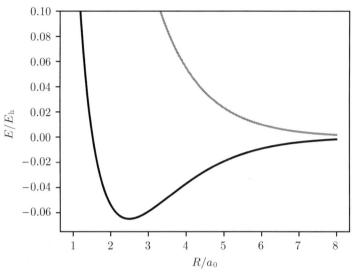

Figure 36.1   Potential energy curves for the bonding (black) and antibonding (gray) states of H_2^+, approximated by the LCAO-MO approach.

We can use SciPy's minimize function to estimate the equilibrium bond length from the minimum of the bonding potential well:

```
from scipy.optimize import minimize
ret = minimize(Ep_lam, 2)          # Initial guess, bond length, R = 2 a_0
ret
```

```
        fun: -0.06483099190652034
   hess_inv: array([[15.3215901]])
        jac: array([-7.63777643e-06])
    message: 'Optimization terminated successfully.'
       nfev: 14
        nit: 6
       njev: 7
     status: 0
    success: True
          x: array([2.49270885])
```

```
# The bond length is currently in atomic units.
bond_length = ret['x'][0]

# Convert to SI units by multiplying by the Bohr radius, a0.
import scipy.constants
a0 = scipy.constants.value('Bohr radius')
bond_length_pm = bond_length * a0 * 1.e12

print(f'Estimated H2+ bond length: {bond_length_pm:.0f} pm')
```

```
Estimated H2+ bond length: 132 pm
```

This compares fairly well with the experimental value of 106 pm.

```
# For the dissociation energy, we would like to convert from Hartree (Eh) to eV.
from scipy.constants import hbar, c, alpha, e
Eh_in_eV = hbar * c * alpha / a0 / e

De = -Ep_lam(bond_length)
De *= Eh_in_eV
print(f'Estimated H2+ dissociation energy: {De:.1f} eV')
```

```
Estimated H2+ dissociation energy: 1.8 eV
```

This compares less well with the experimental value of 2.8 eV.

To plot the molecular wavefunctions, we need to convert the SymPy expressions into NumPy arrays on a suitable grid of spatial coordinates. Adopting a Cartesian coordinate system with nucleus *a* at the origin and the *z*-axis along the internuclear axis, the distances of the electron from the two atoms in the $y = 0$ plane are

$$r_a = \sqrt{x^2 + z^2}, \quad r_b = \sqrt{r_a^2 + R^2 - 2Rz}$$

The code below plots the bonding (Figure 36.2) and antibonding (Figure 36.3) molecular orbitals as Matplotlib surface plots.

```
# We would like a convenient variable name for the numerical value
# of the internuclear distance, but don't want to replace the SymPy
# symbol, R, so use R_.
R_ = bond_length
# Create a grid of points in the xz plane.
x = np.linspace(-4, 4, 80)
z = np.linspace(-2.5, 5, 80)
X, Z = np.meshgrid(x, z)
# Calculate the corresponding distances from the origin (at atom a).
r_a = np.sqrt(X**2 + Z**2)
r_b = np.sqrt(r_a**2 + R_**2 - 2*R_*Z)
```

```
# The coefficients in the LCAO-MO expressions, Psi_+ and Psi_-.
c_p, c_m = 1 / sp.sqrt(2 * (1 + S)), 1 / sp.sqrt(2 * (1 - S))
# SymPy expressions for the basis AOs and the LCAO-MOs.
ra, rb = sp.symbols('r_a r_b', real=True, positive=True)
psia = 1/sp.sqrt(sp.pi) * sp.exp(-ra)
psib = 1/sp.sqrt(sp.pi) * sp.exp(-rb)
Psi_p, Psi_m = c_p * (psia + psib), c_m * (psia - psib)
```

```
# Turn the SymPy expressions into functions.
Psi_p_lam = sp.lambdify((R, ra, rb), Psi_p)
Psi_m_lam = sp.lambdify((R, ra, rb), Psi_m)
# Calculate the wavefunctions on our grid of electron distances.
```

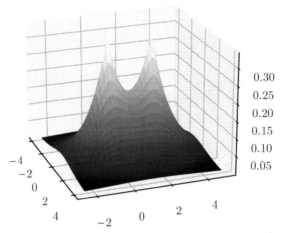

Figure 36.2   The bonding molecular orbital wavefunction of H$_2^+$, approximated by the LCAO-MO approach.

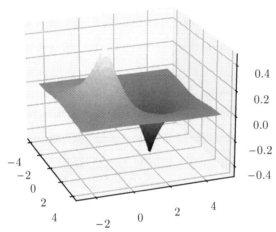

Figure 36.3   The anti-bonding molecular orbital wavefunction of H$_2^+$, approximated by the LCAO-MO approach.

```
Psi_p_ = Psi_p_lam(R_, r_a, r_b)
Psi_m_ = Psi_m_lam(R_, r_a, r_b)
```

```
# Plot the bonding MO as a 3D surface.
from mpl_toolkits.mplot3d import Axes3D
fig = plt.figure()
ax = fig.add_subplot(111, projection='3d')
ax.plot_surface(X, Z, Psi_p_)
# Fix the view (elevation, azimuth) to a suitable orientation.
ax.view_init(20, -20)
```

```
# Plot the anti-bonding MO as a 3D surface.
fig = plt.figure()
ax = fig.add_subplot(111, projection='3d')
ax.plot_surface(X, Z, Psi_m_)
ax.view_init(20, -20)
```

37

Approximations of the Helium Atom Electronic Energy

37.1 Theory

The Hamiltonian operator for the energy of a helium-like atom is, in Hartree atomic units:

$$\hat{H} = -\frac{1}{2}\nabla_1^2 - \frac{1}{2}\nabla_2^2 - \frac{Z}{r_1} - \frac{Z}{r_2} + \frac{1}{r_{12}},$$

where the five terms represent, respectively, the kinetic energy of electron 1, the kinetic energy of electron 2, the potential energy between electron 1 and the nucleus ($Z = 2$ for He), the potential energy between electron 2 and the nucleus and the electron–electron repulsion energy.

Even assuming the nucleus is infinitely heavy, as we have done here, the resulting Schrödinger equation, $\hat{H}\Psi = E\Psi$, is a second-order differential equation in six spatial coordinates (the positions, in three-dimensional space, of the two electrons).

There is no exact, analytical solution to this equation because of the electron-repulsion term. This involves the positions of both electrons and prevents any nice factorization of the wavefunction that might make the differential equation separable:

$$r_{12} = \sqrt{(x_1 - x_2)^2 + (y_1 - y_2)^2 + (z_1 - z_2)^2}.$$

Even in spherical polar coordinates,

$$r_{12} = \sqrt{r_1^2 + r_2^2 - 2r_1 r_2 \cos\theta},$$

where θ is the angle between the electron position vectors r_1 and r_2, the problem is not tractable and approximate methods must be used.

Experimentally, the electronic energy of He is found to be $E_0 = -2.90372$ hartrees, so at least we have something to aim for.

37.2 Attempt 1: Ignore Electron–Electron Repulsion

As a first attempt at solving this problem, we might think to simply ignore the troublesome electron repulsion term:

$$\hat{H}' = -\frac{1}{2}\nabla_1^2 - \frac{1}{2}\nabla_2^2 - \frac{Z}{r_1} - \frac{Z}{r_2}.$$

In this case, the Schrödinger is fully separable, since we can factorize the wavefunction:

$$\Psi'(r_1, r_2) = \psi_1(r_1)\psi_2(r_2)$$

(assuming this spatial wavefunction can be coupled with an antisymmetric spin wavefunction), and obtain two independent hydrogen-like systems:

$$-\frac{1}{2}\nabla_1^2\psi_1 - \frac{Z}{r_1}\psi_1 = E\psi_1 \text{ and } -\frac{1}{2}\nabla_2^2\psi_2 - \frac{Z}{r_2}\psi_2 = E\psi_2.$$

This system *can* be solved exactly, and the resulting functions are the familiar hydrogen-like atomic orbitals. The ground state is therefore given by assigning each electron to a 1s orbital, $\psi_{1s}(r) = (Z^3/\pi)^{1/2}e^{-Zr}$:

$$\Psi'(r_1, r_2) = \frac{Z^3}{\pi}e^{-Zr_1}e^{-Zr_2} = \frac{Z^3}{\pi}e^{-Z(r_1+r_2)}.$$

Each 1s orbital has an energy $-Z^2/2$ so the total electronic energy of He, ignoring electron repulsion, is predicted to be:

```
Z = 2
2 * (-Z**2 / 2)
```

```
-4.0
```

That is, -4 hartrees, which does not compare well with the experimental value of -2.90372 hartrees: clearly, a large error is introduced by neglecting electron repulsion.

```
E0, Ebad = -2.90372, -4
print(f'Error in estimate for non-interacting electrons: '
      f'{(E0 - Ebad)/E0 * 100:.2f} %')
```

```
Error in estimate for non-interacting electrons: -37.75 %
```

37.3 Attempt 2: One-Parameter Variational Approach

The variational theorem assures us that any wavefunction we guess is guaranteed not have an expectation value for its energy, $\tilde{E} = \langle E \rangle$, less than the true ground state energy, E_0, and a sensible first guess wavefunction would be one of the form

$$\tilde{\Psi}(r_1, r_2; \alpha) = \frac{\alpha^3}{\pi} e^{-\alpha(r_1+r_2)} = \tilde{\psi}(r_1)\tilde{\psi}(r_2),$$

where $\tilde{\psi}(r_i) = \sqrt{\alpha^3/\pi}\, e^{-\alpha r_i}$ and α is an adjustable parameter whose optimal value we will seek by minimizing $\langle E \rangle = \langle \tilde{\Psi} | \hat{H} | \tilde{\Psi} \rangle$. To obtain an expression for $\langle E \rangle$, three different integrals are needed:

$$\langle -\tfrac{1}{2}\nabla_1^2 \rangle = \langle -\tfrac{1}{2}\nabla_2^2 \rangle,$$
$$\langle -Z/r_1 \rangle = \langle -Z/r_2 \rangle,$$

and $\langle 1/r_{12} \rangle$.

The first two integrals are relatively straightforward. The first,

$$\langle -\tfrac{1}{2}\nabla_1^2 \rangle = -\frac{1}{2}\int_\tau \tilde{\Psi}^* \nabla_1^2 \tilde{\Psi}\, d\tau$$

$$= -\frac{1}{2}\left(\int_{\tau_2} \tilde{\psi}(r_2)^* \tilde{\psi}(r_2)\, d\tau_2\right)\int_{\tau_1} \tilde{\psi}(r_1)^* \nabla_1^2 \tilde{\psi}(r_1)\, d\tau_1$$

$$= -\frac{1}{2}\int_{\tau_1} \tilde{\psi}(r_1)^* \nabla_1^2 \tilde{\psi}(r_1)\, d\tau_1$$

$$= -\frac{1}{2}\int_0^\infty \tilde{\psi}(r_1)^* \nabla_1^2 \tilde{\psi}(r_1)\, 4\pi r_1^2\, dr_1$$

where we have used (in the second line) the fact that the ∇_1 operator acts only on the coordinates r_1 and in the third line the fact that $\tilde{\psi}(r_2)$ is normalized. Lastly, the integral over the three-dimensional spatial coordinates of the electron is reduced to a single integral over r_1 since the $\tilde{\psi}(r_1)$ is spherically symmetric and hence $d\tau_1 = 4\pi r_1^2\, dr_1$. We need

$$\nabla_1^2 \tilde{\psi}(r_1) = \left[\frac{\partial^2}{\partial r_1^2} + \frac{2}{r_1}\frac{\partial}{\partial r_1}\right]\tilde{\psi}(r_1).$$

The calculus here is not hard, but SymPy can do it for us:

```
import sympy as sp
r1 = sp.symbols('r1', positive=True)
Z, alpha = sp.symbols('Z alpha', positive=True, nonzero=True)

psi1 = sp.sqrt(alpha**3 / sp.pi) * sp.exp(-alpha*r1)

# Check the wavefunction is normalized.
4 * sp.pi * sp.integrate(psi1 * psi1 * r1**2, (r1, 0, sp.oo))
```

1

```
D2_psi1 = psi1.diff(r1, r1) + 2/r1 * psi1.diff(r1)
```

```
4 * sp.pi * sp.integrate(-psi1 / 2 * D2_psi1 * r1**2, (r1, 0, sp.oo))
```

$$\frac{\alpha^2}{2}$$

The second of the necessary integrals also reduces to one dimension under the spherical symmetry of the central potential:

$$\langle -Z/r_1 \rangle = \int_0^\infty \tilde{\psi}(r_1)^* \left(-\frac{Z}{r_1} \right) \tilde{\psi}(r_1)\, 4\pi\, r_1^2\, dr_1$$

```
4 * sp.pi * sp.integrate(psi1 * psi1 * -Z / r1 * r1**2, (r1, 0, sp.oo))
```

$$-Z\alpha$$

Unfortunately, this is where our luck runs out. It would be nice to evaluate the remaining, electron repulsion integral symbolically over three coordinates (r_1, r_2 and θ):

```
r2 = sp.symbols('r2', positive=True)
theta = sp.symbols('theta', positive=True)
psi2 = sp.sqrt(alpha**3 / sp.pi) * sp.exp(-alpha*r2)
Psi = (psi1 * psi2).simplify()
8 * sp.pi**2 * sp.integrate(
    Psi * Psi / sp.sqrt(r1**2 + r2**2 - 2*r1*r2*sp.cos(theta)),
    (r1, 0, sp.oo), (r2, 0, sp.oo), (theta, 0, sp.pi)
    )
```

$$8\alpha^6 \int_0^\pi \int_0^\infty e^{-2\alpha r_2} \int_0^\infty \frac{e^{-2\alpha r_1}}{\sqrt{r_1^2 - 2r_1 r_2 \cos(\theta) + r_2^2}}\, dr_1\, dr_2\, d\theta$$

SymPy returns a modified version of the unevaluated integral, indicating that it couldn't make any progress with it.

Nonetheless, the integral does have an analytical solution:[1]

$$\langle 1/r_{12} \rangle = \frac{5}{8}\alpha$$

---

[1] A derivation of this solution is provided on the website accompanying this book. See E21.1(c) for the numerical evaluation of this integral.

Adding these integrals together gives

$$\tilde{E} = \langle E \rangle = \alpha^2 - 2Z\alpha + \frac{5}{8}\alpha$$

The value of α which minimizes $\tilde{E}$ is obtained by solving

$$\frac{d\tilde{E}}{d\alpha} = 2\alpha - \left(2Z - \frac{5}{8}\right) = 0,$$

that is, $\alpha = Z - \frac{5}{16}$. If α is interpreted as an effective nuclear charge, this expression indicates that each electron shields the other from the nucleus by an amount equal to $\frac{5}{16}$ of an electron charge.

The estimated energy itself is obtained by substituting this value of α into the above expression for $\tilde{E}$:

```python
def calc_variational_E():
    """Calculate the best variational estimate of the He energy."""
    Z = 2
    alpha = Z - 5/16
    Ebest = alpha**2 - (2*Z*alpha - 5 * alpha / 8)
    return Ebest
Ebest_var = calc_variational_E()
print(Ebest_var)
```

```
-2.84765625
```

This approximate energy, $\tilde{E} = -2.84766$ hartree, is much closer to the true value of -2.90372 hartree.

```python
E0 = -2.90372
print(f'Error in variational estimate: {(E0 - Ebest_var)/E0 * 100:.2f} %')
```

```
Error in variational estimate: 1.93 %
```

37.4 Attempt 3: Hartree–Fock with a Minimal Basis

In the Hartree–Fock method, the approximate wavefunction is described by a suitably antisymmetrized product of one-electron wavefunctions, usually considered to be constructed from a Slater determinant, in which the spatial parts of these wavefunctions are linear combinations of some basis functions.

For the case of the helium ground-state, we can start the analysis from a simple two-electron wavefunction:

$$\Psi = \psi(r_1)\psi(r_2)\frac{1}{\sqrt{2}}(\alpha_1\beta_2 - \beta_1\alpha_2),$$

where $\frac{1}{\sqrt{2}}(\alpha_1\beta_2 - \beta_1\alpha_2)$ is the (antisymmetric) singlet spin function.

The simplest possible expansion series for $\psi(r_i)$ is:

$$\psi(r_i) = c_1\phi_1(r_i) + c_2\phi_2(r_i),$$

where the basis functions are the hydrogenic 1s and 2s orbitals:

$$\phi_1(r_i) = \sqrt{\frac{Z^3}{\pi}}\exp(-Zr_i)$$

$$\phi_2(r_i) = \sqrt{\frac{Z^3}{8\pi}}\left(1 - \frac{Zr_i}{2}\right)\exp(-Zr_i/2).$$

which have energies $\epsilon_1 = -Z^2/2$ and $\epsilon_2 = -Z^2/8$.

We need to use at least two basis functions because of the requirement for Ψ to be normalized as we vary c_1 and c_2: This minimal basis actually leaves us with only one degree of freedom with respect to which we will minimize the expectation value of the energy, $\langle E \rangle = \langle\Psi|H|\Psi\rangle$.

To proceed with this simplified Hartree–Fock approach, we need an expression for $\langle\Psi|H|\Psi\rangle$. First, we will collect together the terms of the Hamiltonian operator as follows:

$$\hat{H} = \hat{h}_1 + \hat{h}_2 + \frac{1}{r_{12}}, \quad \text{where } \hat{h}_i = -\frac{1}{2}\nabla_i^2 - \frac{Z}{r_i}.$$

The one-electron parts are straightforward, since Ψ is an eigenfunction of the relevant operators, and the basis is orthonormal:

$$\langle\Psi|\hat{h}_1|\Psi\rangle = 2c_1^2\epsilon_1 = 2c_1^2\left(-\frac{Z^2}{2}\right)$$

$$\langle\Psi|\hat{h}_2|\Psi\rangle = 2c_2^2\epsilon_2 = 2c_2^2\left(-\frac{Z^2}{8}\right)$$

The electron repulsion part is more complicated and might be expected to consist of 16 terms arising from the expansion of each side of the expression

$$\langle\psi(r_1)\psi(r_2)|\frac{1}{r_{12}}|\psi(r_1)\psi(r_2)\rangle = \langle[c_1\phi_1(r_1) + c_2\phi_2(r_1)][c_1\phi_1(r_2) + c_2\phi_2(r_2)]|$$

$$\frac{1}{r_{12}}|[c_1\phi_1(r_1) + c_2\phi_2(r_1)][c_1\phi_1(r_2) + c_2\phi_2(r_2)]\rangle.$$

There is a convenient notation in which to express the terms that derive from this formula. We write:

$$\langle pq|rs \rangle = \langle\phi_p(r_1)\phi_q(r_2)|\frac{1}{r_{12}}|\phi_r(r_1)\phi_s(r_2)\rangle$$

This version of the notation is the *physicists' convention*: The electrons are identified by their order in each side of the expression (electron 1 is associated with ϕ_p
and ϕ_r, electron 2 with ϕ_q and ϕ_s and the $1/r_{12}$ operator is implied but not explicitly
written).

With this in place, one can write:

$$\langle \psi(r_1)\psi(r_2)|\frac{1}{r_{12}}|\psi(r_1)\psi(r_2)\rangle = c_1^4\langle 11|11\rangle + 4c_1^3 c_2\langle 11|12\rangle$$

$$+ 4c_1^2 c_2^2\langle 11|22\rangle + 2c_1^2 c_2^2\langle 12|12\rangle$$

$$+ 4c_1 c_2^3\langle 12|22\rangle + c_2^4\langle 22|22\rangle.$$

There are 16 possible integrals of the form $\langle pq|rs\rangle$ for p, q, r, s each independently
equal to 1 or 2. However, several are equal to each other: Electrons are identical,
so one can permute the electron indices without changing the value of the integral
(e.g., $\langle 12|11\rangle = \langle 21|11\rangle$). Also, since the basis functions are real, the left- and right-
hand sides can be swapped: $\langle 12|11\rangle = \langle 11|12\rangle$ and so on.

The integrals themselves are somewhat tricky, but only need to be evaluated
once. Their values are given in the code below.

```
# Define expressions for the 1s and 2s orbital energies.
epsilon1, epsilon2 = -Z**2 / 2, -Z**2 / 8
# Define dictionary of the necessary electron repulsion integrals.
I = {
    '<11|11>': 5 * Z / 8,
    '<11|12>': 4096 * sp.sqrt(2) * Z / 64827,
    '<11|22>': 16 * Z / 729,
    '<12|12>': 17 * Z / 81,
    '<12|22>': 512 * sp.sqrt(2) * Z / 84375,
    '<22|22>': 77 * Z / 512
}
```

```
# Create a SymPy expression for the energy expectation value, <Psi|H|Psi>.
c1, c2 = sp.symbols('c1 c2', real=True)
Eexpr = (2 * epsilon1 * c1**2 + 2 * epsilon2 * c2**2
         + c1**4 * I['<11|11>'] + 4*c1**3*c2 * I['<11|12>']
         + 2 * (c1*c2)**2 * (2 * I['<11|22>'] + I['<12|12>'])
         + 4 * c1*c2**3 * I['<12|22>'] + c2**4 * I['<22|22>'])

# Turn it into a function that can be used to produce numerical results with NumPy.
E = sp.lambdify((c1, c2), Eexpr.evalf(subs={Z: 2}), 'numpy')
```

With a function to evaluate $\tilde{E} = \langle \Psi|H|\Psi\rangle$ in place, it can be visualized as a func-
tion of the expansion coefficients c_1 and c_2. As a function of these two coordinates, $\tilde{E}$ can be considered as a surface. However, as noted previously, because the
wavefunction must be normalized, we are only interested in the points on this surface that also lie on the circle $c_1^2 + c_2^2 = 1$.

```python
import numpy as np
import matplotlib.pyplot as plt

# Create a mesh of (c1, c2) points, each coordinate between cmin and cmax.
cmin, cmax = -1.2, 1.2
c1, c2 = np.linspace(cmin, cmax, 30), np.linspace(cmin, cmax, 30)
C1, C2 = np.meshgrid(c1, c2)

# Calculate the value of <E> for every point on this mesh.
Elandscape = E(C1, C2)
```

```python
# Create a Matplotlib Figure with a 3D projection.
fig = plt.figure()
ax = fig.add_subplot(projection='3d')
# Plot the energy landscape, <E>(c1, c2) as a wireframe surface.
ax.plot_wireframe(C1, C2, Elandscape, color='k', lw=1)

# Also plot a red cicle indicating the values of <E> we are interested
# in searching for the minimum amongst.
alpha = np.linspace(0, 2*np.pi, 100)
c1, c2 = np.cos(alpha), np.sin(alpha)
ax.plot(c1, c2, E(c1, c2), c='r', lw=2)
ax.set_xlabel(r'$c_1$')
ax.set_ylabel(r'$c_2$')
ax.set_zlabel(r'$E\,/\mathrm{hartree}$')
ax.view_init(45, 130)
```

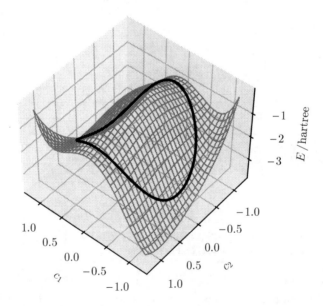

SciPy's minimize method can be used to find the necessary minimum. As described in Section 34.3, this method has an argument, constraints, that allows the user to specify conditions that the fitted parameters must meet. In this case, the constraint itself, $c_1^2 + c_2^2 = 1$, can be expressed as a function that returns 0 when this equality is met: $f_{con}(c_1, c_2) = c_1^2 + c_2^2 - 1$. It will be passed the parameters as a tuple, which we will call p.

```python
from scipy.optimize import minimize

def con(p):
    c1, c2 = p
    return c1**2 + c2**2 - 1

# The constraint itself must be packed into a dictionary that defines its
# type (equality) and the function that evaluates it.
cons = {'type':'eq', 'fun': con}

# The function, E, expects the parameters to be passed to it as separate
# arguments, so unpack the tuple p with the *p notation; the initial
# guess for the parameters c1 and c0 is (1,0).
res = minimize((lambda p: E(*p)), (1,0), constraints=cons)
res
```

```
       fun: -2.823635223051534
       jac: array([-3.45336533,  0.68267366])
   message: 'Optimization terminated successfully'
      nfev: 23
       nit: 7
      njev: 7
    status: 0
   success: True
         x: array([ 0.98101524, -0.19393065])
```

```python
(c1_best, c2_best), Ebest_HF2 = res['x'], res['fun']
print(f'Optimum parameters: c1 = {c1_best:.3f}, c2 = {c2_best:.3f}')
print(f'Optimized Hartree-Fock energy, E = {Ebest_HF2:.5f} hartree')
print(f'Error in Hartree-Fock estimate: {(E0 - Ebest_HF2)/E0 * 100:.2f} %')
```

```
Optimum parameters: c1 = 0.981, c2 = -0.194
Optimized Hartree-Fock energy, E = -2.82364 hartree
Error in Hartree-Fock estimate: 2.76 %
```

Unfortunately, this minimal Hartree–Fock approximation is slightly worse than the one-parameter variational approach taken in attempt 2.

The normalization constraint can also be expressed by writing the parameters c_1 and c_2 in terms of an angle, α:

$$(c_1, c_2) = (\cos \alpha, \sin \alpha),$$

(since $\cos^2 \alpha + \sin^2 \alpha = 1$), reducing the problem to a one-dimensional one.

```python
def E_1d(alpha):
    c1, c2 = np.cos(alpha), np.sin(alpha)
    return E(c1, c2)
```

```
fig = plt.figure()
ax = fig.add_subplot()
alpha = np.linspace(-np.pi/2, np.pi/2, 100)
Ecurve = E_1d(alpha)
ax.plot(alpha, Ecurve)
ax.set_xlabel(r'$\alpha$')
ax.set_ylabel(r'$E\,/\mathrm{hartree}$')
```

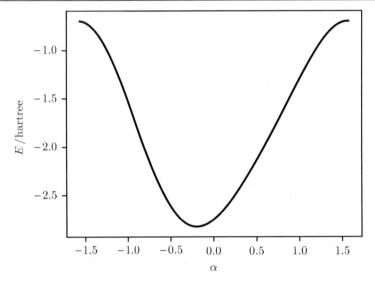

```
res = minimize(E_1d, 0)
res
```

```
       fun: -2.8236352230143815
 hess_inv: array([[0.23542635]])
       jac: array([3.57627869e-07])
  message: 'Optimization terminated successfully.'
      nfev: 12
       nit: 4
      njev: 6
    status: 0
  success: True
         x: array([-0.19516717])
```

This gives the same result in fewer function evaluations and leads to the same parameters:

```
alpha_best = res['x'][0]
Ebest_HF2 = res['fun']
c1_best, c2_best = np.cos(alpha_best), np.sin(alpha_best)
print(f'Optimum parameters: c1 = {c1_best:.3f}, c2 = {c2_best:.3f}')
print(f'Optimized Hartree-Fock energy, E = {Ebest_HF2:.5f} hartree')
```

```
Optimum parameters: c1 = 0.981, c2 = -0.194
Optimized Hartree-Fock energy, E = -2.82364 hartree
```

A more formal and general approach to the Hartree–Fock method (i.e., one that can be more easily used with much larger basis set expansions) starts with the method of *Lagrange multipliers*: the task of minimizing a set of equations $f(c_i)$ with respect to the n parameters c_i whilst simultaneously demanding that some function $g(c_i)$ is a constant is the same as finding the solution to the equations

$$\frac{\partial}{\partial c_i}\left[f(c_i)+\lambda g(c_i)\right]=0,$$

for a constant Lagrange multiplier, λ.

It turns out that for $f(c_i)=\langle\Psi|H|\Psi\rangle$ and $g(c_i)=\langle\psi|\psi\rangle=1$, these equations can always be written in matrix form as the *Hartree–Fock equation*:

$$F(c)c=\epsilon c,$$

where ϵ is a constant related to λ, the column vector c holds the parameters c_i and the elements of the *Fock matrix*, $F(c)$, depend on the derivatives of $f(c_i)$ with respect to c_i. For example, in the two-parameter case, we have been considering so far, the elements of the Fock matrix are:

$$F_{11}(c)=\epsilon_1+\langle 11|11\rangle c_1^2+2\langle 11|12\rangle c_1 c_2+\langle 12|12\rangle c_2^2$$
$$F_{12}(c)=\langle 11|12\rangle c_1^2+2\langle 11|22\rangle c_1 c_2+\langle 12|22\rangle c_2^2$$
$$F_{12}(c)=F_{21}(c)$$
$$F_{22}(c)=\epsilon_2+\langle 12|12\rangle c_1^2+2\langle 12|22\rangle c_1 c_2+\langle 22|22\rangle c_2^2$$

Since the Fock matrix depends on the solution vector c itself, the Hartree–Fock equation cannot be simply inverted to yield the coefficients: This is a nonlinear problem which is usually solved with an iterative approach, as follows.

First, make an initial guess of the solution eigenvector (as for the above mini-mization algorithm), say $c^{(0)}=(1, 0)$. Use this to calculate the elements of the Fock Matrix, $F^{(1)}=F(c^{(0)})$, and form the equation:

$$F^{(1)}c^{(1)}=\epsilon^{(1)}c^{(1)},$$

This can be solved with the usual methods of linear algebra: The goal is to find the eigenvectors and eigenvalues, of $F^{(1)}$, select the smallest eigenvalue and use its corresponding eigenvector as $c^{(1)}$; we then repeat the process with $F^{(2)}=F(c^{(1)})$ and so on, in the hope that the process will converge on the optimum coefficients c which solve the nonlinear Hartree–Fock equation.

```
# From now on, we will deal with numerical values instead of SymPy Symbols.
eps1 = float(epsilon1.evalf(subs={Z: 2}))
```

```
eps2 = float(epsilon2.evalf(subs={Z: 2}))
Irep = dict((k, float(v.evalf(subs={Z: 2}))) for k, v in I.items())
```

```
def Fock(c):
    """Calculate the Fock matrix with coefficient vector, c."""
    c1, c2 = c
    F11 = (eps1 + Irep['<11|11>']*c1**2 + 2* Irep['<11|12>']*c1*c2
              + Irep['<12|12>']*c2**2)
    F12 = Irep['<11|12>']*c1**2 + 2*Irep['<11|22>']*c1*c2 + Irep['<12|22>']*c2**2
    F22 = (eps2 + Irep['<12|12>']*c1**2 + 2*Irep['<12|22>']*c1*c2
              + Irep['<22|22>']*c2**2)
    return np.array(((F11, F12), (F12, F22)))
```

```
def solve_HF(maxit=20, tol=1.e-5):
    """Solve the Hartree-Fock equation iteratively.

    Returns the optimum energy, coefficients, convergence flag and number
    of iterations performed. If the procedure does not converge to a tolerance
    of tol within maxit iterations, the first two of these will be None.

    """

    # Initial guess of coefficient array.
    c = np.array((1,0))
    converged, nit, epsilon_last = False, 0, None
    while not converged and nit < maxit:
        # Set up the Fock matrix and find its eigenvalues and eigenvectors.
        F = Fock(c)
        vals, vecs = np.linalg.eigh(F)
        # Get the coefficient vector, c, corresponding to the smallest eigenvalue.
        i = np.argmin(vals)
        epsilon, c = vals[i], vecs[:,i]
        # Have we converged on a value for epsilon?
        if nit and abs((epsilon - epsilon_last)/epsilon_last) < tol:
            converged = True
        epsilon_last = epsilon
        nit +=1

    if converged:
        return E(*c), c, True, nit

    # No convergence.
    return None, None, False, nit

Ebest_HF2, c, converged, nit = solve_HF()
if converged:
    print(f'Convergence achieved in {nit} iterations.')
    print(f'Optimum parameters: c1 = {c1_best:.3f}, c2 = {c2_best:.3f}')
    print(f'Optimized Hartree-Fock energy, E = {Ebest_HF2:.5f} hartree')
    print(f'Error in Hartree-Fock estimate: {(E0 - Ebest_HF2)/E0 * 100:.2f} %')
else:
    print(f'Failure to converge after {nit} iterations.')
```

```
Convergence achieved in 10 iterations.
Optimum parameters: c1 = 0.981, c2 = -0.194
Optimized Hartree-Fock energy, E = -2.82364 hartree
Error in Hartree-Fock estimate: 2.76 %
```

The Hartree–Fock approach for a basis of N functions involves calculating a larger set of integrals $\langle pq|rs \rangle$. These integrals only need to be evaluated once, however, and can be stored for use in constructing the corresponding Fock matrix on each iteration.

38

Computational Chemistry with Psi4 and Python

Psi4 is a free and open source computational chemistry package for calculating the energies, geometries, vibrational frequencies, and other properties of molecular systems. It is written in C++ but comes with a Python API enabling these calculations to be carried out using a Python script or Jupyter Notebook.

The field of computational chemistry is huge and Psi4 is a mature application with a great deal of functionality; this chapter will present some examples of its use on simple systems to demonstrate the Python Application Programming Interface (API) – for more details the reader is directed to the online manual.[1]

38.1 Installing Psi4

There is a guide to installing Psi4 on different platforms on the project's website (https://www.psicode.org/). It can also be installed as part of the Anaconda package (see https://anaconda.org/psi4/psi4). A full installation includes various numerical libraries and may take up a fair amount of disk space. For the simple molecular systems considered in this chapter, however, not a great deal of computing power is necessary.

38.2 Examples

E38.1

Optimizing the Energy and Geometry of Ammonia

Modern computers have no problem performing quite accurate calculations of the structure and energetics of simple systems, such as small molecules consisting of light (e.g., first-row) atoms. This example will consider the ammonia molecule, NH_3.

---

[1] https://psicode.org/psi4manual/master/index.html.

The Python Psi4 API is imported into the working environment like any other module. Here, we will also import NumPy and Matplotlib for use later.

```
import psi4
import numpy as np
import matplotlib.pyplot as plt
```

To specify the molecule, an initial guess for its atoms' positions can be provided in Cartesian coordinates (`.xyz` format – see Section 15.1.2) or as a "Z-matrix" of bonds and angles. Here, we use the former and choose to place the nitrogen atom at the origin. By default, distances are measured in angstroms.

Pretending to know a little (but not much) about the likely structure of ammonia, we will set up the three hydrogen atoms 1 Å along each of the x-, y- and z-axes.

```
# Cartesian (xyz) format for the atom locations.
NH3 = psi4.geometry("""
N 0 0 0
H 1 0 0
H 0 1 0
H 0 0 1
""")
```

Alternatively, the geometry may be specified in Z-matrix notation, which locates the atoms through distances and angles. One form this could take for our guess at the ammonia structure is in the four-line string below:

```
# An alternative is to use a Z-matrix notation:
NH3_Z = psi4.geometry("""
N
H 1 1.0
H 1 1.0 2 90
H 1 1.0 2 90 3 90
""")
```

Each line of this string corresponds to the location of an atom:

- First, place a nitrogen atom (atom 1);
- Locate a hydrogen (atom 2) a distance 1.0 Å from it;
- Atom 3 is another hydrogen atom, also 1.0 Å from atom 1 (the N), and at an angle 90 degrees from the initial N–H direction;
- Finally, atom 4 is also a hydrogen atom, also 90 degrees from the initial N–H direction, and assigned a dihedral angle of 90 degrees: the plane defined by the location of the atoms 2-1-4 intersects that defined by the location of atoms 2-1-3 at 90 degrees.

If our goal were to use the structure of ammonia in some further calculation rather than to optimize it, we could simply read in the geometry from the PubChem online database.

```
NH3_pubchem = psi4.geometry('pubchem:ammonia')
```

```
Searching PubChem database for ammonia (single best match returned)
Found 1 result(s)
```

(We will use this later to assess how well our own geometry optimization went.) Using our initial guess for the ammonia structure, we can inspect the geometry

```
# NH3.geometry() returns a Psi4 Matrix object: cast it into a regular NumPy array.
arr = NH3.geometry().to_array()
arr
```

```
array([[-0.15818758,  0.11185551,  0.        ],
       [ 1.17805058,  0.11185551, -1.33623816],
       [ 1.17805058,  0.11185551,  1.33623816],
       [-0.15818758, -1.77787062,  0.        ]])
```

Psi4 has performed some transformations to put our ammonia molecule into what it considers to be a standard orientation and converted the distances to atomic units, but the initial-guess bond lengths remain the same at this stage. One way to verify this is to calculate the distances between the bonded atoms. Considering each row of the atomic positions array, arr, as a vector, this can be achieved by calculating the difference between pairs of position vectors and determining their magnitude with linalg.norm. We need to set the argument axis=1 here to return the magnitudes along each column of the difference matrix, for each row:

```
# The atomic unit of distance is the bohr, about 0.53 A:
bohr2angstroms = 0.52917720859

def get_bond_lengths(arr):
    """Return the lengths of the three N-H bonds in Angstroms."""
    # Calculate the bond lengths as the distances between the
    # nitrogen atom (arr[0]) and the hydrogen atoms (arr[1:]).
    return np.linalg.norm(arr[1:] - arr[0], axis=1) * bohr2angstroms

get_bond_lengths(arr)
```

```
array([1., 1., 1.])      # all bond lengths 1 angstrom, as defined earlier
```

To calculate the energy of the molecule (in this geometry, which we understand is distorted from equilibrium), we need to pick a method and a basis set. Psi4 provides a large range of computational methods and basis sets; here, we will use a few of them.

Starting with a basic Hartree–Fock self-consistent field (HF-SCF) calculation using a minimal STO-3G basis (see Example E34.2), calling psi4.energy will produce a fair amount of output, including the part we are interested in:

```
E = psi4.energy('scf/STO-3g', molecule=NH3)
```

```
...

@DF-RHF Final Energy:    -55.42943171301930

   => Energetics <=

     Nuclear Repulsion Energy =           12.2352758064124174
     One-Electron Energy =               -99.3203090257318024
     Two-Electron Energy =                31.6556015063000871
     Total Energy =                      -55.4294317130192979

Computation Completed

...
```

Now, we will optimize the geometry using the same method and basis. The function call to psi4.optimize again produces a lot of informative output; the most important parts are shown below.

```
E = psi4.optimize('scf/STO-3g', molecule=NH3)
```

```
...


Writing optimization data to binary file.
     Final energy is      -55.4555640683953
     Final (previous) structure:
     Cartesian Geometry (in Angstrom)
          N      -0.0617550981     0.0436674486     0.0000000000
          H       0.5575118462     0.1817099879    -0.8144885754
          H       0.5575118462     0.1817099879     0.8144885754
          H      -0.2569767292    -0.9701508019     0.0000000000

 -55.45556406839534

     Saving final (previous) structure.
     Cleaning optimization helper files.
     ------------------------
          OPTKING Finished Execution
     ------------------------


     Final optimized geometry and variables:
     Molecular point group: cs
     Full point group: C3v

     Geometry (in Angstrom), charge = 0, multiplicity = 1:

     N          -0.061755098302     0.043667448780     0.000000000000
     H           0.557511848378     0.181709988575    -0.814488578604
     H           0.557511848378     0.181709988575     0.814488578604
     H          -0.256976730255    -0.970150805711     0.000000000000
```

The optimization has reduced the energy from -55.429 to -55.456 hartree by shifting the atoms' relative positions:

```
arr_opt = NH3.geometry().to_array()
arr_opt
```

```
array([[-0.11670022,  0.08251952,  0.        ],
       [ 1.05354471,  0.34338211, -1.53916035],
       [ 1.05354471,  0.34338211,  1.53916035],
       [-0.48561564, -1.83331932,  0.        ]])
```

```
# Print the three bond lengths in the optimized geometry.
get_bond_lengths(arr_opt)
```

```
array([1.03244319, 1.03244319, 1.03244319])
```

It is helpful to compare this with the reference structure retrieved previously from PubChem:

```
arr_pubchem = NH3_pubchem.geometry().to_array()
get_bond_lengths(arr_pubchem)
```

```
array([1.01900071, 1.01902404, 1.01900415])
```

This indicates that the scf/STO-3g optimized bond lengths are a little too long (1.032 Å compared with 1.019 Å), but given the simplicity of the basis, we didn't do too badly. We can also compare the H–N–H bond angles:

```
def get_bond_angle(arr):
    """Return the H-N-H bond angle, in degrees."""
    NH1 = arr[1] - arr[0]
    NH2 = arr[2] - arr[0]
    rNH1 = np.linalg.norm(NH1)
    rNH2 = np.linalg.norm(NH2)
    # Calculate the angle between vectors a and b from a.b = |a||b|cos(theta)
    return np.degrees(np.arccos(NH1 @ NH2 / rNH1 / rNH2))

print(f'Initialised H-N-H bond angle: {get_bond_angle(arr):.2f} deg')
print(f'scf/STO-3g H-N-H bond angle: {get_bond_angle(arr_opt):.2f} deg')
print(f'Reference H-N-H bond angle: {get_bond_angle(arr_pubchem):.2f} deg')
```

```
Initialised H-N-H bond angle: 90.00 deg
scf/STO-3g H-N-H bond angle: 104.16 deg
Reference H-N-H bond angle: 106.00 deg
```

Bond lengths, angles and rotational constants are such commonly required quantities, that there are methods dedicated to outputting them:

```
NH3.print_bond_angles()
```

```
Removing binary optimization data file.
Cleaning optimization helper files.
    Bond Angles (degrees)

    Angle 2-1-3:  104.165
    Angle 2-1-4:  104.165
    Angle 3-1-4:  104.165
    Angle 1-2-3:   37.918
    Angle 1-2-4:   37.918
    Angle 3-2-4:   60.000
    Angle 1-3-2:   37.918
    Angle 1-3-4:   37.918
    Angle 2-3-4:   60.000
    Angle 1-4-2:   37.918
    Angle 1-4-3:   37.918
    Angle 2-4-3:   60.000
```

```
NH3.print_distances()
```

```
    Interatomic Distances (Angstroms)

    Distance 1 to 2 1.032
    Distance 1 to 3 1.032
    Distance 1 to 4 1.032
    Distance 2 to 3 1.629
    Distance 2 to 4 1.629
    Distance 3 to 4 1.629
```

```
NH3.print_rotational_constants()
```

```
Removing binary optimization data file.
Cleaning optimization helper files.
Rotational constants: A =      9.42676  B =      9.42676  C =      6.30349 [cm^-1]
Rotational constants: A = 282607.23955  B = 282607.23954  C = 188974.00596 [MHz]
```

Finally, for this example, we will compare the calculated energies using different basis sets.[2] For each optimization, the wavefunction computed can be returned along with the energy by setting the argument `return_wfn=True` in the call to `psi4.optimize`. This wavefunction object stores a reference to the basis set used and the number of functions in that basis set, which can be retrieved by calling `wfn.basisset().nbf()`.

---

[2] A description of the various basis sets used in computational chemistry is beyond the scope of this book.

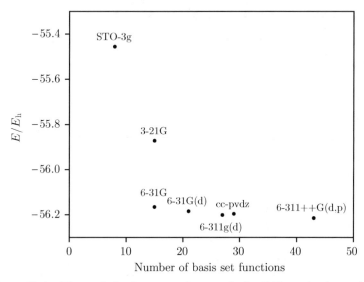

Figure 38.1   The optimized energy of ammonia for different basis set sizes.

```
basis_sets = ['STO-3g', '3-21G', '6-31G', '6-31G(d)', '6-311g(d)',
              'cc-pvdz', '6-311++G(d,p)']
```

```
# The calculated energy and number of basis functions used for each set.
E, nbf = {}, {}
# The N-H bond lengths and H-N-H bond angles.
r, theta = {}, {}
for basis in basis_sets:
    # Cartesian (xyz) format for the atom locations.
    NH3 = psi4.geometry("""
    N 0 0 0
    H 1.1 0 0
    H 0 1.1 0
    H 0 0 1.1
    """)
    E[basis], wfn = psi4.optimize(f'scf/{basis}', molecule=NH3, return_wfn=True)
    arr = NH3.geometry().to_array()
    r[basis] = get_bond_lengths(arr)[0]
    theta[basis] = get_bond_angle(arr)
    nbf[basis] = wfn.basisset().nbf()
```

```
plt.scatter(nbf.values(), E.values())
plt.ylabel(r'$E /E_\mathrm{h}$')
plt.xlabel(r'Number of basis set functions')
for label, x, y in zip(basis_sets, nbf.values(), E.values()):
    plt.annotate(label, (x, y), ha='center', va='bottom')
```

The resulting plot is shown in Figure 38.1.

E38.2

Vibrational Modes of Ethyne

In this example, we will use Psi4 to analyze the vibrations of the ethyne molecule, C_2H_2. First, some imports:

```
import psi4
import numpy as np
import pandas as pd
import matplotlib.pyplot as plt
```

Now, guess and optimize a geometry for the molecule, using some suitable method and basis.

```
# Cartesian (xyz) format for the initial-guess atom locations.
C2H2 = psi4.geometry("""
C -0.6 0 0
C  0.6 0 0
H -1.5 0 0
H  1.5 0 0
""")

# Optimize the geometry.
E = psi4.optimize('scf/cc-pvdz', molecule=C2H2)
```

```
print(f'Energy = {E} Eh')
print(f'Point group:', C2H2.get_full_point_group())
C2H2.print_distances()
C2H2.print_bond_angles()
C2H2.print_rotational_constants()
```

```
Energy = -76.82593102771146 Eh
Point group: D_inf_h
        Interatomic Distances (Angstroms)

        Distance 1 to 2 1.192
        Distance 1 to 3 1.064
        Distance 1 to 4 2.255
        Distance 2 to 3 2.255
        Distance 2 to 4 1.064
        Distance 3 to 4 3.319


        Bond Angles (degrees)

        Angle 2-1-3:  180.000
        Angle 2-1-4:    0.000
        Angle 3-1-4:  180.000
        Angle 1-2-3:    0.000
        Angle 1-2-4:  180.000
        Angle 3-2-4:  180.000
        Angle 1-3-2:    0.000
```

```
        Angle 1-3-4:    0.000
        Angle 2-3-4:    0.000
        Angle 1-4-2:    0.000
        Angle 1-4-3:    0.000
        Angle 2-4-3:    0.000



Rotational constants: A = ********  B =       1.19785  C =       1.19785 [cm^-1]
Rotational constants: A = ********  B =   35910.57745  C =   35910.57745 [MHz]
```

This looks about right: The values given by the useful NIST Computational Chemistry Comparison and Benchmark DataBase are $r_{CC} = 1.203$ Å and $r_{CH} = 1.063$ Å. C_2H_2 is linear, so the rotational constant A is not defined and $B = C$.

The function psi4.frequencies computes harmonic vibrational frequencies (and other molecular properties). A lot of information is output, and can be retrieved from the wavefunction by inspecting the dictionary wfn.frequency_analysis after the calculation has completed. To visualize the normal modes, we can also instruct Psi4 to write a file in MOLDEN format that can be read in later to produce an interactive image.

```
# We want to visualize the normal modes later, so set the normal_modes_write
# flag to True to write a file in MOLDEN format.
psi4.set_options({"normal_modes_write": True})
# The file will have the suffix extension .molden_normal_modes; here set filename
# prefix.
psi4.set_options({"writer_file_label":"C2H2"})

E, wfn = psi4.frequencies('scf/cc-pvdz', molecule=C2H2, return_wfn=True)
```

```
vibinfo = wfn.frequency_analysis
```

The vibinfo dictionary returned contains a lot of information about the frequency analysis, including the following, identified by string keys:

- 'TRV': type of motion: 'TR' for translations and rotations, 'V' for vibrations;
- 'gamma': the irreducible representation ("irrep") defining the vibrational symmetry;
- 'omega': the vibrational frequency as a wavenumber in cm^{-1};
- 'IR_intensity': the infrared intensity of the vibration, in $km\ mol^{-1}$;
- 'degeneracy': the degree of degeneracy of the motion.

For example:

```
vibinfo['TRV'], vibinfo['gamma'], vibinfo['omega']
```

```
(Datum(numeric=False, label='translation/rotation/vibration', units='',
     data=['TR', 'TR', 'TR', 'TR', 'TR', 'V', 'V', 'V', 'V', 'V', 'V', 'V'],
```

```
          comment='', doi=None, glossary=''),
 Datum(numeric=False, label='irreducible representation', units='',
          data=[None, None, None, None, None, 'B2g', 'B1g', 'B1u', 'B2u', 'Ag',
                'B3u', 'Ag'], comment='', doi=None, glossary=''),
 Datum(numeric=True, label='frequency', units='cm^-1',
          data=array([0.00000000e+00+4.02168548e-05j, 0.00000000e+00+2.40321176e-05j,
                      5.66713514e-06+0.00000000e+00j, 1.09034203e-05+0.00000000e+00j,
                      1.15021755e-05+0.00000000e+00j, 7.81951585e+02+0.00000000e+00j,
                      7.81951586e+02+0.00000000e+00j, 8.65090961e+02+0.00000000e+00j,
                      8.65090961e+02+0.00000000e+00j, 2.22468544e+03+0.00000000e+00j,
                      3.57868554e+03+0.00000000e+00j, 3.69109164e+03+0.00000000e+00j]),
          comment='', doi=None, glossary=''))
```

There are a few things to note here. First, the irreducible representations returned are those for the D_{2h} point group instead of the $D_{\infty h}$: This turns out to be the best way to carry out the computation to avoid numerical problems in the kinetic energy operator associated with the "missing" rotational degree of freedom (about the internuclear axis). We can correlate them with the correct $D_{\infty h}$ irreps later (see Table 38.2).

Second, the frequency wavenumbers are returned as complex numbers: Since we are close to the equilibrium geometry, however, the imaginary parts are small or zero, and so we will cast these to real values. (In contrast, configuration of atoms at a reaction transition state would correspond to a saddle point on the potential energy surface and lead to imaginary components to the frequency through an effectively negative force constant: Instead of being pulled back to an equilibrium structure, they are being pushed to a new configuration in the product molecules.)

Finally, as expected for a linear molecule, there are five translational and rotational motions; since we're not that interested in these, they can be filtered out with a method called psi4.driver.qcdb.vib.filter_nonvib:

```
vibonly = psi4.driver.qcdb.vib.filter_nonvib(vibinfo)
```

It will be convenient to pack the relevant dictionary data into a pandas DataFrame:

```
import pandas as pd
d = {k: vibonly[k].data for k in ('gamma', 'degeneracy', 'omega', 'IR_intensity')}
vibs_df = pd.DataFrame(d)
# Convert the frequencies into real dtypes.
vibs_df['omega'] = vibs_df['omega'].values.real
vibs_df
```

	gamma	degeneracy	omega	IR_intensity
0	B2g	2	781.951585	1.621856e-15
1	B1g	2	781.951586	1.053937e-15
2	B1u	2	865.090961	9.504694e+01
3	B2u	2	865.090961	9.504694e+01
4	Ag	1	2224.685440	2.388628e-17
5	B3u	1	3578.685545	1.042737e+02
6	Ag	1	3691.091645	8.240972e-17

Table 38.1   Calculated and benchmark values for the normal mode vibrational frequencies of ethyne, C_2H_2

Symmetry	Calculated /cm^{-1}	Benchmark /cm^{-1}
Π_g	782	612
Π_u	865	730
Σ_g^+	2,225	1,974
Σ_u^+	3,579	3,289
Σ_g^+	3,691	3,374

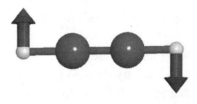

2: Normal mode (782.0 cm^-1}

Select:
1: Normal mode (782.0 cm^-1)
2: Normal mode (782.0 cm^-1)
3: Normal mode (865.1 cm^-1)
4: Normal mode (865.1 cm^-1)
5: Normal mode (2224.7 cm^-1)

Figure 38.2   The interactive widget provided by `fortecubeview` to visualize the normal modes of molecules.

The IR-inactive modes (of gerade-symmetry) have very low intensities, as expected. In a centrosymmetric molecule like ethyne, these will be Raman-active. The D_{2h} irreps correlate with those of $D_{\infty h}$ in the table given below (for the axes orientation used in our calculation), which compares our frequencies with their benchmark values (Table 38.1).

The `fortecubeview` package can be used to visualize the normal modes by reading in the `.molden_normal_modes` file created earlier. This file will contain some identifying number, so inspect the Notebook directory to find out what it is called. In a Jupyter Notebook, a drop-down menu appears from which the mode of interest can be selected (Figure 38.2).

```
import fortecubeview
# Inspecting the directory, this is the filename of our MOLDEN normal modes file.
fortecubeview.vib('C2H2.40259.molden_normal_modes')
```

E38.3

Visualizing the Molecular Orbitals and Energies of Water

Psi4 can calculate molecular orbital energies and wavefunctions which can be visualized using a suitable viewer. In this example, we will optimize the geometry of the water molecule, output its orbital energies, and use the `fortecubeview` library to visualize the HOMO and LUMO.

```
import psi4
import numpy as np
import matplotlib.pyplot as plt
import fortecubeview
```

```
# Z-matrix format for the initial-guess atom locations.
H20 = psi4.geometry("""
O
H 1 1.0
H 1 1.0 2 100
""")

# Optimize the geometry.
E, wfn = psi4.optimize('scf/cc-pvdz', molecule=H20, return_wfn=True)
```

```
...

Final optimized geometry and variables:
    Molecular point group: c2v
    Full point group: C2v

    Geometry (in Angstrom), charge = 0, multiplicity = 1:

    O
    H           1    0.946220
    H           1    0.946220        2   104.629204
```

`fortecubeview` can be used to view the geometry of the molecule (Figure 38.3)

```
fortecubeview.geom(molecule=H20)
```

```
# Print the calculated energies (in hartree).
wfn.print_energies()
```

```
# Print information about the molecular orbitals.
wfn.print_orbitals()
```

H_2O has 10 electrons, so the lowest five molecular orbitals are doubly occupied. The orbitals are labeled by their symmetry irrep preceded by a running index number per irrep.

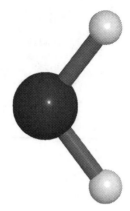

Figure 38.3   The optimized geometry of water.

Figure 38.4   The highest-occupied molecular orbital (HOMO) of water.

To visualize the orbitals, we need to output a set of cube files describing the electron density and value of the wavefunction as a function of three-dimensional space. Here, we will instruct Psi4 to save them to a directory called H2O-cubes (which must exist!) We don't particularly want to calculate these files for *all* the orbitals, so identify them below using the option 'CUBEPROP_ORBITALS'.

```
psi4.set_options({"writer_file_label":"H2O"})
psi4.set_options({
    'CUBEPROP_TASKS': ['ORBITALS', 'DENSITY'],
    'CUBEPROP_ORBITALS': [5, 6],          # HOMO, LUMO
    'CUBEPROP_FILEPATH': 'H2O-cubes'
})
psi4.cubeprop(wfn)
```

The fortecubeview.plot function can be used to render a three-dimensional image depicting the orbital wave function and electron density of a molecule. It will examine a provided file path for cube files, and plot isosurfaces at a provided relative threshold level (sumlevel, which defaults to 0.85). Here, this is done for the highest-occupied and lowest-unoccupied molecular orbitals (Figures 38.4 and 38.5).

MO 6a (4-A1)

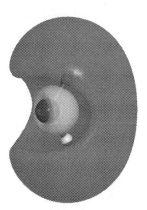

Figure 38.5   The lowest-unoccupied molecular orbital (LUMO) of water.

```
# Possible color schemes are 'emory', 'national', 'bright', 'electron', 'wow'.
# Set the opacity to some value less than 1 to "see through" the isolevel contours.
fortecubeview.plot(path='H2O-cubes', width=500, height=500, colorscheme='emory',
                   sumlevel=0.7, opacity=0.7)
```

38.3 Exercise

P38.1   Use Psi4 to optimize the bond length for the molecules N_2 and F_2. Store the wavefunction object created as wfn and use it to plot a molecular orbital energy level diagram. Are the orderings of the energy levels as expected? Visualize the orbitals with fortecubeview.

Hints: The 'b3lyp/cc-pvdz' method and basis would be a good choice for this exercise. The symmetry labels for the calculated energy levels can be obtained as

```
molecule.irrep_labels()
```

Table 38.2   Descent in symmetry to D_{2h} for some irreducible representations of the $D_{\infty h}$ point group

$D_{\infty h}$	D_{2h}
σ_g	A_g
σ_u	B_{1u}
π_g	$B_{2g} + B_{3g}$
π_u	$B_{2u} + B_{3u}$

These will be for the D_{2h} point group but can be matched up to those of the correct $D_{\infty h}$ group using Table 38.2 (other symmetries are not important) (Table 38.2).

The molecules are closed-shell, so we can inspect wfn for the list of energy levels for, say, the α electrons only:

```
wfn.epsilon_a().to_array()
```

This will return a tuple of arrays, with each array corresponding, in the same order, to each irrep returned by molecule.irrep_labels().

39

Atomic Structure

39.1 One-Electron Atoms

A one-electron (*hydrogenic*) atom with nuclear charge Z may be described by the Hamiltonian operator

$$\hat{H} = -\frac{\hbar^2}{2\mu}\nabla^2 - \frac{Ze^2}{4\pi\epsilon_0 r}, \qquad (39.1)$$

where r is the electron–nucleus distance and

$$\mu = \frac{m_N m_e}{m_N + m_e}$$

is the *reduced mass* of the nucleus–electron pair. Note that $\mu \approx m_e$ because the nucleus, with mass m_N, is much heavier than the electron (mass m_e). In writing the Hamiltonian this way, we have separated the motion of the atom as a whole from its internal structure (relative positions of the nucleus and electron about their mutual centre of mass). The Laplacian operator, ∇^2, is most helpfully expressed in spherical polar coordinates, r, θ and ϕ:

$$\nabla^2 = \frac{1}{r}\frac{\partial^2}{\partial r^2}r + \frac{1}{r^2\sin\theta}\frac{\partial}{\partial\theta}\left(\sin\theta\frac{\partial}{\partial\theta}\right) + \frac{1}{r^2\sin^2\theta}\frac{\partial^2}{\partial\phi^2}.$$

39.1.1 Solving the Schrödinger Equation

The Schrödinger equation, $\hat{H}\psi = E\psi$, resulting from Equation 39.1 is further separable by factoring the wavefunction $\psi(r, \theta, \phi)$ into a radial part, $R(r)$, and an angular part, $Y(\theta, \phi)$:

$$-\frac{\hbar^2}{2\mu r}\frac{\partial^2}{\partial r^2}rR - \frac{Ze^2}{4\pi\epsilon_0 r}R + \frac{l(l+1)\hbar^2}{2\mu r^2}R = ER$$

$$\frac{1}{\sin\theta}\frac{\partial}{\partial\theta}\left(\sin\theta\frac{\partial Y}{\partial\theta}\right) + \frac{1}{\sin^2\theta}\frac{\partial^2 Y}{\partial\phi^2} = -l(l+1)Y$$

These equations can be solved exactly and the resulting wavefunctions, $\psi = R(r)Y(\theta, \phi)$, are the one-electron *atomic orbitals*. The angular wavefunctions, Y_{l,m_l}, are the *spherical harmonics*, described in terms of the quantum numbers, l and m_l. The first few are given below:

$$Y_{0,0} = \frac{1}{2\sqrt{\pi}}$$

$$Y_{1,0} = \frac{1}{2}\sqrt{\frac{3}{\pi}}\cos\theta$$

$$Y_{1,\pm 1} = \mp\frac{1}{2}\sqrt{\frac{3}{2\pi}}\sin\theta e^{\pm i\phi}$$

$$Y_{2,0} = \frac{1}{4}\sqrt{\frac{5}{\pi}}(3\cos^2\theta - 1)$$

$$Y_{2,\pm 1} = \mp\frac{1}{2}\sqrt{\frac{15}{2\pi}}\sin\theta\cos\theta e^{\pm i\phi}$$

$$Y_{2,\pm 2} = \frac{1}{4}\sqrt{\frac{15}{2\pi}}\sin^2\theta e^{\pm 2i\phi}$$

The radial wavefunctions are defined in terms of the quantum numbers n and l, and are best expressed in dimensionless form by taking

$$\rho = \frac{2Z}{na}r, \quad \text{where } a = \frac{m_e}{\mu}a_0 \text{ and } a_0 = \frac{4\pi\epsilon_0\hbar^2}{m_e e^2} \approx 52.9 \text{ pm}$$

is the *Bohr radius*. Using this coordinate, the radial wavefunctions take the form:

$$R(\rho) = N_{n,l}\rho^l L_{n-l-1}^{(2l+1)}(\rho)e^{-\rho/2},$$

where $L_{n-l-1}^{(2l+1)}(\rho)$ is the *associated Laguerre polynomial* of degree $n - l - 1$ and order $2l + 1$ and

$$N_{n,l} = \sqrt{\left(\frac{2Z}{na}\right)^3 \frac{(n-l-1)!}{2n(n+l)!}}$$

is a normalization constant chosen so that

$$\int_0^\infty |R_{n,l}(r)|^2 r^2 \, dr = 1.$$

The first few radial wavefunctions are:

$$R_{1,0} = N_{1,0}e^{-\rho/2},$$
$$R_{2,0} = N_{2,0}(2 - \rho)e^{-\rho/2},$$
$$R_{2,1} = N_{2,1}\rho e^{-\rho/2}.$$

The energies associated with ψ are:

$$E_n = -\frac{\mu Z^2 e^4}{32\pi^2 \epsilon_0^2 \hbar^2 n^2} = R_\infty hc \left(\frac{Z}{n}\right)^2 \left(\frac{\mu}{m_e}\right),$$

where the Rydberg constant (in wavenumbers),

$$R_\infty = \frac{m_e e^4}{8\epsilon_0^2 h^3 c} \approx 109\,737.316\,\text{cm}^{-1}.$$

39.1.2 Radial Wavefunctions and Distribution Functions

SciPy provides a function, `assoc_laguerre`, which calculates the associated Laguerre functions and can be used directly to calculate the radial wavefunctions and hence the radial distribution functions (rdfs), $r^2|R(r)|^2$, of the hydrogenic atom.

```python
import numpy as np
import matplotlib.pyplot as plt
from scipy.constants import m_e, m_p, epsilon_0, hbar, e
from scipy.special import assoc_laguerre
factorial = np.math.factorial
# The Bohr radius, m
a_0 = 4 * np.pi * epsilon_0 * hbar**2 / m_e / e**2
```

After these imports and definitions, we will define a function to calculate the radial wavefunction, $R_{n,l}(r)$ for the atom. r can be a single nucleus–electron distance or an array of such distance values.

```python
def get_R(r, n, el, Z, mu):
    """Return the radial part of the hydrogenic wavefunction at r.

    The wavefunction is that defined by principal quantum number n and
    azimuthal quantum number el for an atom with charge number Z and reduced
    mass mu = m_e.m_N/(m_e + m_N).

    """

    a = m_e / mu * a_0
    f = 2 * Z / n / a
    rho = f * r
    # Normalization constant.
    N = np.sqrt(f**3 * factorial(n - el - 1) / 2 / n /factorial(n + el))
    # The appropriate associated Laguerre polynomial function.
    L = assoc_laguerre(rho, n - el - 1, 2 * el + 1)
    return N * rho**el * L * np.exp(-rho/2)
```

The radial distribution function is straightforward to code:

```python
def get_rdf(r, R):
    """Return the radial distribution function for R(r) at r."""
    return (r * R)**2
```

It might also be useful to have a function that returns the conventional atomic orbital label, given *n* and *l*:

```
def get_orbital_label(n, el):
    """
    Return the label (1s, 2s, ...) for an atomic orbital given by n, el.
    """

    return '{}{:1s}'.format(n, 'spdfg'[el])
```

To compare wavefunctions for different atomic states, we will define a function to pre-calculate them for a range of values of *n* and *l*:

```
def get_R_and_P(r, Z, mu, nmax=3):
    """
    Return a dictionary of radial wavefunctions, R, and radial distribution
    functions, P, for all valid values of quantum numbers l and n <= nmax.
    r is the nucleus-electron distance or array of distances, Z the charge
    nucleus charge number and mu the electron-nucleus reduced mass.

    The returned dictionaries are keyed by atomic orbital label: 1s, 2s, ...

    """

    R, P = {}, {}
    for n in range(1, nmax+1):
        for el in range(0, n):
            orbital_label = get_orbital_label(n, el)
            R[orbital_label] = get_R(r, n, el, Z, mu)
            P[orbital_label] = get_rdf(r, R[orbital_label])
    return R, P
```

First we will plot a comparison of the 1s, 2s and 3s orbital wavefunctions and their radial distribution functions (Figure 39.1).

```
def compare_s_orbitals():
    """Plot the wavefunctions and rdfs for 1s, 2s and 3s orbitals."""

    fig, ax = plt.subplots(nrows=2, ncols=1)
    ax_psi, ax_rdf = ax
    for ao in ('1s', '2s', '3s'):
            ax_psi.plot(r, R[ao], label=ao)
            ax_rdf.plot(r, P[ao], label=ao)

    # A thin (1 pt) black line at y=0, separating positive and negative values.
    ax_psi.axhline(lw=1, c='k')
    ax_psi.set_xlim(0, rmax)
    ax_psi.set_yticks([])
    ax_psi.set_title('Wavefunctions')
    ax_psi.set_xlabel(r'$r$ /m')
    ax_psi.legend()

    ax_rdf.set_xlim(0, rmax)
    ax_rdf.set_ylim(0)
```

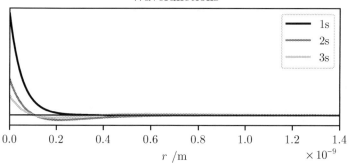

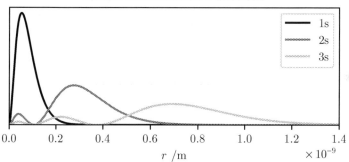

Figure 39.1   Comparison of 1s, 2s and 3s orbitals.

```
ax_rdf.set_yticks([])
ax_rdf.set_xlabel(r'$r$ /m')
ax_rdf.set_title('Radial Distribution Functions')
ax_rdf.legend()

plt.show()
```

```
# Maximum distance from nucleus to consider, in m
rmax = 1.4e-9
r = np.linspace(0, rmax, 1000)
Z, mu = 1, m_e * m_p / (m_e + m_p)
R, P = get_R_and_P(r, Z, mu)

compare_s_orbitals()
```

In a similar way, the three $n = 3$ orbitals can be compared (Figure 39.2):

```
def compare_3el_orbitals():
    fig, ax = plt.subplots(nrows=2, ncols=1)
    ax_psi, ax_rdf = ax
    for ao in ('3s', '3p', '3d'):
        ax_psi.plot(r, R[ao], label=ao)
        ax_rdf.plot(r, P[ao], label=ao)
```

Wavefunctions

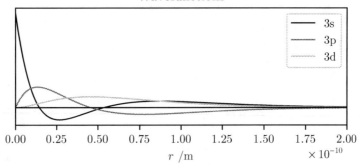

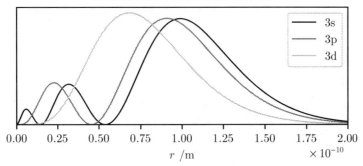

Figure 39.2   Comparison of the 3s, 3p and 3d orbitals.

```
    # A thin (1 pt) black line at y=0.
    ax_psi.axhline(lw=1, c='k')
    ax_psi.set_xlim(0, rmax)
    ax_psi.set_xlabel(r'$r$ /m')
    ax_psi.set_yticks([])
    ax_psi.set_title('Wavefunctions')
    ax_psi.legend()

    ax_rdf.set_xlim(0, rmax)
    ax_rdf.set_xlabel(r'$r$ /m')
    ax_rdf.set_ylim(0)
    ax_rdf.set_yticks([])
    ax_rdf.set_title('Radial Distribution Functions')
    ax_rdf.legend()

    plt.show()

compare_3el_orbitals()
```

Finally, a comparison of the radial wavefunctions and distribution functions for the H atom and He$^+$ ion (Figure 39.3):

```
def compare_H_and_Heplus():
    fig, ax = plt.subplots(nrows=2, ncols=1)
    ax_psi, ax_rdf = ax
```

Wavefunctions

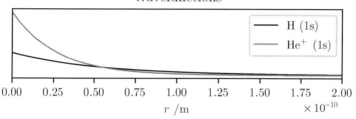

Radial Distribution Functions

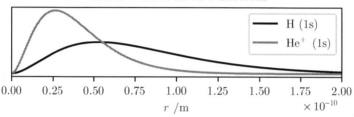

Figure 39.3   Comparison of H and He$^+$ 1s orbitals. The electron is more likely to be found closer to the nucleus for He$^+$ than for H because of the higher nuclear charge.

```
ax_psi.plot(r, R_H['1s'], label='H (1s)')
ax_psi.plot(r, R_He['1s'], label=r'He$^+$ (1s)')
ax_psi.set_xlim(0, rmax)
ax_psi.set_xlabel(r'$r$ /m')
ax_psi.set_yticks([])
ax_psi.set_title('Wavefunctions')
ax_psi.legend()

ax_rdf.plot(r, P_H['1s'], label='H (1s)')
ax_rdf.plot(r, P_He['1s'], label=r'He$^+$ (1s)')
ax_rdf.set_xlim(0, rmax)
ax_rdf.set_xlabel(r'$r$ /m')
ax_rdf.set_yticks([])
ax_rdf.set_title('Radial Distribution Functions')
ax_rdf.legend()

plt.tight_layout()
plt.show()
```

```
rmax = 0.2e-9
r = np.linspace(0, rmax, 1000)

Z, mu = 1, m_e * m_p / (m_e + m_p)
R_H, P_H = get_R_and_P(r, Z, mu)

m_4He = 6.64424e-27
Z, mu = 2, m_e * m_4He / (m_e + m_4He)
R_He, P_He = get_R_and_P(r, Z, mu)

compare_H_and_Heplus()
```

39.1.3 Angular Wavefunctions

SciPy's function, sph_harm, calculates the value of a spherical harmonic with order m and degree n, Y_n^m, at angular coordinates θ and ϕ. We can compute the values of Y_n^m on a grid of such angles, and then map their coordinates to the Cartesian axis system to plot using Matplotlib's 3-D plotting routine, plot_surface.

First, the imports:

```
import numpy as np
import matplotlib.pyplot as plt
# The following import configures Matplotlib for 3D plotting.
from mpl_toolkits.mplot3d import Axes3D
from scipy.special import sph_harm
```

The array of angular values is realized as a meshgrid: a pair of two-dimensional arrays: one with columns of increasing θ values duplicated down its rows and one with rows of increasing ϕ values. These angles are then converted to Cartesian coordinates on the unit sphere with the usual transformation:

$$x = \sin\theta \sin\phi$$

$$y = \sin\theta \cos\phi$$

$$z = \cos\phi$$

```
# Grids of azimuthal and polar angles
theta = np.linspace(0, np.pi, 100)
phi = np.linspace(0, 2*np.pi, 100)
# Create a 2-D meshgrid of (theta, phi) angles.
theta, phi = np.meshgrid(theta, phi)
# Calculate the Cartesian coordinates of each point in the mesh.
xyz = np.array([np.sin(theta) * np.sin(phi),
                np.sin(theta) * np.cos(phi),
                np.cos(theta)])
```

To plot a spherical harmonic, we calculate its values on the grid of angular coordinates, then take its real part, and transform to Cartesian coordinates. The plot_surface function takes these Cartesian coordinates and plots a three-dimensional surface as patches in the correct orientation (see Figure 39.4). In order to depict the sign of the function, a color map is chosen that maps the function values to two colors: blue for negative values and red for positive values.

```
def show_Y(el, mel):
    """Plot the spherical harmonic of degree el and order mel."""

    # NB In SciPy's sph_harm function the azimuthal coordinate, theta,
    # comes before the polar coordinate, phi.
```

Figure 39.4   The $l = 3$, $m_l = 1$ f-orbital.

```
Y = sph_harm(mel, el, phi, theta)

# Colour the plotted surface according to the sign of Y.
cmap = plt.cm.ScalarMappable(cmap=plt.get_cmap('bwr'))
cmap.set_clim(-0.5, 0.5)

fig = plt.figure(figsize=plt.figaspect(1.))
ax = fig.add_subplot(projection='3d')
# We choose to plot the real part of Y: retrieve its x, y and z components
Yx, Yy, Yz = np.abs(Y.real) * xyz
ax.plot_surface(Yx, Yy, Yz,
                facecolors=cmap.to_rgba(Y.real),
                rstride=2, cstride=2)
ax.axis('off')
plt.show()
```

```
el, mel = 3, 1        # One of the f-orbitals
show_Y(el, mel)
```

39.1.4 Atomic Spectra

The Rydberg constant, R_∞ is available within the `scipy.constants` package. We can use it to calculate the energy levels of the hydrogen atom (in the clamped-nucleus approximation, in wavenumber units, cm^{-1}) as $E_n = -R_\infty/n^2$, which we store in a dictionary, keyed by n.

```
from scipy.constants import Rydberg as R
# Convert the Rydberg constant from m-1 to cm-1
R /= 100

nmax = 20
ngrid = np.arange(1, nmax+1, 1, dtype=int)
energy_level = dict((n, -R/n**2) for n in ngrid)
```

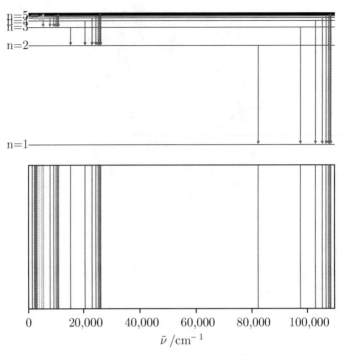

Figure 39.5   The atomic spectrum of hydrogen.

The emission spectrum of the hydrogen atom is often considered in terms of *spectral series*; each series consists of transitions which terminate on a common lower level, n_f. The transition wavenumbers within a series are given by

$$\tilde{\nu}_{i \to f} = R_\infty \left(\frac{1}{n_f^2} - \frac{1}{n_i^2} \right).$$

The set of transitions with $n_f = 1, 2, 3, 4, 5$ are known as the Lyman, Balmer, Paschen, Brackett and Pfund series, respectively.

The following function depicts the hydrogen atom transitions on an energy level diagram and as a spectrum (Figure 39.5).

```
def show_series(wavelengths=False):
    """Plot a diagram indicating the transitions within a hydrogenic atom.

    Set wavelengths=True to show the spectrum as a function of wavelength
    in nm; otherwise wavenumbers (cm-1) will be used.

    """

    # Factor for the conversion cm to nm.
    FAC = 1.e7
```

```python
# Ensure the figure is large enough to resolve its important features.
xsize = ysize = 800     # figure dimensions in pixels
DPI = 200               # figure resolution (dots per inch)
fig = plt.figure(figsize=(xsize // DPI, ysize // DPI), dpi=DPI)
plt.rcParams.update({'font.size': 6})

# The top Axes object will be the energy level diagram, the bottom will be
# the spectrum. sharex is True so that zooming one also zooms the other.
ax_E, ax_spec = fig.subplots(nrows=2, ncols=1, sharex=True)
ax_spec.set_facecolor('k')

for n in ngrid:
    E = energy_level[n]
    # Draw a horizontal line for the energy level E.
    ax_E.axhline(E, c='k', lw=0.5)
    # Add labels for the lowest 3 energy levels.
    if n < 4:
        ax_E.text(0, E, f'n={n} ', va='center', ha='right')

def plot_transition_series(nf):
    """Plot the spectral series terminating on the level n = nf."""

    # Series will be plotted in different colours according to this scheme.
    colour = [None, 'm', 'b', 'g', 'y', 'r']

    for ni in range(nf+1, 10):
        # The transition wavenumber, freq, is the difference in energy levels.
        dy = freq = energy_level[ni] - energy_level[nf]
        if wavelengths:
            dy = FAC / freq
        # Annotate the energy level diagram with an arrow between the levels.
        ax_E.annotate('', xy=(dy, energy_level[nf]),
                    xytext=(dy, energy_level[ni]),
                    arrowprops=dict(arrowstyle='-|>', fc=colour[nf],
                            ec=colour[nf], shrinkA=0, shrinkB=0,
                            mutation_scale=4, lw=0.8)
                    )
        # Add a vertical line to the spectrum plot.
        ax_spec.axvline(dy, c=colour[nf], lw=0.8)

# Plot the series n->1, n->2, ..., n->5.
for nf in range(1, 6):
    plot_transition_series(nf)

# Tidy the plot Axes a little.
ax_E.axis('off')
ax_spec.yaxis.set_visible(False)

if wavelengths:
    # Plot the wavelength spectrum on a log plot.
    ax_spec.set_xscale('log')
    ax_spec.set_xlabel(r'$\lambda\,/\mathrm{nm}$')
    ax_E.set_xlim(10000, FAC / R)
else:
```

```
        ax_spec.set_xlabel(r'$\tilde{\nu}\,/\mathrm{cm^{-1}}$')
        ax_E.set_xlim(0, R)
    plt.tight_layout()

show_series(wavelengths=False)
```

39.2 Many-Electron Atoms

39.2.1 Effective Nuclear Charge

In an atom with more than one electron, the electrons may be considered to occupy atomic orbitals of a similar form to the hydrogenic, one-electron wavefunctions considered previously. In the many-electron atom, each electron experiences
a Coulombic attraction for the nucleus, but is also repelled by the other electrons.

In the simplest approximation, the balance of these effects is captured by assigning to each orbital an *effective nuclear charge*, Z_{eff}, which accounts for the
shielding of the full nuclear charge by other electrons as well as the possible *pen-
etration* effect due to lobes of lower angular momentum orbitals occurring within
(closer to the nucleus than) the maxima of inner orbitals.

The effective nuclear charges often used by chemists are those calculated by
Clementi et al.[1] These were calculated for "Slater type" orbitals (see Example E34.2)
rather than the hydrogenic orbitals,[2] but, nonetheless, they can be used to make
some qualitative predictions.

This section will use the same imports and the functions get_R and get_rdf previously defined. The Clementi effective nuclear charges are provided at the URL
https://scipython.com/chem/egjd/ in the Python file Zeff.py as a dictionary, which can be imported as:

```
from Zeff import Zeff
# Output some examples.
print('Zeff for lithium orbitals:', Zeff['Li'])
print('Zeff for flourine orbitals:', Zeff['F'])
```

```
Zeff for lithium orbitals: {'1s': 2.691, '2s': 1.279}
Zeff for flourine orbitals: {'1s': 8.65, '2s': 5.128, '2p': 5.1}
```

It is helpful to define a function to return the quantum numbers n and l from the
orbital label (e.g., 3d implies $n = 3$, $l = 2$) and another to retrieve the radial distribution functions for all orbitals in an atom, using the Zeff dictionary.

---

[1] E. Clementi, D. L. Raimondi, "Atomic Screening Constants from SCF Functions". *J. Chem. Phys.* **38**(11):
2686–2689 (1963); E. Clementi, D. L. Raimondi, W. P. Reinhardt, "Atomic Screening Constants from SCF
Functions. II. Atoms with 37 to 86 Electrons". *J. Chem. Phys.* **47**, 1300–1307 (1967).

[2] Slater-type orbitals omit the oscillatory associated Laguerre function factor and are not orthogonal to one
another, and so the Z_{eff} derived from them cannot be considered independently for each wavefunction.

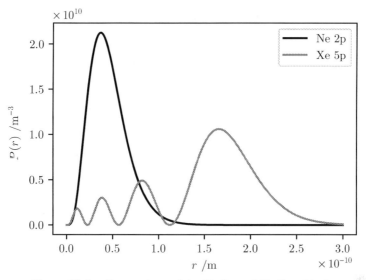

Figure 39.6   Comparison of the Ne 2p and Xe 5p orbitals.

```
def get_n_el(orbital_label):
    """Return the quantum numbers n and l for a given atomic orbital."""
    n = int(orbital_label[0])
    el = 'spdf'.index(orbital_label[1])
    return n, el

def get_orbital_rdfs(atom_symbol, r):
    """Return all atomic orbital radial distribution functions for an atom.

    The atom is identified by its element symbol, and a dictionary of rdfs,
    keyed by orbital label ('1s', '2s', '2p', ...) is returned.

    """

    P = {}
    for orbital in Zeff[atom_symbol]:
        n, el = get_n_el(orbital)
        R = get_R(r, n, el, Zeff[atom_symbol][orbital], m_e)
        P[orbital] = get_rdf(r, R)
    return P
```

First, we will plot the valence orbitals of the noble gas atoms neon and xenon, 2p
and 5p, respectively (Figure 39.6). A (very crude) estimate of the atom size can be
established from the position of the maximum in its valence rdf.

```
def compare_Ne_Xe_valence_orbitals():
    # A grid of distances from the nucleus, in m.
    rmax = 3.e-10
    r = np.linspace(0, rmax, 5000)
```

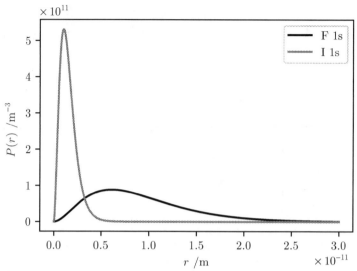

Figure 39.7   Comparison of the 1s orbitals for fluorine and iodine: The 1s electrons are held much closer to the nucleus for iodine.

```
P_Ne = get_orbital_rdfs('Ne', r)
P_Xe = get_orbital_rdfs('Xe', r)

# np.argmax returns the index of the maximum value in an array: use this index
# to find the corresponding distance from the nucleus in the r array.
rmax_Ne = r[np.argmax(P_Ne['2p'])]
rmax_Xe = r[np.argmax(P_Xe['5p'])]
print('Estimated atomic radius of Ne: {:.0f} pm'.format(rmax_Ne * 1.e12))
print('Estimated atomic radius of Xe: {:.0f} pm'.format(rmax_Xe * 1.e12))

fig, ax = plt.subplots()
ax.plot(r, P_Ne['2p'], label='Ne 2p')
ax.plot(r, P_Xe['5p'], label='Xe 5p')
ax.set_xlabel(r'$r$ /m')
ax.set_ylabel(r'$P(r)$ /m$^{-3}$')
ax.legend()
```

```
compare_Ne_Xe_valence_orbitals()
```

We might expect the higher nuclear charge of heavier atoms to hold the core electrons (those in the innermost orbitals, 1s, 2s, etc.) closer to the nucleus than that of lighter atoms, and that is indeed the case (Figure 39.7):

```
def compare_1s_orbitals():
    # A grid of distances from the nucleus, in m.
    rmax = 3.e-11
    r = np.linspace(0, rmax, 5000)

    P_F = get_orbital_rdfs('F', r)
```

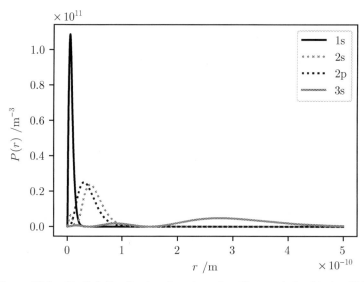

Figure 39.8   Radial distribution functions for all occupied orbitals of Na.

```
    P_I = get_orbital_rdfs('I', r)
    fig, ax = plt.subplots()
    ax.plot(r, P_F['1s'], label='F 1s')
    ax.plot(r, P_I['1s'], label='I 1s')
    ax.set_xlabel(r'$r$ /m')
    ax.set_ylabel(r'$P(r)$ /m$^{-3}$')
    ax.legend()
    plt.show()

compare_1s_orbitals()
```

Finally, we can plot all the radial distribution functions for every filled orbital in an atom (Figure 39.8).

```
def plot_all_rdfs(atom_symbol):
    """Plot radial distribution functions for all occupied orbitals in an atom."""
    # A grid of distances from the nucleus, in m.
    rmax = 5.e-10
    r = np.linspace(0, rmax, 5000)

    P_Na = get_orbital_rdfs(atom_symbol, r)
    fig, ax = plt.subplots()
    for orbital in P_Na:
        ax.plot(r, P_Na[orbital], label=orbital)
    ax.set_xlabel(r'$r$ /m')
    ax.set_ylabel(r'$P(r)$ /m$^{-3}$')
    ax.legend()
    plt.show()
plot_all_rdfs('Na')
```

Here, the outermost 3s electron, despite having a maximum in its rdf well outside the $n = 1$ and $n = 2$ shells, has lobes which penetrate these shells to some extent, and so its Z_{eff} is appreciably greater than 1:

```
print(Zeff['Na']['3s'])
```

```
2.507
```

Excited states in which this electron is promoted to higher energy orbitals with larger angular-momentum would lead to more effective screening of the +11 nuclear charge by the other 10 electrons and a Z_{eff} closer to 1.

39.2.2 Ionization Energies and Polarizability

In this section, the pandas library will be used to read and plot atomic data from the comma-separated values file atomic_data.csv, which is available from https://scipython.com/chem/egje/.

```
import numpy as np
import matplotlib.pyplot as plt
import pandas as pd
```

```
# Read the data into a DataFrame and inspect the first few rows.
atomic_data = pd.read_csv('atomic_data.csv', index_col=0)
atomic_data.head()
```

Atom	Ionization energy /eV	Polarizability /au	Atomic radius /pm	Melting point /K
H	13.598434	4.50711	53.0	14.01
He	24.587389	1.38375	31.0	NaN
Li	5.391715	164.11250	167.0	453.69
Be	9.322699	37.74000	112.0	1560.00
B	8.298019	20.50000	87.0	2349.00

The polarizability, α, of an atom may be thought of as a measure of the ease with which its electron cloud can be distorted. More precisely, the electric dipole, μ, induced in an atom by an electric field E as that field causes the position of the average center of negative electronic charge to move away from the nucleus is $\mu = \alpha E$. It might be expected that electrons that are less tightly held to the nucleus would be easier to push around than those which are more strongly attracted to it, and that larger atoms with more such electrons would also be more polarizable. This is shown in the following scatter plot (Figure 39.9).

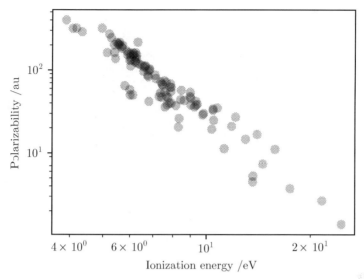

Figure 39.9   A scatter plot of atomic ionization energy against polarizability.

```python
def plot_alpha_ip():
    """Produce a scatter plot of polarizability against ionization energy."""
    fig, ax = plt.subplots()
    # Note that the marker specified in points-squared (an area) so we need
    # the square of the atomic radius (suitably scaled).
    ax.scatter(atomic_data['Ionization energy /eV'],
               atomic_data['Polarizability /au'], alpha=0.3)
    # Plot the data on log-log axes and label.
    ax.set_yscale('log')
    ax.set_xscale('log')
    ax.set_xlabel('Ionization energy /eV')
    ax.set_ylabel('Polarizability /au')

plot_alpha_ip()
```

The argument `alpha=0.3` sets the opacity of the scatter plot markers to 30% so that they can be resolved to some extent when they overlap.

We can also pick out particular groups of elements for more detailed consideration, and label the element symbols with a text annotation. The size of the scatter plot markers can be set in proportion to the atom sizes (Figure 39.10).

```python
# Four groups of the periodic table that we're interested in.
groups = {'Alkali metals': ['Li', 'Na', 'K', 'Rb', 'Cs'],
          'Noble gases': ['He', 'Ne', 'Ar', 'Kr', 'Xe', 'Rn'],
          'Halogens': ['F', 'Cl', 'Br', 'I', 'At'],
          'Alkaline earth metals': ['Be', 'Mg', 'Ca', 'Sr', 'Ba', 'Ra']
          }


def plot_grouped_elements():
    fig, ax = plt.subplots()
    for group_name in groups:
```

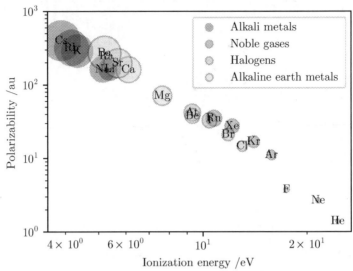

Figure 39.10   A scatter plot of polarizability against ionization energy for groups
of elements.

```
    # Add a scatter plot for each group of elements.
    ax.scatter(atomic_data.loc[groups[group_name]]['Ionization energy /eV'],
            atomic_data.loc[groups[group_name]]['Polarizability /au'],
            atomic_data.loc[groups[group_name]]['Atomic radius /pm']**2/100,
            alpha=0.3, label=group_name)
    # Add text labels of the element symbols in each group at the centre of
    # each marker.
    for symbol, row_data in atomic_data.loc[groups[group_name]].iterrows():
        ax.text(row_data['Ionization energy /eV'],
                row_data['Polarizability /au'],
                symbol, va='center', ha='center')
ax.set_yscale('log')
ax.set_xscale('log')
ax.set_ylim(1, 1000)
ax.set_xlabel('Ionization energy /eV')
ax.set_ylabel('Polarizability /au')
legend = ax.legend()
# We have to manually set the legend marker sizes, otherwise they end up
# all different.
for handle in legend.legendHandles:
    handle.set_sizes([32])

plot_grouped_elements()
```

Finally, we might expect there to be a relationship between the polarizability and
the melting point of element which exists as a molecular solid at low enough temperature, since the cohesive forces holding these solids together are the *dispersion*
(van der Waals) forces which arise from the instantaneous dipole – induced dipole
interaction. The following code demonstrates this relationship for the halogen and
noble gas groups (Figure 39.11).

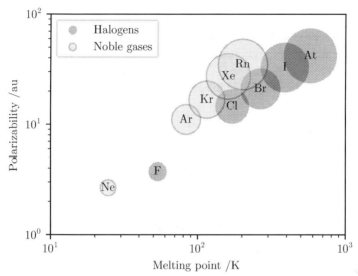

Figure 39.11   A scatter plot of polarizability against melting point for two groups of elements.

```
def plot_alpha_melting_point():
    fig, ax = plt.subplots()
    for group_name in ['Halogens', 'Noble gases']:
        ax.scatter(atomic_data.loc[groups[group_name]]['Melting point /K'],
                   atomic_data.loc[groups[group_name]]['Polarizability /au'],
                   atomic_data.loc[groups[group_name]]['Atomic radius /pm']**2/10,
                   alpha=0.3, label=group_name)
        # Add text labels of the element symbols at the centre of each marker.
        for symbol, row_data in atomic_data.loc[groups[group_name]].iterrows():
            ax.text(row_data['Melting point /K'], row_data['Polarizability /au'],
                    symbol, va='center', ha='center')
    ax.set_yscale('log')
    ax.set_xscale('log')
    ax.set_xlim(10, 1000)
    ax.set_ylim(1, 100)
    ax.set_xlabel('Melting point /K')
    ax.set_ylabel('Polarizability /au')
    legend = ax.legend(loc='upper left')
    # We have to manually set the legend marker sizes, otherwise they end up
    # all different.
    for handle in legend.legendHandles:
        handle.set_sizes([32])

plot_alpha_melting_point()
```

39.2.3 Grotrian Diagrams

```
import numpy as np
import matplotlib.pyplot as plt
```

The file `Na-trans.txt`, available at `https://scipython.com/chem/egjf/`, contains the energy levels and Einstein *A*-factors for some of the stronger transitions in the sodium atom, arranged in columns: lower-state configuration, lower-state energy (cm^{-1}), upper-state configuration, lower-state energy (cm^{-1}) and *A*-factor (s^{-1}). The first few rows of this file can be inspected with the operating system head command:

```
!head 'Na-trans.txt'
```

```
2p6.3s      0.000000 2p6.5p 35042.850000 5.31e+05
2p6.3s      0.000000 2p6.4p 30272.580000 2.73e+06
2p6.3p 16956.170250 2p6.5d 37036.772000 4.88e+06
2p6.3p 16956.170250 2p6.6s 36372.618000 2.27e+06
2p6.3p 16956.170250 2p6.4d 34548.764000 1.21e+07
2p6.3s      0.000000 2p6.3p 16956.170250 6.14e+07
2p6.3p 16956.170250 2p6.5s 33200.673000 4.98e+06
2p6.3p 16956.170250 2p6.3d 29172.887000 5.14e+07
2p6.4s 25739.999000 2p6.5p 35042.850000 7.24e+05
2p6.3p 16956.170250 2p6.4s 25739.999000 1.76e+07
```

We will read the transition data into a dictionary of energy levels (with the configuration strings for keys) and a dictionary of *A*-factors (with a tuple of (upper state, lower state) strings for keys):

```python
filename = 'Na-trans.txt'
Elevels, transA = {}, {}
with open(filename) as fi:
    for line in fi:
        fields = line.split()
        confi, Ei, confk, Ek, A = fields
        Elevels[confi] = float(Ei)
        Elevels[confk] = float(Ek)
        transA[(confi, confk)] = float(A)
```

A *Grotrian diagram* indicates the transitions between energy levels allowed for an atom. The levels are depicted in separate columns for states with different orbital angular momentum quantum number, *l*. The selection rule $\Delta l = \pm 1$ predicts that only transitions between neighboring columns of levels will be strong. The following code plots Figure 39.12, a Grotrian diagram for sodium using the energy levels from the above file and colors the transitions according to their historical groupings: *sharp*, *principal*, *diffuse* and *fundamental*.

```python
def plot_grotrian():
    fig, ax = plt.subplots()
    # Position the energy levels horizontally according to their angular momentum.
    xpos = {'s': 0.2, 'p': 0.4, 'd': 0.6, 'f': 0.8}
    # Each level is indicated by a line this long.
    levlen = 0.1
    # Colour lines belonging to different series as follows.
```

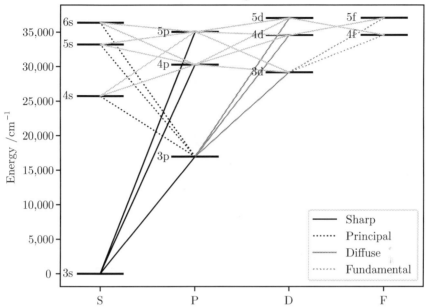

Figure 39.12 A Grotrian diagram for Na.

```
colours = {('p', '3s'): 'tab:blue',      # Sharp
           ('s', '3p'): 'tab:orange',    # Principal
           ('d', '3p'): 'tab:green',     # Diffuse
           ('f', '3d'): 'tab:red'}       # Fundamental

# Add the levels and label them by the valence electron's atomic orbital.
for conf, E in Elevels.items():
    el = conf[-1]
    ax.hlines(E, xpos[el]-levlen/2, xpos[el]+levlen/2)
    ax.text(xpos[el]-levlen/2-0.03, E, conf[-2:], va='center')

# Add lines for the transitions, colour-coded by series.
for states, A in transA.items():
    confi, confk = states
    eli, elk = confi[-1], confk[-1]
    colour = colours.get((elk, confi[-2:]), 'gray')
    ax.plot([xpos[eli], xpos[elk]], [Elevels[confi], Elevels[confk]],
            c=colour, lw=1)

# Label the total orbital electronic angular momentum on the x-axis.
ax.set_xticks(list(xpos.values()))
ax.set_xticklabels([e.upper() for e in xpos.keys()])
ax.set_ylabel(r'Energy /cm$^{-1}$')
ax.set_title('Grotrian diagram for Na')

# We have to add the legend manually to avoid duplicates:
# Create (but don't plot) some fake lines and label them.
```

```
    legend_lines = [plt.Line2D([0], [0], color=c, lw=1) for c in colours.values()]
    legend_labels = ['Sharp', 'Principal', 'Diffuse', 'Fundamental']
    ax.legend(legend_lines, legend_labels)

plot_grotrian()
```

The famous sodium "D-lines" are an intense doublet of transitions in the "sharp" series at around 589 nm (16,980 cm^{-1}) from $3p \rightarrow 3s$, the upper state being split by spin–orbit coupling into $^2P_{1/2}$ and $^2P_{3/2}$ levels. These are not resolved in this diagram.

40

Solutions

P2.1

Assuming Arrhenius-type behavior, the temperature-dependence of the reaction rate coefficient is

$$k = A \exp\left(-\frac{E_a}{RT}\right) \quad \Rightarrow \ln k = \ln A - \frac{E_a}{RT}.$$

Assuming A is approximately constant, changing the temperature from T_1 to T_2 causes a change in k from k_1 to k_2 given by

$$\ln\left(\frac{k_2}{k_1}\right) = -\frac{E_a}{R}\left(\frac{1}{T_2} - \frac{1}{T_1}\right)$$

increasing the rate from k_1 to $10k_1$ therefore requires an increase in temperature from T_1 to T_2 given by

$$\ln 10 = \frac{E_a}{R}\left(\frac{1}{T_1} - \frac{1}{T_2}\right),$$

which is readily rearranged to give the activation energy, E_a.

```
import numpy as np
# The gas constant in J.K-1.mol-1 (4 s.f.).
R = 8.314
# The two temperatures, in K.
T1, T2 = 25 + 273.15, 47 + 273.15
```

```
# Calculate the activation energy, in J.mol-1.
Ea = R * np.log(10) / (1/T1 - 1/T2)
```

```
Ea
```

```
83059.88928684713
```

479

The activation energy is therefore approximately 83 kJ mol^{-1}.

P2.2

The number of moles of $C_{55}H_{104}O_6$ in 1 kg of fat is:

```
M_H, M_C, M_O = 1, 12, 16
# Molar mass of average fat molecule, in g.mol-1.
M_fat = 55 * M_C + 104 * M_H + 6 * M_O
# Number of moles of fat molecules in 1 kg.
n_fat = 1000 / M_fat
n_fat
```

```
1.1627906976744187
```

Complete combustion of each mole of fat would proceed through the overall reaction

$$C_{55}H_{104}O_6 + 78O_2 \rightarrow 52H_2O + 55CO_2,$$

producing 55 mol of CO_2. Therefore:

```
n_CO2 = n_fat * 55
n_H2O = n_fat * 52
print(f'Amount of CO2 produced when 1 kg of fat is metabolized: {n_CO2:.2f} mol')
print(f'Amount of H2O produced when 1 kg of fat is metabolized: {n_H2O:.2f} mol')
```

```
Amount of CO2 produced when 1 kg of fat is metabolized: 63.95 mol
Amount of H2O produced when 1 kg of fat is metabolized: 60.47 mol
```

The corresponding masses of CO_2 and H_2O are therefore:

```
M_CO2, M_H2O = M_C + 2 * M_O, 2 * M_H + M_O
m_CO2 = n_CO2 * M_CO2
m_H2O = n_H2O * M_H2O
print(f'Mass of CO2 produced: {m_CO2:.0f} g')
print(f'Mass of H2O produced: {m_H2O:.0f} g')
```

```
Mass of CO2 produced: 2814 g
Mass of H2O produced: 1088 g
```

Roughly two-thirds of the mass of the products comes from the oxygen used in the metabolism rather than the fat itself.

To calculate the percentage of the $C_{55}H_{104}O_6$ that ends up as each product, consider that all of the carbon and two-thirds of its oxygen atoms end up in CO_2, whilst all of the hydrogen and one-third of the oxygen atoms produce H_2O (the oxygen atoms in these two molecules are rapidly exchanged in the body).

```
perc_CO2 = (55 * M_C + 2/3 * 6 * M_O) / M_fat * 100
perc_H2O = (104 * M_H + 1/3 * 6 * M_O) / M_fat * 100
print(f'Proportion of fat converted to CO2: {perc_CO2:.0f}%')
print(f'Proportion of fat converted to H2O: {perc_H2O:.0f}%')
```

```
Proportion of fat converted to CO2: 84%
Proportion of fat converted to H2O: 16%
```

That is, when fat is metabolized, by far the largest part of it is exhaled as carbon dioxide.[1]

P2.3

```
import numpy as np

# The gas constant in J.K-1.mol-1 (4 s.f.).
R = 8.314
# Molar enthalpy of vaporization of water, kJ.mol-1.
DvapH = 44
# Normal boiling point of water at sea level, K.
T0 = 100 + 273.15

# Atmosphere scale height, H, in m.
H = 8000
# Height of Everest, m.
z = 8849
# Atmospheric pressure at sea level, atm.
p0 = 1
```

```
# Atmospheric pressure at the summit of Everest, atm.
p1 = p0 * np.exp(-z/H)
p1
```

```
0.3308384795152694
```

The air pressure at the summit is about a third that at sea level.

Integrating the Clausius–Clapeyron equation yields

$$\ln\left(\frac{p_1}{p_0}\right) = -\frac{\Delta_{vap}H_m}{R}\left(\frac{1}{T_0} - \frac{1}{T_1}\right),$$

which we need to rearrange in terms of the boiling point, T_1, at air pressure p_1:

$$T_1 = \left[\frac{1}{T_0} - \frac{R}{\Delta_{vap}H_m}\ln\left(\frac{p_1}{p_0}\right)\right]^{-1}$$

[1] R. Meerman and A. J. Brown, *BMJ* **349**, g7257 (2014); https://doi.org/10.1136/bmj.g7782; N. Lifson et al., *J. Biol. Chem.* **180**, 803 (1949).

```
# NB Convert DvapH from kJ.mol-1 to K.mol-1.
T1 = 1 / (1 / T0 - R / 1000 / DvapH * np.log(p1 / p0))
T1
```

```
346.1531389115387
```

This is approximately 73°C. In fact, the effective scale height, H, is not constant but itself depends on temperature and so a figure of about 70°C would be about right.

P3.1

Here is one solution:

```
# Triple point of CO2 (K, Pa).
T3_CO2, p3_CO2 = 216.58, 5.185e5
# Enthalpy of fusion of CO2 (kJ.mol-1).
DfusH_CO2 = 9.019
# Entropy of fusion of CO2 (J.K-1.mol-1).
DfusS_CO2 = 40
# Enthalpy of vaporization of CO2 (kJ.mol-1).
DvapH_CO2 = 15.326
# Entropy of vaporization of CO2 (J.K-1.mol-1).
DvapS_CO2 = 70.8

# Triple point of H2O (K, Pa).
T3_H2O, p3_H2O = 273.16, 611.73
# Enthalpy of fusion of H2O (kJ.mol-1).
DfusH_H2O = 6.01
# Entropy of fusion of H2O (J.K-1.mol-1).
DfusS_H2O = 22.0
# Enthalpy of vaporization of H2O (kJ.mol-1).
DvapH_H2O = 40.68
# Entropy of vaporization of H2O (J.K-1.mol-1).
DvapS_H2O = 118.89
```

```
print(' '*22 + 'CO2' + ' '*8 + 'H2O')
print('-'*40)
print('p3    /Pa            ', f'{p3_CO2:6.0f}      {p3_H2O:10.2f}')
print('T3    /K             ', f'{T3_CO2:9.2f}      {T3_H2O:9.2f}')
print('DfusH /kJ.mol-1      ', f'{DfusH_CO2:10.3f}   {DfusH_H2O:10.3f}')
print('DfusS /J.K-1.mol-1', f'{DfusS_CO2:8.1f}      {DfusS_H2O:8.1f}')
print('DvapH /kJ.mol-1      ', f'{DvapH_CO2:10.3f}   {DvapH_H2O:10.3f}')
print('DvapS /J.K-1.mol-1', f'{DvapS_CO2:8.1f}      {DvapS_H2O:8.1f}')
```

```
                        CO2       H2O
----------------------------------------
p3    /Pa            518500       611.73
T3    /K             216.58       273.16
DfusH /kJ.mol-1       9.019         6.010
```

```
DfusS  /J.K-1.mol-1     40.0      22.0
DvapH  /kJ.mol-1        15.326    40.680
DvapS  /J.K-1.mol-1     70.8      118.9
```

P5.1

First, simply concatenate the element symbols and convert them all to the same case:

```python
element_symbols = [
  'H', 'He', 'Li', 'Be', 'B', 'C', 'N', 'O', 'F', 'Ne', 'Na', 'Mg', 'Al',
  'Si', 'P', 'S', 'Cl', 'Ar', 'K', 'Ca', 'Sc', 'Ti', 'V', 'Cr', 'Mn', 'Fe',
  'Co', 'Ni', 'Cu', 'Zn', 'Ga', 'Ge', 'As', 'Se', 'Br', 'Kr', 'Rb', 'Sr', 'Y',
  'Zr', 'Nb', 'Mo', 'Tc', 'Ru', 'Rh', 'Pd', 'Ag', 'Cd', 'In', 'Sn', 'Sb', 'Te',
  'I', 'Xe', 'Cs', 'Ba', 'La', 'Ce', 'Pr', 'Nd', 'Pm', 'Sm', 'Eu', 'Gd', 'Tb',
  'Dy', 'Ho', 'Er', 'Tm', 'Yb', 'Lu', 'Hf', 'Ta', 'W', 'Re', 'Os', 'Ir', 'Pt',
  'Au', 'Hg', 'Tl', 'Pb', 'Bi', 'Po', 'At', 'Rn', 'Fr', 'Ra', 'Ac', 'Th', 'Pa',
  'U', 'Np', 'Pu', 'Am', 'Cm', 'Bk', 'Cf', 'Es', 'Fm', 'Md', 'No', 'Lr', 'Rf',
  'Db', 'Sg', 'Bh', 'Hs', 'Mt', 'Ds', 'Rg', 'Cn', 'Nh', 'Fl', 'Mc', 'Lv', 'Ts',
  'Og'
]

all_element_letters = ''.join(element_symbols).lower()
all_element_letters
```

```
'hhelibebcnofnenamgalsipsclarkcasctivcrmnfeconicuzngageassebrkrrbsryzrnbmotcrur
hpdagcdinsnsbteixecsbalaceprndpmsmeugdtbdyhoertmybluhftawreosirptauhgtlpbbipoa
trnfrraacthpaunppuamcmbkcfesfmmdnolrrfdbsgbhhsmtdsrgcnnhflmclvtsog'
```

Then loop over the letters of the alphabet and count:

```python
for letter in 'abcdefghijklmnopqrstuvwxyz':
    print(letter, ':', all_element_letters.count(letter))
```

```
a : 16
b : 14
c : 16
d : 8
e : 13
f : 8
g : 9
h : 10
i : 8
j : 0
k : 3
l : 9
m : 12
n : 15
o : 8
p : 10
q : 0
```

```
r : 19
s : 16
t : 12
u : 7
v : 2
w : 1
x : 1
y : 3
z : 2
```

'r' is the most common letter, followed by 'a', 'c', and 's'. The letters 'j' and 'q' do not appear at all.

P7.1

First create sets for each crystal structure:

```
fcc = set(['Cu', 'Co', 'Fe', 'Mn', 'Ni', 'Sc'])
bcc = set(['Cr', 'Fe', 'Mn', 'Ti', 'V'])
hcp = set(['Co', 'Ni', 'Sc', 'Ti', 'Zn'])
```

(a) To find which metals only exist in a single structure, take the difference between the sets:

```
print('fcc only:', fcc - bcc - hcp)
print('bcc only:', bcc - fcc - hcp)
print('hcp only:', hcp - fcc - bcc)
```

```
fcc only: {'Cu'}
bcc only: {'V', 'Cr'}
hcp only: {'Zn'}
```

(b) There are three ways in which a metal can exist in two of these structures. (Note that the set intersection operator is &):

```
print('fcc and bcc:', fcc & bcc - hcp)
print('fcc and hcp:', fcc & hcp - bcc)
print('bcc and hcp:', bcc & hcp - fcc)
```

```
fcc and bcc: {'Mn', 'Fe'}
fcc and hcp: {'Sc', 'Co', 'Ni'}
bcc and hcp: {'Ti'}
```

(c) To find the metals which do not form an hcp structure, take the union of the fcc and bcc metals, and then the difference with the hcp set. The union operator, |, has lower precedence than the difference operator, -, so we need parentheses:

```
print('No hcp:', (fcc | bcc) - hcp)
```

```
No hcp: {'Fe', 'Cr', 'Cu', 'Mn', 'V'}
```

Do any of these metals exist in all three structures?

```
print('fcc and bcc and hcp:', fcc & bcc & hcp)
```

```
fcc and bcc and hcp: set()
```

No – the intersection of all three sets is the empty set.

P7.2

Here is one solution:

```python
def get_stoichiometry(formula):
    """Return all the element symbols in the string formula."""

    n = len(formula)

    # First get all of the "subformulas": each element symbol
    # and its stoichiometry as given in the formula.
    i = 0
    subformulas = []
    for j in range(1, n):
        # The sentinel for a new element symbol is a capital letter.
        if formula[j].isupper():
            subformulas.append(formula[i:j])
            i = j
    # The last one needs special treatment because it has no other
    # one following it.
    subformulas.append(formula[i:])

    # Now build a dictionary of element symbol: stoichiometry pairs.
    stoichiometry = {}
    for s in subformulas:
        # Get the element symbol
        symbol = s[0]
        if len(s) > 1:
            if s[1].islower():
                symbol = s[:2]

        # What's left after the symbol?
        s_n = s[len(symbol):]
        if not s_n:
            # Nothing? Then the stoichiometry is 1
            n = 1
        else:
            # Convert the stoichiometry string into an int.
            n = int(s_n)
        # Add to the stoichiometry dictionary.
```

```
        if symbol not in stoichiometry:
            # This is the first time we've encountered this symbol.
            stoichiometry[symbol] = n
        else:
            # Add to an existing stoichiometry entry.
            stoichiometry[symbol] += n
    return stoichiometry

def get_stoichiometric_formula(formula):
    """Parse a chemical formula and return its stoichiometric formula."""

    stoichiometry = get_stoichiometry(formula)
    # Build the stoichiometric formula string.
    stoichiometric_formula = ''
    for symbol, stoich in stoichiometry.items():
        stoichiometric_formula += symbol
        # Only add the stoichiometry number if it is different from 1.
        if stoich > 1:
            stoichiometric_formula += str(stoich)
    return stoichiometric_formula
```

Some examples:

```
formula = 'CClH2CBr2COH'
print(get_stoichiometry(formula))
print(get_stoichiometric_formula(formula))
```

```
{'C': 3, 'Cl': 1, 'H': 3, 'Br': 2, 'O': 1}
C3ClH3Br2O
```

```
get_stoichiometric_formula('CH3CH2OH')
```

```
'C2H6O'
```

```
get_stoichiometric_formula('C6H5CH3')
```

```
'C7H8'
```

```
get_stoichiometric_formula('HOCH2CH2OH')
```

```
'H6O2C2'
```

P8.1

The provided data, in the file `thermo-data.csv` are:

```
    Formula,     DfHm /kJ.mol-1,     Sm /J.K-1.mol-1
    H2O(g),      -241.83,            188.84
```

```
CO2(g),      -393.52,           213.79
LiOH(s),     -484.93,           42.81
Li2CO3(s),   -1216.04,          90.31
```

This file can be read in with a simple function:

```python
import math

def read_data(filename):
    DfHm, Sm = {}, {}
    with open(filename) as fi:
        fi.readline()
        for line in fi.readlines():
            fields = line.split(',')
            compound = fields[0].strip()
            DfHm[compound] = float(fields[1])
            Sm[compound] = float(fields[2])
    return DfHm, Sm

filename = 'thermo-data.csv'
DfHm, Sm = read_data(filename)
print('Enthalpies of formation:', DfHm)
print('Molar entropies:', Sm)
```

```
Enthalpies of formation: {'H2O(g)': -241.83, 'CO2(g)': -393.52,
                          'LiOH(s)': -484.93, 'Li2CO3(s)': -1216.04}
Molar entropies: {'H2O(g)': 188.84, 'CO2(g)': 213.79,
                  'LiOH(s)': 42.81, 'Li2CO3(s)': 90.31}
```

For the reaction

$$2\text{LiOH(s)} + \text{CO}_2\text{(g)} \rightarrow \text{Li}_2\text{CO}_3\text{(s)} + \text{H}_2\text{O(g)}$$

The relevant reaction enthalpy and entropy are:

```python
DrH = DfHm['Li2CO3(s)'] + DfHm['H2O(g)'] - 2 * DfHm['LiOH(s)'] - DfHm['CO2(g)']
DrS = Sm['Li2CO3(s)'] + Sm['H2O(g)'] - 2 * Sm['LiOH(s)'] - Sm['CO2(g)']
print(f'DrH = {DrH:.2f} kJ.mol-1')
print(f'DrS = {DrS:.2f} J.K-1.mol-1')
```

```
DrH = -94.49 kJ.mol-1
DrS = -20.26 J.K-1.mol-1
```

The standard Gibbs free energy of reaction and equilibrium constant are:

$$\Delta_r G^{\ominus} = \Delta_r H^{\ominus} - T\Delta_r S^{\ominus}, \quad K = \exp\left(-\frac{\Delta_r G^{\ominus}}{RT}\right).$$

```python
R = 8.314        # Gas constant, J.K-1.mol-1
T = 298          # Temperature, K

DrG = DrH - T * DrS / 1000
```

```
K = math.exp(-DrG * 1000/ R / T)

print(f'DrG = {DrG:.2f} kJ.mol-1')
print(f'K = {K:.1e}')
```

```
DrG = -88.45 kJ.mol-1
K = 3.2e+15
```

The equilibrium constant is large and the products are strongly favored under standard conditions at room temperature.

P8.2

The data file consists of columns delimited by whitespace, so each line can be split into its fields with split(). There is a header line with the column names that can be skipped.

The entropy of vaporization is calculated as:

$$\Delta_{vap}S = \frac{\Delta_{vap}H}{T_b},$$

and converted into the units $J\,K^{-1}\,mol^{-1}$.

```
filename = 'vap-data.txt'

data = {}
with open(filename) as fi:
    # Skip header line.
    fi.readline()
    for line in fi.readlines():
        fields = line.split()
        compound, Tb, DvapH = fields[0], float(fields[1]), float(fields[2])
        DvapS = DvapH * 1000 / Tb
        # Data for each compound is stored in this dictionary.
        data[compound] = {'Tb': Tb, 'DvapH': DvapH, 'DvapS': DvapS}
        print(f'{compound:>12s}: DvapS = {DvapS:5.1f} J.K-1.mol-1')
```

```
      CH3Cl: DvapS =   87.0 J.K-1.mol-1
 CH2CHCHCH2: DvapS =   83.7 J.K-1.mol-1
     C2H5OH: DvapS =  110.1 J.K-1.mol-1
        H2O: DvapS =  109.1 J.K-1.mol-1
        CH4: DvapS =   73.1 J.K-1.mol-1
      HCO2H: DvapS =   60.7 J.K-1.mol-1
      CH3OH: DvapS =  104.2 J.K-1.mol-1
      C8H18: DvapS =   86.3 J.K-1.mol-1
    C6H5CH3: DvapS =   86.5 J.K-1.mol-1
```

Some of these compounds show significant deviation from Trouton's rule. Those with higher values of $\Delta_{vap}S$ have extensive hydrogen-bonding in the liquid (and hence greater than normal ordering in this phase).

Formic acid (HCO_2H) forms dimers in the gas phase, which is, therefore, less disordered than the ideal gas assumed by Trouton's rule, so $\Delta_{vap}S < 85\,J\,K^{-1}\,mol^{-1}$.

Methane (CH_4) has a rather low boiling point: Trouton's rule is known to break down substances with $T_b < 150\,K$ which have very weak intermolecular interactions (and hence greater disorder) in the liquid phase; In addition, quantum mechanical effects must be taken into account when considering the entropy of the gas phase of methane close to its boiling point: its widely spaced rotational energy levels are not populated the way a classical theory predicts.

To write a new file, open one with the mode `'w'`:

```
with open('vap-data2.txt', 'w') as fo:
    print('Compound      Tb /K    DvapH /kJ.mol-1'
          'DvapS /J.K-1.mol-1', file=fo)
    for compound, vap_data in data.items():
        print('{:10s}    {:6.1f}    {:12.1f}    {:12.1f}'
              .format(compound,
                      vap_data['Tb'],
                      vap_data['DvapH'],
                      vap_data['DvapS']
                      ),
              file=fo)
```

P9.1

```
import numpy as np
a = np.linspace(1,48,48).reshape(3,4,4)
a
```

```
array([[[ 1.,  2.,  3.,  4.],
        [ 5.,  6.,  7.,  8.],
        [ 9., 10., 11., 12.],
        [13., 14., 15., 16.]],

       [[17., 18., 19., 20.],
        [21., 22., 23., 24.],
        [25., 26., 27., 28.],
        [29., 30., 31., 32.]],

       [[33., 34., 35., 36.],
        [37., 38., 39., 40.],
        [41., 42., 43., 44.],
        [45., 46., 47., 48.]]])
```

(a) 20.0 is located in the last column of the first row, of the second block:

```
a[1,0,3]       # or a[1,0,-1]
```

```
20.0
```

(b) This is the whole of the third row of the first block:

```
a[0,2,:]
```

```
array([ 9.,  10.,  11.,  12.])
```

(c) This is the whole of the final block:

```
a[-1,:,:]     # or a[-1,...]: use ellipsis for repeated :
```

```
array([[33.,  34.,  35.,  36.],
       [37.,  38.,  39.,  40.],
       [41.,  42.,  43.,  44.],
       [45.,  46.,  47.,  48.]])
```

(d) The first two elements of the second row from every block:

```
a[:,1,:2]
```

```
array([[ 5.,   6.],
       [21.,  22.],
       [37.,  38.]])
```

(e) The last two elements of every row in the final block, but backward:

```
a[-1,:,:1:-1]
```

```
array([[36.,  35.],
       [40.,  39.],
       [44.,  43.],
       [48.,  47.]])
```

(f) The first element of each row (taken backwards) from each block:

```
a[:,::-1,0]
```

```
array([[13.,   9.,   5.,   1.],
       [29.,  25.,  21.,  17.],
       [45.,  41.,  37.,  33.]])
```

P9.2

The file containing the students' yield data starts like this:

```
Student ID   KCl yield (g)

  1 25.0
  2 23.5
  3 261.2
  4 16.4
  5 19.0
  6 *
  7 18.7
  8 18.5
```

Missing data are apparently indicated by an asterisk, and there are two header lines. We don't really need the Student ID column for this analysis, so we can read only the second column with `genfromtxt`:

```python
import numpy as np
masses = np.genfromtxt('KCl-yields.txt', skip_header=2, missing_values='*',
                       usecols=1)
masses
```

```
array([ 25. ,   23.5, 261.2,   16.4,  19. ,    nan,  18.7,  18.5,  28.6,
         21.4,  19.3,  29.7,    nan,  21.2,  18.4,  23.7,  24.9,  15.6,
         22.9,   0. ,  18.8,  20.8,    nan,  23.9,  16.8,  24.6,  23.8,
         27.4,  17.9,  27.5])
```

Calculating the median and standard deviation of the raw data (whilst ignoring missing data points):

```python
median, std = np.nanmedian(masses), np.nanstd(masses)
median, std
```

```
(21.4, 45.688600061536036)
```

We can identify invalid data with a boolean index and set these entries equal to NaN:

```python
masses[(abs(masses) > median+3*std) | (masses == 0)] = np.nan
masses
```

```
array([25. ,  23.5,   nan, 16.4, 19. ,   nan, 18.7, 18.5, 28.6, 21.4, 19.3,
       29.7,   nan, 21.2, 18.4, 23.7, 24.9, 15.6, 22.9,   nan, 18.8, 20.8,
         nan, 23.9, 16.8, 24.6, 23.8, 27.4, 17.9, 27.5])
```

The report on the processed data is then:

```python
print(f'Minimum yield: {np.nanmin(masses):.1f} g')
print(f'Maximum yield: {np.nanmax(masses):.1f} g')
print(f'Mean yield: {np.nanmean(masses):.1f} g')
print(f'Median yield: {np.nanmedian(masses):.1f} g')
print(f'Yield standard deviation: {np.nanstd(masses):.1f} g')
```

```
Minimum yield: 15.6 g
Maximum yield: 29.7 g
Mean yield: 21.9 g
Median yield: 21.4 g
Yield standard deviation: 3.9 g
```

```
# Molar masses of KClO3 and KCl in g.mol-1.
M_KClO3, M_KCl = 122.6, 74.6
```

```
# Initial mass (g) and amount (mol) of KClO3.
m_KClO3 = 50
n_KClO3 = m_KClO3 / M_KClO3
# Final amount (mol) and mass (g) of KCl, assuming complete
# decomposition of KClO3.
n_KCl = n_KClO3
m_KCl = M_KCl * n_KCl
print(f'Maximum theoretical yield of KCl: {m_KCl:.1f} g')
```

```
Maximum theoretical yield of KCl: 30.4 g
```

```
percentage_yields = masses / m_KCl * 100
print('Reported percentage yields of KCl')
for percentage_yield in sorted(percentage_yields[~np.isnan(percentage_yields)]):
    if not np.isnan(percentage_yield):
        print(f'{percentage_yield:.1f}%')
```

```
Reported percentage yields of KCl
51.3%
53.9%
55.2%
58.8%
60.5%
60.8%
61.5%
61.8%
62.5%
63.4%
68.4%
69.7%
70.3%
75.3%
77.2%
77.9%
78.2%
78.6%
80.9%
81.8%
82.2%
90.1%
90.4%
```

```
94.0%
97.6%
```

P10.1

The following expressions for the different types of average speed can be derived:

$$v_\star = \sqrt{\frac{2k_\mathrm{B}T}{m}} \quad \text{(mode)},$$

$$\langle v \rangle = \sqrt{\frac{8k_\mathrm{B}T}{\pi m}} \quad \text{(mean)},$$

$$\langle v^2 \rangle^{1/2} = \sqrt{\frac{3k_\mathrm{B}T}{m}} \quad \text{(rms speed)}.$$

```python
import numpy as np
import matplotlib.pyplot as plt
# Boltzmann constant, J.K-1, atomic mass unit (kg)
kB, u = 1.381e-23, 1.661e-27
```

```python
# Gas particle masses, in kg.
m_Ar, m_Xe = 40 * u, 131 * u
# Temperature in K.
T = 300
```

```python
def fMB(v, T, m):
    """
    Return value of the Maxwell-Boltzmann distribution for a molecule of
    mass m (in kg) at temperature T (in K), at the speed v (in m.s-1).

    """

    fac = m / 2 / kB / T
    return (fac / np.pi)**1.5 * 4 * np.pi * v**2 * np.exp(-fac * v**2)

def MB_averages(T, m):
    """
    Return the mode, mean and root-mean-square average speeds for the Maxwell-
    Boltzmann distribution for a gas of particles of mass m at temperature T.

    """

    fac = kB * T / m
    mode = np.sqrt(2 * fac)
    mean = np.sqrt(8 * fac / np.pi)
    rms = np.sqrt(3 * fac)
    return mode, mean, rms

def plot_MB(T, m, label, ls='-'):
    """
```

Plot the Maxwell-Boltzmann speed distribution for an ideal gas of
particles with mass m (in kg) at temperature T (in K). Indicate on
the plot the modal, mean and rms speeds. label is the label to
assign to the plotted line and ls the linestyle to use.

```
    """

    # A grid of speeds between 0 and vmax.
    vmax = 1000
    v = np.linspace(0, vmax, 1000)
    # Calculate and plot the Maxwell-Boltzmann distribution.
    f = fMB(v, T, m)
    plt.plot(v, f, c='k', ls=ls, label=label)

    # Calculate the various averages.
    mode, mean, rms = MB_averages(T, m)
    # It will be useful to have the value of the distribution maximum.
    fmax = np.max(f)
    # Plot straight lines from 0 to the corresponding point on the
    # distribution curve for each of the averages.
    plt.plot([mode, mode], [0, fmax], c='gray')
    plt.plot([mean, mean], [0, fMB(mean, T, m)], c='gray')
    plt.plot([rms, rms], [0, fMB(rms, T, m)], c='gray')

    # Add labels: use a bit of trial and error to get the placement right.
    plt.text(mode, fmax*1.05, 'mode', ha='center')
    plt.text(mean, fmax, 'mean')
    plt.text(rms, fmax*0.95, 'rms')
    # Set the y-limits from zero to 10% above the distribution maximum.
    plt.ylim(0, fmax*1.1)
    plt.legend()
```

```
plot_MB(T, m_Ar, 'Ar')
plot_MB(T, m_Xe, 'Xe', '--')
```

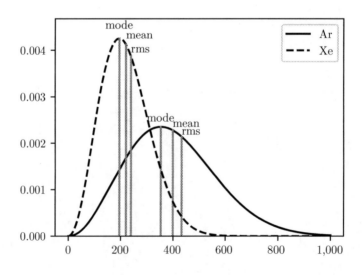

P15.1

1. Cc1ccccc1
2. CCCCC/C=C/CCCCCCCCCC(O)=O
3. C(C(=O)O)N
4. CC1=CC[C@@H](CC1)C(=C)C
5. C1CCC2CCCC2C1

P15.2

1. Phenol, Ph–OH
2. 2-Methyl-1-butanol, $CH_3CH_2CH(CH_3)CH_2OH$
3. 1-Methyl-cyclopentene,

4. Proline,

P17.1

The code below plots a $p - V$ diagram for the Carnot Cycle.

```python
import numpy as np
from scipy.constants import R
import matplotlib.pyplot as plt

def get_p(V, T):
    """Return the pressure of 1 mol of an ideal gas at (V, T)."""
    return R * T / V
```

```python
def plot_isotherm(Vmin, Vmax, T, c='k'):
    """Plot a single isotherm for temperature T between Vmin and Vmax.

    Returns the arrays of V, p values defining the isotherm.

    """
```

```
    V = np.linspace(Vmin, Vmax, 1000)
    p = get_p(V, T)
    plt.plot(V, p, c=c, label=f'{T:.1f} K', lw=1)
    return V, p

def plot_adiabat(Vmin, Vmax, p0, V0, c='k'):
    """Plot a single adiabat for between Vmin and Vmax.

    p0, V0 is a fixed point on the adiabat; the function returns the arrays
    of V, p values defining the adiabat.

    """
    V = np.linspace(Vmin, Vmax, 1000)
    p = p0 * (V/V0)**(-5/3)
    plt.plot(V, p, c=c, lw=1)
    return V, p

def carnot_cycle(V1, Thot, r_i, r_a):
    """
    Plot a p-V diagram for a Carnot cycle for an ideal gas starting at
    a compressed state (V1, Thot) defined by isothermal compression ratio
    r_i = V2/V1 and adiabatic compression ratio, r_a = V3/V2.

    """

    # Constant-volume heat capacity, "gamma" parameter (Cp/Cv).
    C_V, gamma = 3/2 * R, 5/3
    # Calculate the state variables at each stage of the cycle.
    V2 = V1 * r_i
    V3 = V2 * r_a
    V4 = V3 * V1 / V2
    p1, p2 = get_p(V1, Thot), get_p(V2, Thot)
    Tcold = Thot * (V2 / V3) ** (gamma - 1)
    p3, p4 = get_p(V3, Tcold), get_p(V4, Tcold)

    # Plot the isotherms and adiabats for the cycle.
    VA, pA = plot_isotherm(V1, V2, Thot, c='r')
    VB, pB = plot_adiabat(V2, V3, p2, V2)
    VC, pC = plot_isotherm(V4, V3, Tcold, c='b')
    VD, pD = plot_adiabat(V1, V4, p1, V1)
    plt.xlabel('$V\;/\mathrm{m^3}$')
    plt.ylabel('$p\;/\mathrm{Pa}$')

    # Step 1: isothermal expansion, V1 -> V2 at Thot.
    w1 = - n * R * Thot * np.log(V2 / V1)
    q1 = -w1
    # Step 2: adiabatic expansion, (V2, Thot) -> (V3, Tcold).
    w2 = n * C_V * (Tcold - Thot)
    q2 = 0
    # Step 3: isothermal compression, V3 -> V4, at Tcold.
    w3 = - n * R * Tcold * np.log(V4 / V3)
    q3 = -w3
    # Step 4: adiabatic compression, (V4, Tcold -> V1, Thot).
```

```
        w4 = n * C_V * (Thot - Tcold)
        q4 = 0

        # Label the states.
        states = [(p1, V1), (p2, V2), (p3, V3), (p4, V4)]
        for i, (p, V) in enumerate(states):
            plt.text(V, p, 'ABCD'[i])
        plt.legend()

        # Total energy input through heating:
        q = q1
        # Total work done *on* the gas:
        w = w1 + w2 + w3 + w4
        # Efficiency.
        eta = -w / q
        print('Efficiency = {:.1f} %'.format(eta * 100))
```

```
Efficiency = 37.0 %
```

```
n = 1
V1, Thot = 0.2, 800
r_i, r_a = 2, 2
carnot_cycle(V1, Thot, r_i, r_a)
```

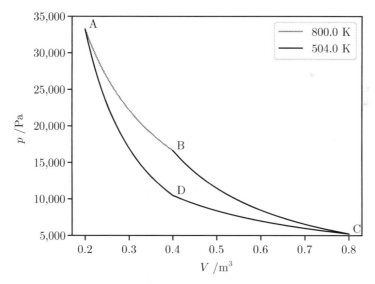

The efficiency reported above is, indeed, found to be given by $\eta = 1 - \frac{T_{\text{cold}}}{T_{\text{hot}}}$:

```
(1 - 504 / 800) * 100
```

```
37.0
```

P18.1

(a) Balancing a redox reaction

The provided reaction, in basic solution, needs hydroxide ions and water molecules added before it can be balanced:

$$a\text{MnO}_4{}^-(\text{aq}) + b\text{I}^-(\text{aq}) + c\text{H}_2\text{O(l)} \longrightarrow d\text{MnO}_2(\text{s}) + e\text{I}_2(\text{aq}) + f\text{OH}^-(\text{aq}).$$

The equations governing the balance of the reaction can be written as

$$\begin{aligned}
\text{Mn}: &\quad a + d = 0 \\
\text{O}: &\quad 4a + c + 2d + f = 0 \\
\text{I}: &\quad b + 2e = 0 \\
\text{H}: &\quad 2c + f = 0 \\
\text{charge}: &\quad a + b + f = 0
\end{aligned}$$

(the unique solution, if it exists will then involve the coefficients d, e and f having the opposite signs from coefficients a, b, and c). As before, we will fix $a = 1$:

```
import numpy as np
M = np.array([[1, 0, 0, 1, 0, 0],
              [4, 0, 1, 2, 0, 1],
              [0, 1, 0, 0, 2, 0],
              [0, 0, 2, 0, 0, 1],
              [1, 1, 0, 0, 0, 1]])
A = np.vstack((M, [1, 0, 0, 0, 0, 0]))
x = np.array([0, 0, 0, 0, 0, 1])
coeffs = np.linalg.solve(A, x)
coeffs
```

```
array([ 1. ,   3. ,   2. ,  -1. ,  -1.5, -4. ])
```

Therefore, the balanced equation is:

$$2\,\text{MnO}_4{}^-(\text{aq}) + 6\,\text{I}^-(\text{aq}) + 4\,\text{H}_2\text{O(l)} \longrightarrow 2\,\text{MnO}_2(\text{s}) + 3\,\text{I}_2(\text{aq}) + 8\,\text{OH}^-(\text{aq})$$

(b) Balancing a complex reaction

This reaction can be balanced in the same way.

$$a\text{Cr}_7\text{N}_{66}\text{H}_{96}\text{C}_{42}\text{O}_{24} + b\text{MnO}_4{}^- + c\text{H}^+ \longrightarrow d\text{Cr}_2\text{O}_7{}^{2-} + e\text{Mn}^{2+} + f\text{CO}_2 + g\text{NO}_3{}^- + h\text{H}_2\text{O}$$

```
import numpy as np
M = np.array([[ 7, 0, 0, 2, 0, 0, 0, 0],
              [66, 0, 0, 0, 0, 0, 1, 0],
              [96, 0, 1, 0, 0, 0, 0, 2],
              [42, 0, 0, 0, 0, 1, 0, 0],
```

```
            [24, 4, 0, 7, 0, 2, 3, 1],
            [ 0, 1, 0, 0, 1, 0, 0, 0],
            [ 0, -1, 1, -2, 2, 0, -1, 0]])
Mrank = np.linalg.matrix_rank(M)
print(f'rank of M = {Mrank}; full rank? {Mrank==M.shape[0]}')
A = np.vstack((M, [0, 0, 0, 0, 0, 0, 0, 1]))
x = np.array([0, 0, 0, 0, 0, 0, 0, 1])
coeffs = np.linalg.solve(A, x)
coeffs
```

```
rank of M = 7; full rank? True

array([-0.00532198, -0.62586482, -1.48908994,  0.01862693,  0.62586482,
        0.22352315,  0.35125067,  1.         ])
```

This doesn't look very encouraging, but if we divide by the smallest value, we can work out the integer coefficients:

```
coeffs /= np.min(np.abs(coeffs))
coeffs
```

```
array([  -1.  , -117.6, -279.8,    3.5,  117.6,   42. ,   66. ,  187.9])
```

The balanced reaction is therefore:

$$10\,Cr_7N_{66}H_{96}C_{42}O_{24} + 1176\,MnO_4^- + 2798\,H^+ \longrightarrow 35\,Cr_2O_7^{2-} + 1176\,Mn^{2+}$$
$$+ 420\,CO_2 + 660\,NO_3^- + 1879\,H_2O$$

(c) The reaction between copper and nitric acid

For the reaction:

$$aCu(s) + bHNO_3(aq) \longrightarrow cCu(NO_3)_2(aq) + dNO(g) + eNO_2(g) + fH_2O(l)$$

The chemical composition matrix is:

$$\mathbf{M} = \begin{pmatrix} -1 & 0 & 1 & 0 & 0 & 0 \\ 0 & -1 & 0 & 0 & 0 & 2 \\ 0 & -1 & 2 & 1 & 1 & 0 \\ 0 & -3 & 6 & 1 & 2 & 1 \end{pmatrix}$$

```
M = np.array([[-1,  0, 1, 0, 0, 0],
              [ 0, -1, 0, 0, 0, 2],
              [ 0, -1, 2, 1, 1, 0],
              [ 0, -3, 6, 1, 2, 1]])
n = M.shape[1]
r = np.linalg.matrix_rank(M)
print(f'n = {n}, r = {r}, nullity = {n-r}')
```

```
n = 6, r = 4, nullity = 2
```

The nullity of **M** is 2, so there is no unique way to balance the reaction as written.

Given the additional information about the gaseous product in the limit of strong and weak nitric acid solution, the reactions can be balanced, however. With concentrated HNO_3, we can eliminate the column corresponding to NO since NO_2 is observed to be the product:

```
M = np.array([[-1,   0,  1,  0,  0],
              [ 0,  -1,  0,  0,  2],
              [ 0,  -1,  2,  1,  0],
              [ 0,  -3,  6,  2,  1]])
n = M.shape[1]
r = np.linalg.matrix_rank(M)
print(f'n = {n}, r = {r}, nullity = {n-r}')
A = np.vstack((M, [1, 0, 0, 0, 0]))
x = np.array([0, 0, 0, 0, 1])
coeffs = np.linalg.solve(A, x)
coeffs
```

```
n = 5, r = 4, nullity = 1

array([1., 4., 1., 2., 2.])
```

The balanced reaction in this case is therefore:

$$Cu(s) + 4\,HNO_3(aq, conc) \longrightarrow Cu(NO_3)_2(aq) + 2\,NO_2(g) + 2\,H_2O(l)$$

In dilute nitric acid, we can eliminate the column corresponding to NO_2 instead:

```
M = np.array([[-1,   0,  1,  0,  0],
              [ 0,  -1,  0,  0,  2],
              [ 0,  -1,  2,  1,  0],
              [ 0,  -3,  6,  1,  1]])
n = M.shape[1]
r = np.linalg.matrix_rank(M)
print(f'n = {n}, r = {r}, nullity = {n-r}')
A = np.vstack((M, [1, 0, 0, 0, 0]))
x = np.array([0, 0, 0, 0, 1])
coeffs = np.linalg.solve(A, x)
coeffs
```

```
n = 5, r = 4, nullity = 1

array([1.        , 2.66666667, 1.        , 0.66666667, 1.33333333])
```

To get integer coefficients, multiply by 3:

```
coeffs * 3
```

```
array([3., 8., 3., 2., 4.])
```

The balanced reaction in this case is

$$3\,\text{Cu(s)} + 8\,\text{HNO}_3(\text{aq, dil}) \longrightarrow 3\,\text{Cu(NO}_3)_2(\text{aq}) + 2\,\text{NO(g)} + 4\,\text{H}_2\text{O(l)}$$

P18.2

```
import numpy as np
s = [None] * 6
names = ['e', '(12)', '(13)', '(23)', '(123)', '(132)']
s[0] = np.eye(3, dtype=int)   # e
s[1] = np.array([[0, 1, 0], [1, 0, 0], [0, 0, 1]], dtype=int) # (12)
s[2] = np.array([[0, 0, 1], [0, 1, 0], [1, 0, 0]], dtype=int) # (13)
s[3] = np.array([[1, 0, 0], [0, 0, 1], [0, 1, 0]], dtype=int) # (23)
s[4] = np.array([[0, 1, 0], [0, 0, 1], [1, 0, 0]], dtype=int) # (123)
s[5] = np.array([[0, 0, 1], [1, 0, 0], [0, 1, 0]], dtype=int) # (132)
```

(a) First demonstrate that the permutation matrices are orthogonal:

```
for i, m in enumerate(s):
    minv = np.linalg.inv(m)
    if np.allclose(minv, m.T):
        print(f's[{i}] is orthogonal')
```

```
s[0] is orthogonal
s[1] is orthogonal
s[2] is orthogonal
s[3] is orthogonal
s[4] is orthogonal
s[5] is orthogonal
```

(b) Show that some power of each of s is equal to s[0], the identity matrix:

```
for i, m in enumerate(s):
    n = 1
    while not np.allclose(np.linalg.matrix_power(m, n), s[0]):
        n += 1
    print(f's[{i}]^{n} = e')
```

```
s[0]^1 = e
s[1]^2 = e
s[2]^2 = e
s[3]^2 = e
s[4]^3 = e
s[5]^3 = e
```

(c) From the Cayley table for the S_3 group, the parities of the permutations are as follows:

```
parities = [1, -1, -1, -1, 1, 1]
for name, parity in zip(names, parities):
    print(f'{name:>5s}: {parity:2d}')
```

```
      e:   1
  (12):  -1
  (13):  -1
  (23):  -1
 (123):   1
 (132):   1
```

The relation between these parities and the matrix determinants can be verified as follows:

```
for i, m in enumerate(s):
    det = np.linalg.det(m)
    print(f'{names[i]:>5s}: {parities[i]:2d} == {det:4.1f}')
```

```
      e:   1 ==   1.0
  (12):  -1 ==  -1.0
  (13):  -1 ==  -1.0
  (23):  -1 ==  -1.0
 (123):   1 ==   1.0
 (132):   1 ==   1.0
```

P19.1

The data file provided is:

```
# Reaction rates, v (in mM.s-1), for enzymatic kinetics of pepsin with
# bovine serum albumin (S, in mM).)
S /mM  v /mM.s-1
 0.1 0.00339
 0.2 0.00549
 0.5 0.00875
 1.0 0.01090
 5.0 0.01358
10.0 0.01401
20.0 0.01423
```

and can be read in with `np.genfromtxt`:

```
import numpy as np
import matplotlib.pyplot as plt

E0 = 0.028
S, v = np.genfromtxt('pepsin-rates.txt', skip_header=3, unpack=True)
S, v
```

```
(array([ 0.1,   0.2,   0.5,   1. ,   5. ,  10. ,  20. ]),
 array([0.00339, 0.00549, 0.00875, 0.0109 , 0.01358, 0.01401, 0.01423]))
```

The Lineweaver–Burk method involves taking the inverse of the Michaelis–Menten equation:

$$\frac{1}{v} = \frac{1}{v_{max}} + \frac{K_M}{v_{max}} \frac{1}{[S]}.$$

The best-fit parameters can be found by linear least-squares regression of $1/v$ against $1/[S]$.

```
X = np.vstack((np.ones(len(S)), 1/S)).T
res = np.linalg.lstsq(X, 1/v, rcond=None)
print('Fit parameters: {}\nSquared residuals: {}\n'
      'Rank of X: {}\nSingular values of X:{}'.format(*res))
```

```
Fit parameters: [69.13781054 22.58790909]
Squared residuals: [0.00799023]
Rank of X: 2
Singular values of X:[11.52087785  2.07891166]
```

```
coeffs = res[0]
vmax_fit = 1/coeffs[0]
KM_fit = coeffs[1] * vmax_fit
v_fit = 1 / (KM_fit / vmax_fit / S + 1 / vmax_fit)
k2_fit = vmax_fit / E0
print(f'Fitted KM = {KM_fit:.3f} mM')
print(f'Fitted vmax = {vmax_fit:.4f} s-1')
print(f'Fitted k2 = {k2_fit:.2f} s-1')
plt.plot(1/S, 1/v, 'x')
plt.plot(1/S, 1 / v_fit)
plt.xlabel('$1/\mathrm{[S]}\;\/\mathrm{mM^{-1}}$')
plt.ylabel('$1/v\;/\mathrm{s\,mM^{-1}}$')
```

```
Fitted KM = 0.327 mM
Fitted vmax = 0.0145 s-1
Fitted k2 = 0.52 s-1
```

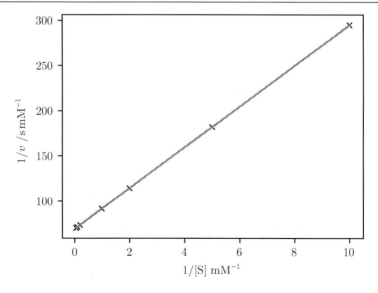

P19.2

For "simple" linear regression on n points using a straight line model, $y = a + bx$, Equation 19.4,

$$\hat{\beta} = \begin{pmatrix} a \\ b \end{pmatrix} = (\mathbf{X}^T \mathbf{X})^{-1} \mathbf{X}^T \mathbf{y},$$

defines the best-fit coefficients a and b where the matrices $\mathbf{y}$ and $\mathbf{X}$

$$\mathbf{y} = \begin{pmatrix} y_1 \\ y_2 \\ \vdots \\ y_n \end{pmatrix}, \quad \text{and} \quad \mathbf{X} = \begin{pmatrix} 1 & x_1 \\ 1 & x_2 \\ \vdots \\ 1 & x_n \end{pmatrix}. \tag{40.1}$$

We have

$$\mathbf{X}^T \mathbf{X} = \begin{pmatrix} 1 & 1 & 1 & \cdots & 1 \\ x_1 & x_2 & x_3 & \cdots & x_n \end{pmatrix} \begin{pmatrix} 1 & x_1 \\ 1 & x_2 \\ 1 & x_3 \\ \vdots & \vdots \\ 1 & x_n \end{pmatrix} = \begin{pmatrix} n & \sum x_i \\ \sum x_i & \sum x_i^2 \end{pmatrix}. \tag{40.2}$$

Taking the inverse,

$$(\mathbf{X}^T \mathbf{X})^{-1} = \frac{1}{\Delta} \begin{pmatrix} \sum x_i^2 & -\sum x_i \\ -\sum x_i & n \end{pmatrix}, \tag{40.3}$$

where the determinant,

$$\Delta = n \sum x_i^2 - \left(\sum x_i \right)^2 = n S_{xx} - S_x^2.$$

Next,

$$(\mathbf{X}^T \mathbf{X})^{-1} \mathbf{X}^T = \frac{1}{\Delta} \begin{pmatrix} \sum x_i^2 & -\sum x_i \\ -\sum x_i & n \end{pmatrix} \begin{pmatrix} 1 & 1 & 1 & \cdots & 1 \\ x_1 & x_2 & x_3 & \cdots & x_n \end{pmatrix}$$

$$= \frac{1}{\Delta} \begin{pmatrix} \sum x_i^2 - x_1 \sum x_i & \sum x_i^2 - x_2 \sum x_i & \cdots & \sum x_i^2 - x_n \sum x_i \\ -\sum x_i + n x_1 & -\sum x_i + n x_2 & \cdots & -\sum x_i + n x_n \end{pmatrix}.$$

Finally,

$$\hat{\beta} = (\mathbf{X}^T\mathbf{X})^{-1}\mathbf{X}^T\mathbf{y}$$

$$= \frac{1}{\Delta}\begin{pmatrix} \sum x_i^2 - x_1\sum x_i & \sum x_i^2 - x_2\sum x_i & \cdots & \sum x_i^2 - x_n\sum x_i \\ -\sum x_i + nx_1 & -\sum x_i + nx_2 & \cdots & -\sum x_i + nx_n \end{pmatrix}\begin{pmatrix} y_1 \\ y_2 \\ \vdots \\ y_n \end{pmatrix} = \begin{pmatrix} a \\ b \end{pmatrix},$$

where

$$a = \frac{1}{\Delta}\left[y_1\left(\sum x_i^2 - x_1\sum x_i\right) + y_2\left(\sum x_i^2 - x_2\sum x_i\right) + \cdots + y_n\right.$$
$$\left.\left(\sum x_i^2 - x_n\sum x_i\right)\right]$$
$$= \frac{1}{\Delta}\left[\left(\sum y_i\right)\left(\sum x_i^2\right) - \left(\sum x_iy_i\right)\left(\sum x_i\right)\right]$$
$$= \frac{S_yS_{xx} - S_{xy}S_x}{nS_{xx} - S_x^2}$$

and

$$b = \frac{1}{\Delta}\left[y_1\left(nx_1 - \sum x_i\right) + y_2\left(nx_2 - \sum x_i\right) + \cdots + y_n\left(nx_n - \sum x_i\right)\right]$$
$$= \frac{1}{\Delta}\left[n\left(\sum x_iy_i\right) - \left(\sum y_i\right)\left(\sum x_i\right)\right]$$
$$= \frac{nS_{xy} - S_xS_y}{nS_{xx} - S_x^2}.$$

P20.1

1. For $y = ap \pm bq$, using

$$\frac{\partial y}{\partial p} = a, \quad \frac{\partial y}{\partial q} = \pm b,$$

$$\sigma_y = \sqrt{\left(\frac{\partial y}{\partial p}\right)^2\sigma_p^2 + \left(\frac{\partial y}{\partial q}\right)^2\sigma_q^2} = \sqrt{a^2\sigma_p^2 + b^2\sigma_q^2}.$$

2. For $y = cpq$,

$$\frac{\partial y}{\partial p} = cq, \quad \frac{\partial y}{\partial q} = cp,$$

and

$$\sigma_y = \sqrt{c^2 q^2 \sigma_p^2 + c^2 p^2 \sigma_q^2} = c\sqrt{p^2 q^2 \left(\frac{\sigma_p}{p}\right)^2 + p^2 q^2 \left(\frac{\sigma_q}{q}\right)^2}$$

$$= |y|\sqrt{\left(\frac{\sigma_p}{p}\right)^2 + \left(\frac{\sigma_q}{q}\right)^2}.$$

3. For $y = a \ln(bx)$,

$$\frac{\partial y}{\partial x} = \frac{a}{x}$$

and

$$\sigma_y = \sqrt{\left(\frac{\partial y}{\partial x}\right)^2 \sigma_x^2} = \sqrt{\left(\frac{a}{x}\right)^2 \sigma_x^2} = \left|\frac{a}{x}\right| \sigma_x$$

4. For $y = ae^{bx}$,

$$\frac{\partial y}{\partial x} = abe^{bx} = by$$

and

$$\sigma_y = \sqrt{b^2 y^2 \sigma_x^2} = |yb| \sigma_x.$$

P20.2

As given in the question,

$$y = p_1 - p_0 = \left[\frac{l_0}{l_0 - (h - l_1)} - 1\right] p_0 = \rho g l_1$$

Assuming that only the uncertainties in the measurements of h and l_1 are significant,

$$\sigma_y = \sqrt{\left(\frac{\partial y}{\partial h}\right)^2 \sigma_h^2 + \left(\frac{\partial y}{\partial l_1}\right)^2 \sigma_{l_1}^2},$$

where

$$\frac{\partial y}{\partial h} = \frac{l_0 p_0}{[l_0 - (h - l_1)]^2}$$

$$\frac{\partial y}{\partial l_1} = -\frac{l_0 p_0}{[l_0 - (h - l_1)]^2}.$$

Therefore,

$$\sigma_y = \sqrt{\frac{l_0^2 p_0^2}{[l_0 - (h - l_1)]^4} \left(\sigma_h^2 + \sigma_{l_1}^2\right)}$$

$$= \frac{l_0 p_0}{[l_0 - (h - l_1)]^2} \sqrt{\sigma_h^2 + \sigma_{l_1}^2}.$$

```python
import numpy as np
import matplotlib.pyplot as plt
```

```python
# Measured distances, h and l1, in cm.
h = [29.9, 35. , 39.9, 45.1, 50. , 55. , 59.9, 65. , 70.1, 74.9, 80.1,
     85.1, 90.1, 95. ]
L1 = [29.3, 34.3, 39.2, 44.5, 49.1, 54. , 58.7, 64. , 68.9, 73.5, 78.6,
      83.7, 88.6, 93.4]
n = len(h)

# Create NumPy arrays and convert units to m.
h = np.array(h) / 100
L1 = np.array(L1) / 100
```

```python
# Define the experimental parameters.
# Air pressure, Pa.
p0 = 1037 * 100
# Water density, kg.m-3.
rho = 1000
# Length of the test tube diving bell, m.
L0 = 0.2
# Measurement uncertainties, m.
sig_L1 = sig_h = 1e-3
```

```python
z = L0 - (h - L1)
x = L1
y = p1_minus_p0 = (L0 / z - 1) * p0

sigma_y = p0 * L0 / z**2 * np.sqrt(sig_L1**2 + sig_h**2)
```

```python
def plot_data():
    plt.errorbar(L1, p1_minus_p0, yerr=sigma_y, xerr=sig_L1, marker='o',
                 ls='', capsize=4)
```

```
    plt.xlabel('$l_1\;/\mathrm{m}$')
    plt.ylabel('$p_1 - p_0\;/\mathrm{Pa}$')
plot_data()
```

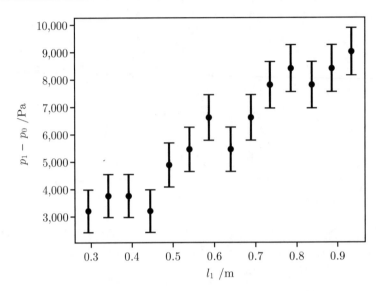

A straight line fit, $y = a + bx$ through these data provides an estimated value for g:

```
Sx, Sy, Sxx, Sxy = np.sum(x), np.sum(y), np.sum(x*x), np.sum(x*y)
Delta = n * Sxx - Sx**2
a = (Sy * Sxx - Sxy * Sx) / Delta
b = (n*Sxy - Sy * Sx) / Delta
print('a, b =', a, b)
g_est = b / rho
print('g-estimated =', g_est)
```

```
a, b = 113.64118138687233 9633.702167054678
g-estimated = 9.633702167054679
```

The uncertainty in the fitted parameter b is

$$\sigma_b = \sigma_y \sqrt{\frac{n}{nS_{xx} - S_x^2}},$$

and so we can estimate the uncertainty in g as

```
sigma_g = np.mean(sigma_y) * np.sqrt(n / Delta) / rho
sigma_g
```

```
1.1051296914381148
```

The result of this experiment is therefore $g = 9.6 \pm 1.1 \text{ m.s}^{-2}$.

P21.1

It is helpful to define some variables for the quantities provided.

```python
import numpy as np
from numpy.polynomial import Polynomial as P
from scipy.integrate import quad

# Gas constant, J.K-1.mol-1
R = 8.314

# Molar masses of H2 and CH4, g.mol 1.
M_H2, M_CH4 = 2, 16
# Number of moles of H2 and CH4 in 1 kg of each gas.
nH2, nCH4 = 1000 / M_H2, 1000 / M_CH4

# Molar combustion enthalpies, J.mol-1
DcHm_H2, DcHm_CH4 = -286e3, -891e3
# Specific combustion enthalpies, J.kg-1
DcHs_H2, DcHs_CH4 = DcHm_H2 * nH2, DcHm_CH4 * nCH4

# Temperature, K
T = 298
# Initial and final pressures, Pa
p1, p2 = 1e5, 2e7

# van der Waals parameters a and b (m6.Pa.mol-2 and m3.mol-1).
aH2, bH2 = 2.476e-2, 2.661e-5
aCH4, bCH4 = 0.2300, 4.301e-5
```

In this case, we know the amount of gas and the final pressure and temperature but not the volume. The van der Waals equation can be rearranged into the following cubic equation in V, which can be solved for its real root if p_2, n, and T are known (along with the van der Waals parameters, a and b).

$$pV^3 - n(pb + RT)V^2 + n^2 aV - n^3 ab = 0.$$

With this information, the work integral can be evaluated analytically or numerically as before:

$$w = -\int_{V_1}^{V_2} p \, dV = -\int_{V_1}^{V_2} \frac{nRT}{V - nb} - \frac{n^2 a}{V^2} \, dV.$$

```python
def get_V(n, p, T, a, b):
    """Invert the van der Waals equation to return V from n, p, T."""
    poly = Polynomial([-n**3 * a * b, n**2 * a, -n*(p*b + R*T), p])
    roots = poly.roots()

    # If we don't have exactly one real root, something is wrong (e.g.
    # the van der Waals equation gives unphysical results for this n, p, T).
    assert sum(np.isreal(roots)) == 1

    return roots[np.isreal(roots)][0].real
```

```
def get_w_per_kg(n, p2, T, a, b):
    """Return the work required to compress the gas to p2."""

    # Ideal gas equation is fine for the initial volume.
    V1 = n * R * T / p1
    # Get the final, compressed, volume from the van der Waals equation.
    V2 = get_V(n, p2, T, a, b)

    # Van der Waals equation of state.
    def p(V, n, R, T, a, b):
        return n * R * T / (V - n * b) - n**2 * a / V**2

    # Numerical integration of p with respect to V.
    wp, err = quad(p, V1, V2, args=(n, R, T, a, b))
    return -wp

wH2 = get_w_per_kg(nH2, p2, T, aH2, bH2)
print(f'Work required for compression of H2 to 200 bar: {wH2/1.e6:.1f} MJ/kg')
print(f'({wH2 / -DcHs_H2 * 100:.1f}% of enthalpy released on combustion)')


wCH4 = get_w_per_kg(nCH4, p2, T, aCH4, bCH4)
print(f'Work required for compression of CH4 to 200 bar: {wCH4/1.e6:.1f} MJ/kg')
print(f'({wCH4 / -DcHs_CH4 * 100:.1f}% of enthalpy released on combustion)')
```

```
Work required for compression of H2 to 200 bar: 6.5 MJ/kg
(4.6% of enthalpy released on combustion)
Work required for compression of CH4 to 200 bar: 0.8 MJ/kg
(1.4% of enthalpy released on combustion)
```

Note that over eight times more energy is required to compress hydrogen than to compress methane (since there are eight times as many H_2 molecules in 1 kg than CH_4 molecules).

P22.1

First define a function to calculate the Planck distribution as a function of wavelength:

$$u(\lambda, T) = \frac{8\pi^2 hc}{\lambda^5} \frac{1}{e^{hc/\lambda k_B T} - 1},$$

```
import numpy as np
from scipy.constants import h, c, k as kB
from scipy.optimize import minimize_scalar
import matplotlib.pyplot as plt
```

```
def planck(lam, T):
    """
```

```
Return the Planck function for wavelength lam (m) and temperature T (K).
"""
    return 8 * np.pi**2 * h * c / lam**5 / (np.exp(h*c/lam/kB/T) - 1)
```

```
# We want to find the maximum in the function u(lam, T), which will be the
# *minimum* in -u(lam, T):
def func(lam, T):
    return -planck(lam, T)
```

Simply calling `minimize_scalar` won't work:

```
minimize_scalar(func, args=(T,))
```

```
--------------------------------------------------------------------
ZeroDivisionError                        Traceback (most recent call last)
Input In [x], in <module>
----> 1 minimize_scalar(func, args=(T,))
...
Input In [x], in planck(lam, T)
      1 def planck(lam, T):
----> 2     return 8 * np.pi**2 * h * c / lam**5 / (np.exp(h*c/lam/kB/T) - 1)

ZeroDivisionError: float division by zero
```

Here, the algorithm has apparently started at `lam = 0` in trying to bracket the minimum and promptly encountered a division by zero error from the definition of $u(\lambda, T)$. We will need to bracket the minimum manually: A good place to start would be to plot the function:

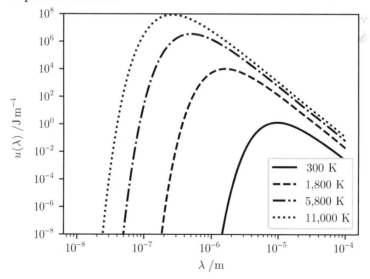

```
Ts = np.array([300, 1800, 5800, 11000])
# The form of the Planck function motivates the use of a logarithmic
```

```
# scale: set up a grid of wavelengths from 10^-8 to 10^-4 m.
lam = np.logspace(-8, -4, 1000)
for T in Ts:
    plt.plot(lam, planck(lam, T), label=f'{T} K')
plt.xscale('log')
plt.yscale('log')
plt.ylim(1.e-8, 1.e8)
plt.xlabel(r'$\lambda \;/\mathrm{m}$')
plt.ylabel(r'$u(\lambda) \;/\mathrm{J\,m^{-4}}$')
plt.legend()
```

With this information, we can choose a bracketing interval of, for example, 10^{-7} − 10^{-3} m: At the temperatures being considered, the maximum in $u(\lambda, T)$ always lies between these wavelengths.

```
lam_max = []
for T in Ts:
    res = minimize_scalar(func, args=(T,), bracket=(1.e-7, 1.e-3))
    if res.success:
        lam_max.append(res.x)
    else:
        print('Something went wrong. Inspect the returned object:')
        print(res)
```

```
lam_max
```

```
[9.659237948074667e-06,
 1.6098731068658823e-06,
 4.996162293070267e-07,
 2.6343372711944905e-07]
```

To demonstrate the linear relation between $1/T$ and λ_{max}, fit a straight line using np.linalg.lstsq.

```
X = np.vstack((np.ones(len(Ts)), 1/Ts)).T
ret = np.linalg.lstsq(X, lam_max, rcond=None)
ret
```

```
(array([2.01226347e-13, 2.89777132e-03]),
 array([1.35844905e-25]),
 2,
 array([2.00000108, 0.00267348]))
```

(Note that the very small (squared) residual error, 1.35844905e-25, indicating a very good fit.) This suggests that, for the Wien displacement law, $\lambda_{max} = b/T$, the parameter $b = 2.8978 \times 10^{-3}$ m K. This is in good agreement with the accepted value:

```
import scipy.constants as pc
pc.value('Wien displacement law constant')
```

```
0.0028977685
```

P22.2

For both approaches, we need the second derivative of the wavefunction:

$$\psi = \exp\left(-\alpha q^2/2\right)$$

$$\Rightarrow \frac{d\psi}{dq} = -\alpha q\psi$$

$$\Rightarrow \frac{d^2\psi}{dq^2} = -\alpha\psi + \alpha^2 q^2\psi = \alpha(\alpha q^2 - 1)\psi.$$

(a) For the numerical solution, define a function to calculate the Rayleigh–Ritz ratio,

$$\langle E'\rangle = \frac{\langle\psi|\hat{H}|\psi\rangle}{\langle\psi|\psi\rangle},$$

where

$$\langle\psi|\hat{H}|\psi\rangle = \int_{-\infty}^{\infty}\psi\left(-\frac{1}{2}\frac{d^2\psi}{dq^2} + \frac{1}{2}q^4\psi\right)dq$$

$$= \frac{1}{2}\int_{-\infty}^{\infty}\left(q^4 - \alpha^2 q^2 + \alpha\right)\psi^2\,dq.$$

```python
import numpy as np
from scipy.integrate import quad
import matplotlib.pyplot as plt

def psi(q, alpha):
    """Return the unnormalized trial wavefunction."""
    return np.exp(-alpha * q**2 / 2)

def psi2(q, alpha):
    """Return the square of the unnormalized trial wavefunction."""
    return psi(q, alpha)**2

def rayleigh_ritz(alpha):
    """Return the Rayleigh-Ritz ratio, <psi|H|psi> / <psi|psi>."""

    def func(q, alpha):
        return 0.5 * (q**4 - (alpha * q)**2 + alpha) * psi2(q, alpha)

    num, _ = quad(func, -np.inf, np.inf, args=(alpha,))
    det, _ = quad(psi2, -np.inf, np.inf, args=(alpha,))
    return num / det
```

We can learn more by plotting this $\langle E' \rangle$ as a function of α:

```
alpha_grid = np.linspace(0.5, 2.5, 25)
Eexp_grid = [rayleigh_ritz(alpha) for alpha in alpha_grid]
plt.plot(alpha_grid, Eexp_grid)
plt.xlabel(r'$q$')
plt.ylabel(r"$\langle E' \rangle$")
```

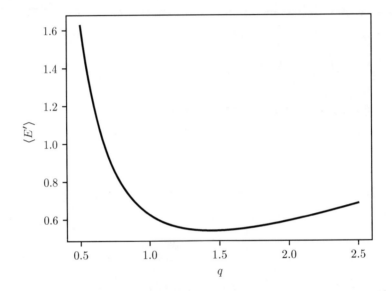

Evidently, we can bracket the minimum in $\langle E' \rangle(\alpha)$ using the values $\alpha_a = 1.0$, $\alpha_b = 1.5$ and $\alpha_c = 2.0$, since $\langle E' \rangle(\alpha_a) > \langle E' \rangle(\alpha_b)$ and $\langle E' \rangle(\alpha_c) > \langle E' \rangle(\alpha_b)$.

```
minimize_scalar(rayleigh_ritz, bracket=(1, 1.5, 2))
```

```
    fun: 0.5408435888652784
message: '\nOptimization terminated successfully;\nThe returned value satisfies
         the termination criteria\n(using xtol = 1.48e-08 )'
   nfev: 14
    nit: 10
success: True
      x: 1.4422495872533527
```

The optimum value of α is 1.442, giving an energy of $\langle E' \rangle = 0.541$ in units of $(k\hbar^4/\mu^2)^{1/3}$.

(b) To confirm the numerical result, we need to evaluate the integrals analytically. Consider the following family of integrals:

$$I_n = \int_{-\infty}^{\infty} q^{2n} e^{-\alpha q^2} \, dq = \int_{-\infty}^{\infty} q^{2n-1} q e^{-\alpha q^2} \, dq$$

$$= \left[q^{2n-1} \left(-\frac{1}{2\alpha} \right) e^{-\alpha q^2} \right]_{-\infty}^{\infty} - \int_{-\infty}^{\infty} (2n-1) q^{2n-2} \left(-\frac{1}{2\alpha} \right) e^{-\alpha q^2} \, dq$$

$$= 0 + \frac{2n-1}{2\alpha} I_{n-1}.$$

The integrals we need are therefore $I_0 = \sqrt{\pi/\alpha}$ (given) and

$$I_1 = \frac{1}{2\alpha} I_0$$

$$I_2 = \frac{3}{2\alpha} I_1 = \frac{3}{4\alpha^2} I_0.$$

The Rayleigh–Ritz ratio is therefore (see above):

$$\langle E' \rangle = \frac{\langle \psi | \hat{H} | \psi \rangle}{\langle \psi | \psi \rangle} = \frac{1}{2I_0} \int_{-\infty}^{\infty} \left(q^4 - \alpha^2 q^2 + \alpha \right) \psi^2 \, dq$$

$$= \frac{1}{2I_0} \left[I_2 - \alpha^2 I_1 + \alpha I_0 \right] = \frac{1}{2I_0} \left[\frac{3}{4\alpha^2} I_0 - \alpha^2 \frac{1}{2\alpha} I_0 + \alpha I_0 \right]$$

$$= \frac{3}{8\alpha^2} + \frac{\alpha}{4}$$

This function has a minimum at

$$\frac{d\langle E' \rangle}{d\alpha} = -\frac{3}{4\alpha^3} + \frac{1}{4} = 0 \Rightarrow \alpha = 3^{1/3}$$

Therefore:

```
alpha = 3**(1/3)
print(f'alpha = {alpha:.3f}')
print(f"optimum <E'> = {rayleigh_ritz(alpha):.3f}")
```

```
alpha = 1.442
optimum <E'> = 0.541
```

These are the same values as those determined numerically in (a).

P23.1

First, the imports and some function definitions as before:

```
import numpy as np
from scipy.special import hermite
from scipy.integrate import quad
import matplotlib.pyplot as plt
```

```
def psi_v(q, v):
    """Return the harmonic oscillator wavefunction, psi_v, at q."""
    Nv = 1 / np.sqrt( np.sqrt(np.pi) * 2**v * np.math.factorial(v))
    Hv = hermite(v)
    return Nv * Hv(q) * np.exp(-q**2 / 2)

def P_v(q, v):
    """Return the probability distribution function for psi_v."""
    return psi_v(q, v)**2

def get_A(v):
    """Return the classical oscillator amplitude for state v."""
    return np.sqrt(2*v + 1)

def P_cl(q, A):
    return 1/np.pi / np.sqrt(A**2 - q**2)
```

For $v = 20$, there will be 21 peaks: a suitable number of regions to average over would be about half this, so as to smooth out these oscillations.

```
v = 20
nregions = v // 2
A = get_A(v)
Dq = 2 * A / nregions
P_qm = np.zeros(nregions)
q = np.zeros(nregions)
for i in range(nregions):
    # Integrate the square of the wavefunction over the ith region
    # between -A + Dq.i and -A + Dq.(i+1)
    a = -A + i*Dq
    P_qm[i] = quad(P_v, a, a + Dq, args=(v,))[0] / Dq
    q[i] = a + Dq / 2
```

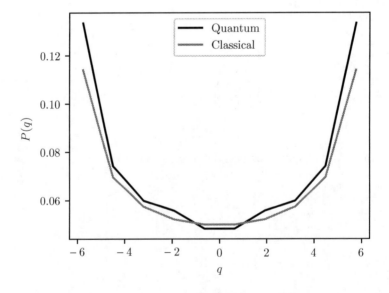

```
# Classical probabilities for the regions
P_cl_v = P_cl(q, A)
plt.plot(q, P_qm, c='k', label='Quantum')
plt.plot(q, P_cl_v, c='#777777', label='Classical')
plt.xlabel(r'$q$')
plt.ylabel(r'$P(q)$')
```

P26.1

Starting from the Brusselator mechanism,

$$A \rightarrow X \qquad\qquad k_1$$
$$2X + Y \rightarrow 3X \qquad\qquad k_2$$
$$B + X \rightarrow Y + D \qquad\qquad k_3$$
$$X \rightarrow E \qquad\qquad k_4$$

write expressions for the rate of change of the intermediate concentrations, [X] and [Y]:

$$\frac{d[X]}{dt} = k_1[A] + k_2[X]^2[Y] - k_3[B][X] - k_4[X],$$
$$\frac{d[Y]}{dt} = -k_2[X]^2[Y] + k_3[B][X].$$

Let $x = p[X]$, $y = q[Y]$ and $\tau = rt$:

$$\frac{r}{p}\frac{dx}{d\tau} = k_1[A] + \frac{k_2}{p^2 q}x^2 y - \frac{k_3[B]}{p}x - \frac{k_4}{p}x,$$
$$\frac{r}{q}\frac{dy}{d\tau} = -\frac{k_2}{p^2 q}x^2 y + \frac{k_3[B]}{p}x.$$

Rearranging,

$$\frac{dx}{d\tau} = \frac{pk_1[A]}{r} + \frac{k_2}{pqr}x^2 y - \frac{1}{r}(k_3[B] + k_4)\,x,$$
$$\frac{dy}{d\tau} = -\frac{k_2}{p^2 r}x^2 y + \frac{k_3[B]q}{pr}x.$$

Choosing $k_4/r = 1$ and $pk_1[A]/r = 1$ fixes $r = k_4$ and $p = k_4/(k_1[A])$. It would be nice if the factors involving [B] were the same. We can define

$$b = \frac{k_3[B]}{r} = \frac{k_3[B]q}{pr} \qquad \Rightarrow q = p = \frac{k_4}{k_1[A]},$$

which leaves

$$a = \frac{k_2}{p^2 r} = \frac{k_2}{pqr} = \frac{k_1^2 k_2 [A]^2}{k_4^3}.$$

Therefore,

$$\frac{dx}{d\tau} = 1 + ax^2 y - (b+1)x,$$
$$\frac{dy}{d\tau} = -ax^2 y + bx.$$

If the values of x and y are constant in time at the fixed point, $(x_\star, y_\star)$, then it must be the case that:

$$\frac{dx}{d\tau} = 0 = 1 + ax_\star^2 y_\star - (b+1)x_\star, \text{ and}$$
$$\frac{dy}{d\tau} = 0 = -ax_\star^2 y_\star + bx_\star.$$

This pair of equations are readily solved to yield $x_\star = 1$, $y_\star = b/a$.

```python
import numpy as np
from scipy.integrate import solve_ivp
from matplotlib import pyplot as plt
```

```python
def deriv(t, X, a, b):
    """Return the derivatives dx/dtau and dy/dtau."""
    x, y = X
    dxdtau = 1 + a * x**2 * y - (b+1)*x
    dydtau = - a * x**2 * y + b * x
    return dxdtau, dydtau
```

```python
def plot_brusselator(a, b, irow, axes):
    """
    Integrate the Brusselator equations for reactant concentrations a, b
    and plot on the row indexed at irow of the figure axes.

    """

    # Initial and final (scaled) time points.
    taui, tauf = 0, 100
    # Initial (scaled) concentrations of the intermediates.
    x0, y0 = 0, 0
    soln = solve_ivp(deriv, (taui, tauf), (x0, y0),
                     dense_output=True, args=(a, b))

    tau = np.linspace(taui, tauf, 1000)
    x, y = soln.sol(tau)
    axes[irow][0].plot(tau, x, lw=1)
    axes[irow][0].plot(tau, y, lw=1)
```

```
        axes[irow][0].legend((r'$x$', r'$y$'), loc='upper right')
        axes[irow][0].set_xlabel(r'$\tau$')
        text_hpos = (tauf - taui) / 2
        axes[irow][0].text(text_hpos, 0.1, f'$a={a}, b={b}$', ha='center')
        axes[irow][1].plot(x, y, lw=1)
        axes[irow][1].set_xlabel(r'$x$')
        axes[irow][1].set_ylabel(r'$y$')
```

```
fig, axes = plt.subplots(nrows=3, ncols=2, figsize=(6, 8))
# Plot the Brusselator solutions for different reactant concentrations, a and b.
plot_brusselator(1, 0.7, 0, axes)
plot_brusselator(1, 1.8, 1, axes)
plot_brusselator(1, 2.05, 2, axes)

fig.tight_layout()
plt.show()
```

The resulting plots are shown in Figure 40.1.

As it turns out, for $(a, b) = (1, 0.7)$ and $(1, 1.8)$, the Brusselator is stable and the intermediate concentrations converge on the fixed point $(x_\star, y_\star) = (1, b/a)$. For $(a, b) = (1, 2.05)$, the intermediate concentrations enter a *limit cycle*.

P31.1

The code below produces simulated spectra for $^1H^{35}Cl$ at two temperatures, 100 K and 600 K.

```
import numpy as np
import matplotlib.pyplot as plt
import pandas as pd

from scipy.constants import physical_constants
# The second radiation constants, c2 = hc/kB, in cm.K
c2 = physical_constants['second radiation constant'][0] * 100
# Speed of light (m.s-1), atomic mass constant (kg) and Boltzmann constant (J.K-1)
c = physical_constants['speed of light in vacuum'][0]
u = physical_constants['atomic mass constant'][0]
kB = physical_constants['Boltzmann constant'][0]
```

```
df = pd.read_csv('diatomic-data.tab', sep='\s+', index_col=0, na_values='*')

def get_E(v, J, molecule):
    """Return the energy, in cm-1, of the state (v, J) for a molecule."""

    # Get the parameters, replacing NaN values with 0
    prms = df[molecule].fillna(0)
    vfac, Jfac = v + 0.5, J * (J+1)
    Bv = prms['Be'] - prms['alpha_e'] * vfac + prms['gamma_e'] * vfac**2
    Dv = prms['De'] - prms['beta_e'] * vfac
    F = Bv * Jfac - Dv * Jfac**2
```

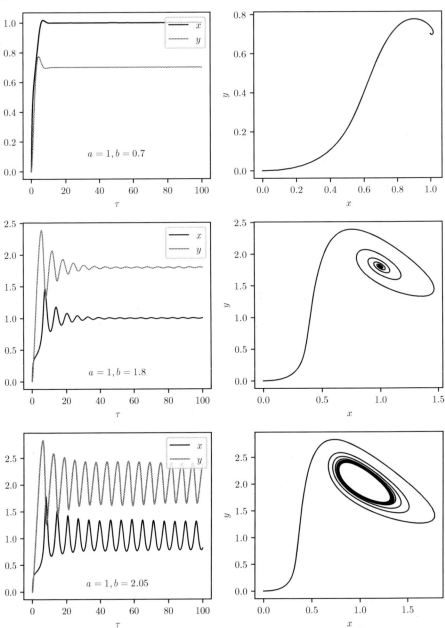

Figure 40.1   Plots of the Brusselator dynamical system for different initial (scaled) reactant concentrations, a and b. The time-dependence of the (scaled) intermediate concentrations is plotted on the left and the corresponding phase space plot on the right.

```
G = prms['we'] * vfac - prms['wexe'] * vfac**2 + prms['weye'] * vfac**3
return G + F
```

```
def get_states(molecule, states_vmax=2, states_Jmax=50):
    index = pd.MultiIndex.from_tuples(
        ((v, J) for v in range(states_vmax+1) for J in range(states_Jmax+1)),
        names=['v', 'J'])
    states = pd.DataFrame(index=index, columns=('E', 'g'), dtype=float)
    for v, J in index:
        # Set the energy and the degeneracy columns.
        states.loc[(v, J), 'E'] = get_E(v, J, molecule)
        states.loc[(v, J), 'g'] = 2 * J + 1
    # Convert the dtype of the degeneracy column to int.
    return states.convert_dtypes()


def get_partition_sum(E0, T):
    E, g = states['E'], states['g']
    # In the partition sum, measure energies from the zero-point level.
    return np.sum(g * np.exp(-c2 * (E-E0) / T))


def get_trans(E0, q, T, trans_Jmax=20):
    Jpp_P = np.arange(trans_Jmax, 0, -1)
    Jpp_R = np.arange(0, trans_Jmax)
    Pbranch = [f'P({Jpp})' for Jpp in Jpp_P]
    Rbranch = [f'R({Jpp})' for Jpp in Jpp_R]
    trans = pd.DataFrame(index=np.concatenate((Pbranch, Rbranch)))
    trans['Jpp'] = np.concatenate((Jpp_P, Jpp_R))
    trans['Jp'] = np.concatenate((Jpp_P-1, Jpp_R+1))
    vp, vpp = 1, 0
    for br, (Jpp, Jp) in trans[['Jpp', 'Jp']].iterrows():
        # Store the transition's upper and lower state energies in the trans table,
        # measured from the zero-point level.
        Epp = states.loc[(vpp, Jpp), 'E'] - E0
        Ep = states.loc[(vp, Jp), 'E'] - E0
        nu = Ep - Epp
        # Also calculate the relative intensity, including a Honl-London factor,
        #
        HLfac = Jpp
        if Jp - Jpp == 1:
            HLfac = Jpp + 1
        rS = nu * HLfac / q * np.exp(-c2 * Epp / T) * (1 - np.exp(-c2 * nu / T))
        trans.loc[br, ['Epp', 'Ep', 'nu0', 'rS']] = Epp, Ep, nu, rS
    return trans


def fG(nu, nu0, alphaD):
    """Normalized Gaussian function with HWHM alphaD, centered on nu0."""
    N = np.sqrt(np.log(2) / np.pi) / alphaD
    return N * np.exp(-np.log(2) * ((nu - nu0) / alphaD)**2)


def fL(nu, nu0, gammaL):
    """Normalized Lorentzian function with HWHM gammaL, centered on nu0."""
    N = 1 / np.pi
    return N * gammaL / ((nu - nu0)**2 + gammaL**2)


def get_spec(trans, HWHM, flineshape=fG):
    """Return the spectrum and a suitable wavenumber grid."""
    # Maximum and minimum transition wavenumbers.
    numin, numax = trans['nu0'].min(), trans['nu0'].max()
```

```
    # Pad the wavenumber grid by 5 cm-1 at each end
    numin, numax = round(numin) - 5, round(numax) + 5
    nu = np.arange(numin, numax, 0.001)

    spec = np.zeros_like(nu)
    for br, row in trans.iterrows():
        spec += row['rS'] * flineshape(nu, row['nu0'], HWHM)
    spec /= spec.max()
    return nu, spec
```

```
def calc_spec(T, HWHM, flineshape=fG):
    """Calculate and return the 1-0 rovibrational spectrum at temperature T."""
    E0 = get_E(0, 0, molecule)
    q = get_partition_sum(E0, T)
    trans = get_trans(E0, q, T)
    nu, spec = get_spec(trans, HWHM, flineshape)
    return nu, spec
```

```
# Process the data for the molecule of interest, (1H)(35Cl).
molecule = '(1H)(35Cl)'
gammaL = 2
states = get_states(molecule)
```

```
# Read in the experimental transition parameters.
expt_df = pd.read_csv('HITRAN_1H-35Cl_1-0.csv')

def calc_expt_spec(nu, T, HWHM, flineshape=fG):
    """Calculate the experimental spectrum at temperature T."""
    Tref = 296
    E0 = get_E(0, 0, molecule)
    qref = get_partition_sum(E0, Tref)
    q = get_partition_sum(E0, T)
    expt_spec = np.zeros_like(nu)
    for br, row in expt_df.iterrows():
        nu0, Epp = row['nu'], row['elower']
        sw = (row['sw'] * qref / q *
              np.exp(-c2 * Epp / T) / np.exp(-c2 * Epp / Tref) *
              (1 - np.exp(-c2 * nu0 / T)) / (1 - np.exp(-c2 * nu0 / Tref))
              )
        expt_spec += sw * flineshape(nu, nu0, HWHM)
    expt_spec /= expt_spec.max()
    return expt_spec
```

```
def plot_spectra(nu, spec, expt_spec, T):
    """Plot spec and expt_spec on the same Axes and label."""
    plt.plot(nu, spec, lw=1, label='theory')
    plt.plot(nu, expt_spec, lw=1, label='experiment')
    plt.xlabel(r'$\tilde{\nu}\;/\mathrm{cm^{-1}}$')
    plt.ylabel(r'$\sigma$ (arb. units)')
    plt.title(f'{T} K')
    plt.legend()
```

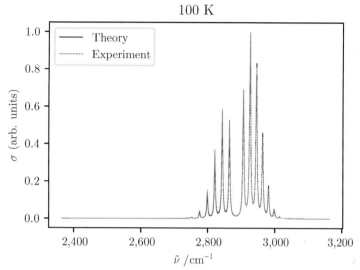

Figure 40.2   Comparison of theoretical and experimental spectra for ^{1}H^{35}Cl at 100 K.

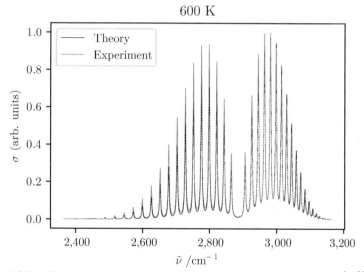

Figure 40.3   Comparison of theoretical and experimental spectra for ^{1}H^{35}Cl at 600 K.

```
# Temperature, K.
T = 100
# Calculate the spectrum using the calculated line list.
nu, spec = calc_spec(T, gammaL, fL)
# Calculate the spectrum using the experimental line list
expt_spec = calc_expt_spec(nu, T, gammaL, fL)

plot_spectra(nu, spec, expt_spec, T)
```

```
# Temperature, K.
T = 600
# Calculate the spectrum using the calculated line list.
nu, spec = calc_spec(T, gammaL, fL)
# Calculate the spectrum using the experimental line list
expt_spec = calc_expt_spec(nu, T, gammaL, fL)

plot_spectra(nu, spec, expt_spec, T)
```

The spectra at 100 K and 600 K are plotted in Figures 40.2 and 40.3. The agreement is generally good, though with greater discrepancies at the higher temperature.

P33.1

The secular matrix can be constructed from the connectivity of the matrix with reference to the atom labeling provided:

```
import numpy as np
import matplotlib.pyplot as plt
```

```
M = np.zeros((10, 10))
M[0,1] = M[1,2] = M[2,3] = M[3,4] = M[4,5] = M[0,5] = 1
M[5,6] = M[6,7] = M[7,8] = M[8,9] = M[0,9] = 1
```

```
# Calculate the eigenvalues and eigenvectors as before.
E, eigvecs = np.linalg.eigh(M, UPLO='U')
```

```
# A quick-and-dirty way to calculate the atom positions.
c = np.sqrt(3) / 2
theta = np.radians(np.arange(-60, 300, 60))
pos = np.empty((2, 10))
pos[:, :6] = np.array((np.sin(theta) + c, np.cos(theta)))
theta = np.radians(np.arange(180, 420, 60))
pos[:, 6:] = np.array((np.sin(theta[:4]) - c, np.cos(theta[:4])))

def plot_structure(ax):
    idx = list(range(10)) + [0, 5]
    for i in range(len(idx)-1):
        p1x, p1y = pos.T[idx[i]]
        p2x, p2y = pos.T[idx[i+1]]
        ax.plot([p1x, p2x], [p1y, p2y], lw=2, c='k')

def plot_MOs(i, ax, pos, eigvecs):
    # Plot the molecular framework.
    ax.plot(*pos, c='k', lw=2)
    for j, coeff in enumerate(eigvecs[:,i]):
        ax.add_patch(plt.Circle(pos.T[j], radius=coeff,
```

```
                              fc=colors[np.sign(coeff)]))
    # Ensure the circles are circular and turn off axis spines and labels
    ax.axis('square')
    ax.axis('off')

colors = {-1: 'tab:red', 1: 'tab:blue'}
fig, axes = plt.subplots(nrows=5, ncols=2, figsize=(6, 8))
for i in range(10):
    ax = axes[i // 2, i % 2]
    plot_structure(ax)
    plot_MOs(i, ax, pos, eigvecs)
plt.show()
```

The resulting plot is shown in Figure 40.4.

```
# Output the energy levels.
print('E / |beta|')
print('-'*18)
for Elevel in E:
    print(f'{-Elevel:>9.3f}')
```

```
E / |beta|
------------------
    2.303
    1.618
    1.303
    1.000
    0.618
   -0.618
   -1.000
   -1.303
   -1.618
   -2.303
```

```
# Calculate the pi-bond orders.
nocc = [0]*5 + [2]*5
idx = np.where(M == 1)
P = np.sum(nocc * eigvecs[idx,:][0] * eigvecs[idx,:][1], axis=1)
for j, (i, k) in enumerate(zip(*idx)):
    print(f'P{i}{k} = {P[j]:.3f}')
```

```
P01 = 0.555
P05 = 0.518
P09 = 0.555
P12 = 0.725
P23 = 0.603
P34 = 0.725
P45 = 0.555
P56 = 0.555
P67 = 0.725
P78 = 0.603
P89 = 0.725
```

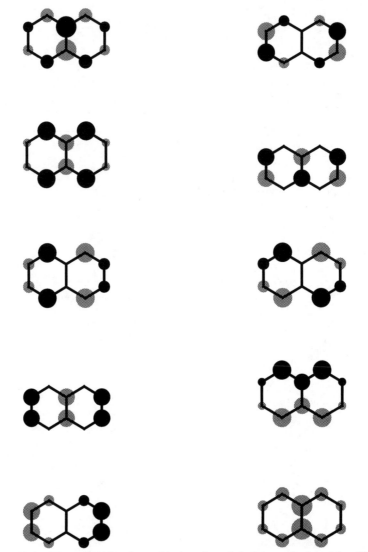

Figure 40.4   The π molecular orbitals of naphthalene calculated by Hückel theory.

```
Q = np.sum(nocc * eigvecs**2, axis=1)
Q
```

```
array([1., 1., 1., 1., 1., 1., 1., 1., 1., 1.])
```

The π electron charge is evenly distributed across the atoms.

P33.2

```
import numpy as np
import matplotlib.pyplot as plt
```

Set up the secular matrix according to the suggested atom numbering for the structure of pyridine:

```
M = np.zeros((6, 6))
M[0,0] = 0.5
M[0,1] = M[0,5] = 0.8
M[1,2] = M[2,3] = M[3,4] = M[4,5] = 1
```

```
# Calculate the eigenvalues and eigenvectors as before.
E, eigvecs = np.linalg.eigh(M, UPLO='U')
```

```
print('E / |beta| | ', ('    c{:1s}    '*6).format(*'N12345'))
for Elevel, coeffs in zip(E, eigvecs.T):
    print(f'{-Elevel:>9.3f}   | ', ('{: 7.3f}'*6).format(*coeffs))
```

```
E / |beta| |      cN      c1      c2      c3      c4      c5
    1.849  |   -0.243   0.357  -0.465   0.503  -0.465   0.357
    1.000  |   -0.000   0.500  -0.500  -0.000   0.500  -0.500
    0.667  |    0.564  -0.412  -0.177   0.530  -0.177  -0.412
   -1.000  |    0.000   0.500   0.500  -0.000  -0.500  -0.500
   -1.062  |    0.665   0.234  -0.284  -0.535  -0.284   0.234
   -1.954  |    0.424   0.386   0.414   0.424   0.414   0.386
```

Finally, a plot of the molecular orbitals (Figure 40.5):

```
colors = {-1: 'tab:red', 1: 'tab:blue'}
# Arrange the atoms equally around a circle to make a hexagon.
# The first point is included at the end as well to close the ring.
theta = np.radians(np.arange(0, 420, 60))
pos = np.array((np.sin(theta), np.cos(theta)))

def plot_MOs(i, ax, pos, eigvecs):
    # Plot the molecular framework.
    ax.plot(*pos, c='k', lw=2)
    for j, coeff in enumerate(eigvecs[:,i]):
        ax.add_patch(plt.Circle(pos.T[j], radius=coeff,
                                fc=colors[np.sign(coeff)]))
    # Ensure the circles are circular and turn off axis spines and labels
    ax.axis('square')
    ax.axis('off')

# Plot the MO wavefunctions on a 2x3 grid of Axes.
fig, axes = plt.subplots(nrows=2, ncols=3)
for i in range(6):
    plot_MOs(i, axes[i // 3, i % 3], pos, eigvecs)
```

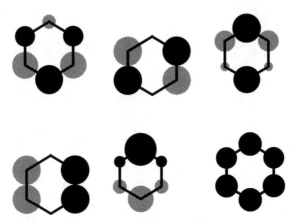

Figure 40.5   The π molecular orbitals of pyridine calculated by Hückel theory.

P34.1

(a) First, import the necessary libraries and define the functions used:

```python
import numpy as np
from scipy.optimize import least_squares, minimize
from scipy.integrate import quad
import matplotlib.pyplot as plt
```

```python
def STO1(r, zeta=1):
    """The normalized 1s Slater-type orbital."""
    return np.sqrt(zeta**3 / np.pi) * np.exp(-zeta * r)

def pGTO(r, alpha):
    """
    The normalized, s-type primitive Gaussian-type orbital with exponent
    parameter alpha.

    """

    return (2 * alpha / np.pi)**0.75 * np.exp(-alpha * r**2)

def GTO(r, prms):
    """
    Return the Gaussian-type orbital psi_NG as a linear expansion
    of primitive Gaussians. The coefficients are packed into prms as
    [c1, alpha1, c2, alpha2, ..., cN, alphaN].

    """

    n = len(prms) // 2
    psi = 0
    for k in range(n):
        c_k, alpha_k = prms[2*k:2*k+2]
        psi += c_k * pGTO(r, alpha_k)
    return psi
```

```
def get_c1(red_prms):
    """

    Determine the value of c1 that normalizes the STO-NG function described
    by the coefficients red_prms = [alpha1, c2, alpha2, ..., cN, alphaN].

    """

    def resid(c1):
        """Return the difference between <Psi_NG|Psi_NG> and unity."""
        prms = np.concatenate((c1, red_prms))
        I = quad(lambda r: GTO(r, prms)**2 * r**2, 0, np.inf)[0] * 4 * np.pi
        return I - 1

    c1guess = 0.1
    fit = least_squares(resid, [c1guess])
    return fit.x[0]
```

To fit the discretized STO, either use SciPy's `curve_fit` method, or define a function that returns the root-mean square difference between the values of ψ^{3G} and those of $\phi_1^S(r_i, 1)$:

```
r = np.linspace(0, 5, 500)
phi_S = STO1(r)

def STO_NG_resid(fit_prms):
    c1 = get_c1(fit_prms)
    prms = np.concatenate(([c1], fit_prms))
    psi_NG = GTO(r, prms)
    return psi_NG - phi_S

def rms_resid(fit_prms):
    resid = STO_NG_resid(fit_prms)
    return np.sqrt(np.mean(resid**2))
```

The fit has a tendency to wander off into unlikely values of its parameters, so start with those values of c_k and α_k fit to maximize the overlap (Example E34.2) and constrain the parameter values to within the interval [0, 15]:

```
# Initial guesses for alpha1, c2, alpha2, c3, alpha3.
x0 = [0.1099509 , 0.53494108, 0.40671275, 0.15398082, 2.23376247]
fit = least_squares(STO_NG_resid, x0, bounds=(0, 15))
#fit = minimize(rms_resid, x0, bounds=[(0, 15)] * len(x0))
```

Now, as before, complete the parameter set by finding c_1 to make ψ^{3G} normalized:

```
c1 = get_c1(fit.x)
prms3 = np.concatenate(([c1], fit.x))
prms3
```

```
array([ 0.78430464,  0.18827274,  0.30768249,  1.15546627,  0.02577338,
        14.36592071])
```

And calculate the expectation value of the energy:

```
def func(r, prms):
    """
    Return the integrand for the expectation value of the energy of the STO-NG
    function defined by prms = [c1, alpha1, c2, alpha2, ..., cN, alphaN],
    <phi_NG | H | phi_NG> as a function of the electron-nucleus distance, r.
    """

    phi_NG = GTO(r, prms)
    N = len(prms) // 2
    rhs = 0
    for k in range(N):
        ck, alphak = prms[2*k:2*k+2]
        rhs += ck * (3 * alphak - 2 * (alphak * r)**2 - 1/r) * pGTO(r, alphak)
    return phi_NG * rhs * 4 * np.pi * r**2

def get_E(prms):
    E = quad(func, 0, np.inf, args=(prms,))[0]
    return E

Eexact = -0.5
E = get_E(prms3)
error = abs((E - Eexact) / Eexact) * 100
print(f'STO-{N}G: {E:.4f} Eh, error = {error:.2f}%')
```

```
STO-3G: -0.4935 Eh, error = 1.30%
```

This is obviously a little worse than the method of maximizing the overlap (which gave an error of 1.02%).

(b) To minimize $\langle E \rangle$, we now have almost everything we need. Define a function to return the energy given a set of fit parameters, and use it with the same initial guess as before:

```
def fit_E(fit_prms):
    c1 = get_c1(fit_prms)
    prms = np.concatenate(([c1], fit_prms))
    return get_E(prms)

x0 = [0.1099509 , 0.53494108, 0.40671275, 0.15398082, 2.23376247]
fit3 = minimize(fit_E, x0)
```

```
fit3
```

```
      fun: -0.4969792526930734
 hess_inv: array([[ 6.30944422e-01, -1.72680979e+00,  3.33980741e+00,
            -4.24532064e-01,  2.39658646e+01],
```

```
   [-1.72680979e+00,  6.27166742e+00, -1.07781939e+01,
     1.22715725e+00, -6.48771540e+01],
   [ 3.33980741e+00, -1.07781939e+01,  2.69793989e+01,
    -4.02071355e+00,  2.27087973e+02],
   [-4.24532064e-01,  1.22715725e+00, -4.02071355e+00,
     8.56241240e-01, -4.59201975e+01],
   [ 2.39658646e+01, -6.48771540e+01,  2.27087973e+02,
    -4.59201975e+01,  3.03750558e+03]])
    jac: array([5.59538603e-06, 4.42564487e-06, 2.30595469e-06, 4.26545739e-06,
     4.43309546e-07])
message: 'Optimization terminated successfully.'
   nfev: 186
    nit: 28
   njev: 31
 status: 0
success: True
      x: array([0.15137657, 0.40788764, 0.68128952, 0.07047645, 4.50030797])
```

This time we can read $\langle E \rangle$ directly from the `fit3` object:

```
E = fit3.fun
error = abs((E - Eexact) / Eexact) * 100
print(f'STO-{N}G: {E:.4f} Eh, error = {error:.2f}%')
```

```
STO-3G: -0.4970 Eh, error = 0.60%
```

This time the error in the energy is smaller, and presumably the best we can do with the STO-3G contracted Gaussian function.

P34.2

First, calculate the Gibbs free energy of formation of the species involved in the processes

$$CO(g) + H_2O(g) \rightleftharpoons H_2(g) + CO_2(g)$$

$$CH_4(g) + H_2O(g) \rightleftharpoons CO(g) + 3\,H_2(g)$$

using the thermodynamic data read in from `thermo-data.csv` and the Shomate Equations given in Example E31.2.

```
import numpy as np
from scipy.constants import R     # Gas constant, J.K-1.mol-1
import pandas as pd
from scipy.optimize import minimize

# Standard pressure, Pa.
pstd = 1.e5

# Read in the thermodynamic data, strip stray spaces from the column names.
df = pd.read_csv('thermo-data.csv', comment='#', index_col=0)
```

```
df.columns = df.columns.str.strip()

def get_DfH(formula, T):
    """Return the standard molar enthalpy of formation of formula at T K."""
    s = df.loc[formula]
    t = T / 1000
    return (s.DfHo_298 + s.A * t + s.B * t**2 / 2 + s.C * t**3 / 3
            + s.D * t**4 / 4 - s.E / t + s.F - s.H)

def get_S(formula, T):
    """Return the standard molar entropy of formula at T K."""
    s = df.loc[formula]
    t = T / 1000
    return (s.A * np.log(t) + s.B * t + s.C * t**2 / 2 + s.D * t**3 / 3
            - s.E / 2 / t**2 + s.G)

def get_DfG(formula, T):
    DfHo = get_DfH(formula, T)
    So = get_S(formula, T)
    DfGo = DfHo - T * So / 1000
    return DfGo
```

```
# The conditions of the reaction mixture.
T = 1000          # K
ptot = 4.e5       # Pa
species = ('CO', 'H2O', 'H2', 'CO2', 'CH4')

# Calculate the standard Gibbs free energy of formation at temperature T.
DfGo = {}
for formula in species:
    DfGo[formula] = get_DfG(formula, T)
```

The function to be minimized is the total Gibbs free energy of the reaction mixture,

$$nG = \sum_J n_J \Delta_f G_J^{\ominus} + RT \sum_J n_J \ln\left(x_J \frac{p_{tot}}{p^{\ominus}}\right).$$

```
def nG(narr):
    """Return the total Gibbs free energy of the reaction mixture."""
    ntot = sum(narr)
    ret = 0
    for i, n in enumerate(narr):
        formula = species[i]
        x = n / ntot
        ret += n * (DfGo[formula] * 1000 + R * T * np.log(x * ptot/pstd))
    return ret
```

Now carry out the fit, with suitable constraints on the amounts of each atom (which must remain constant) and species (which must be nonnegative).

```
# Initial amounts
n0 = {'CO': 0, 'H2O': 1.5, 'H2': 0, 'CO2': 0, 'CH4': 2}
# We start with this number of carbon, oxygen and hydrogen atoms
# and must keep them constant.
nC = n0['CO'] + n0['CO2'] + n0['CH4']
nO = n0['CO'] + n0['H2O'] + 2*n0['CO2']
nH = 2*n0['H2O'] + 2*n0['H2'] + 4*n0['CH4']

# The constraints for the fit: conserve the number of moles of each
# atom and ensure that the amounts of each compound is positive.
# n[0]    n[1]    n[2]    n[3]    n[4]
# CO      H2O     H2      CO2     CH4
cons = [{'type': 'eq', 'fun': lambda n: n[0] + n[1] + 2*n[3] - nO},     # O atoms
        {'type': 'eq', 'fun': lambda n: n[0] + n[3] + n[4] - nC},       # C atoms
        {'type': 'eq', 'fun': lambda n: 2*n[1] + 2*n[2] + 4*n[4] - nH},# H atoms
        {'type': 'ineq', 'fun': lambda n: n[0]},       # nCO  > 0
        {'type': 'ineq', 'fun': lambda n: n[1]},       # nH2O > 0
        {'type': 'ineq', 'fun': lambda n: n[2]},       # nH2  > 0
        {'type': 'ineq', 'fun': lambda n: n[3]},       # nCO2 > 0
        {'type': 'ineq', 'fun': lambda n: n[4]},       # nCH4 > 0
        ]

# Use the Sequential Least Squares Programming (SLSQP) algorithm, which allows
# constraints to be specified.
# The initial guess composition is somewhat arbitrary, but should be
# consistent with the constraints.
ret = minimize(nG, (0.5, 0.25 , 1, 0.375 , 1.125), method="SLSQP",
               constraints=cons)
ret
```

```
      fun: -1260147.0829021323
      jac: array([-327859.5625   , -458495.46875 , -138946.      , -647409.296875,
          -286201.78125 ])
  message: 'Optimization terminated successfully'
     nfev: 70
      nit: 11
     njev: 11
   status: 0
  success: True
        x: array([0.7840544 , 0.40727639, 2.96950161, 0.15433461, 1.061611  ])
```

```
print(f'T = {T} K, ptot = {ptot/1.e5} bar')
print('Equilibrium composition (mol)')
for i, formula in enumerate(species):
    print(f'{formula:>3s}: {ret.x[i]:.3f}')
```

```
T = 1000 K, ptot = 4.0 bar
Equilibrium composition (mol)
 CO: 0.784
H2O: 0.407
 H2: 2.970
```

```
CO2: 0.154
CH4: 1.062
```

P35.1

SymPy can handle these algebra and calculus problems directly.

```
import sympy as sp
x = sp.Symbol('x', real=True)
```

(a)

```
expr = (x**3 - 1) / x**2 / (x - 2)**3
expr
```

$$\frac{x^3 - 1}{x^2 (x - 2)^3}$$

```
sp.apart(expr)
```

$$-\frac{3}{16(x-2)} + \frac{5}{4(x-2)^2} + \frac{7}{4(x-2)^3} + \frac{3}{16x} + \frac{1}{8x^2}$$

(b)

```
expr = (x**3 + 14) / (x**3 - 3*x**2 + 8*x)
expr
```

$$\frac{x^3 + 14}{x^3 - 3x^2 + 8x}$$

```
sp.apart(expr)
```

$$\frac{5x - 11}{4(x^2 - 3x + 8)} + 1 + \frac{7}{4x}$$

(c)

```
expr = 1 / (4 - x)**2
expr
```

$$\frac{1}{(4 - x)^2}$$

```
sp.integrate(expr, (x, -sp.oo, 2))
```

$$\frac{1}{2}$$

(d) In the usual applications of this integral, *a* and *b* are assumed to be real and positive, and *c* is assumed to be real:

```
a, b = sp.symbols('a b', real=True, positive=True)
c = sp.symbols('c', real=True)
sp.integrate(sp.exp(-a * x**2) * sp.exp(-b * (x-c)**2),
             (x, -sp.oo, sp.oo)).simplify()
```

$$\frac{\sqrt{\pi}\,e^{-\frac{abc^2}{a+b}}}{\sqrt{a+b}}$$

Note that because the product of two Gaussian functions (on different centres) can be written as another Gaussian function, this is proportional to

$$\int_{-\infty}^{\infty} e^{-(a+b)x^2}\,dx.$$

(e)

```
n = sp.Symbol('n', integer=True)
sp.summation(1/n**2, (n, 1, sp.oo))
```

$$\frac{\pi^2}{6}$$

P35.2

First, import SymPy, and define some symbols:

```
import sympy as sp
x = sp.Symbol('x', real=True)
n = sp.symbols('n', integer=True, positive=True)
L = sp.symbols('L', real=True, positive=True)
```

```
psi = sp.sin(n * sp.pi * x / L)
I = sp.integrate(psi**2, (x, 0, L)).simplify()
N = sp.sqrt(1/I)
N
```

$$\frac{\sqrt{2}}{\sqrt{L}}$$

```
psi = N * psi
psi
```

$$\frac{\sqrt{2}\sin\left(\frac{\pi n x}{L}\right)}{\sqrt{L}}.$$

The values of $\langle x \rangle$ and $\langle p_x \rangle$ are given:

```
Ex, Ep = L/2, 0
```

The value of $\langle x^2 \rangle$ can be found by direct integration:

$$\langle x^2 \rangle = \int_0^L \psi^*(x)x^2\psi(x)\,\mathrm{d}x$$

```
Ex, Ep = L/2, 0
Ex2 = sp.integrate(x**2 * psi**2, (x, 0, L))
Ex2.simplify()
```

$$\frac{L^2}{3} - \frac{L^2}{2\pi^2 n^2}$$

The operator $\hat{p}_x^2 = -\hbar^2 \frac{\mathrm{d}^2}{\mathrm{d}x^2}$ and so

$$\langle \hat{p}_x^2 \rangle = -\hbar^2 \int_0^L \psi^*(x)\frac{\mathrm{d}^2\psi(x)}{\mathrm{d}x^2}\,\mathrm{d}x$$

```
hbar = sp.Symbol('hbar', real=True, positive=True)
Ep2 = -hbar**2 * sp.integrate(psi * sp.diff(psi, x, 2), (x, 0, L))
Ep2
```

$$\frac{\pi^2\hbar^2 n^2}{L^2}.$$

The standard deviation uncertainties in x and p_x are:

```
Delta_x = sp.sqrt(Ex2 - Ex**2).simplify()
Delta_p = sp.sqrt(Ep2 - Ep**2).simplify()
print(Delta_x.simplify())
print(Delta_p.simplify())
```

```
L*sqrt(3*pi**2*n**2 - 18)/(6*pi*n)
pi*hbar*n/L
```

```
unc_prod = (Delta_x * Delta_p).simplify()
print(unc_prod)
```

```
hbar*sqrt(3*pi**2*n**2 - 18)/6
```

This is the product of the uncertainties, $\Delta x \Delta p_x$. Its value increases monotonically with n, and has its minimal value for $n = 1$:

```
unc_prod.subs(n, 1)
```

$$\frac{\hbar\sqrt{-18 + 3\pi^2}}{6}.$$

Since π is a little bit larger than 3, this is clearly greater than $\hbar/2$ but to verify, divide by $\hbar/2$ and evaluate (expecting a value greater than 1):

```
(unc_prod.subs(n, 1) * 2 / hbar).evalf()
```

$$1.13572361677322$$

and so the Heisenberg Uncertainty Principle is satisfied.

P38.1

The required imports for this exercise are as follows:

```
import psi4
import numpy as np
import pandas as pd
import matplotlib.pyplot as plt
import fortecubeview
```

It will be helpful to define a function to return the wavefunction and orbital energies for a diatomic molecule identified by its two atom symbols:

```
def get_wfn(atom1, atom2):
    # Initial-guess atom locations with the bond length 1 A.
    molecule = psi4.geometry(
        f"""
        {atom1} 0 0 -0.5
        {atom2} 0 0  0.5
        """)
    # Optimize the geometry.
    E, wfn = psi4.optimize('b3lyp/cc-pvdz', molecule=molecule, return_wfn=True)
    return molecule, wfn

def get_wfn_and_energies(atom1, atom2):
```

```
    """
    Calculate the wavefunction and list of orbitals for the diatomic
    molecule with atoms atom1 and atom2.
    """

    molecule, wfn = get_wfn(atom1, atom2)

    irrep_map = {'Ag': 'sigma_g', 'B1u': 'sigma_u', 'B3u': 'pi_u',
                 'B2u': 'pi_u', 'B3g': 'pi_g', 'B2g': 'pi_g'}
    # Get the irreducible representations and arrays of orbital
    # energies corresponding to each of them.
    irreps = molecule.irrep_labels()
    MO_energies = wfn.epsilon_a().to_array()
    # Build a list of (irrep, energy) tuples.
    energy_list = []
    for irrep, Earray in zip(irreps, MO_energies):
        # Relabel the irrep from the D2h to Dinfh point group.
        Dinfh_irrep = irrep_map.get(irrep)
        if Dinfh_irrep:
            # LaTeX format for orbital symmetry label, preceeded by running index.
            energy_list.extend([[f'{i+1}\\{Dinfh_irrep}', E]
                                for i,E in enumerate(Earray)])
    # Sort the list into ascending order of energy.
    energy_list.sort(key=lambda t: t[1])
    return molecule, wfn, energy_list
```

```
# Perform the calculation for N2.
N2, N2wfn, N2energies = get_wfn_and_energies('N', 'N')
```

```
# Perform the calculation for F2.
F2, F2wfn, F2energies = get_wfn_and_energies('F', 'F')
```

```
def plot_MO_energies(ax, wfn, MO_energies):
    """Plot an MO diagram on Axes ax."""

    # total number of electrons (alphas and betas).
    nelec = int(sum(nocc.sum() for nocc in wfn.occupation_a().to_array())
                + sum(nocc.sum() for nocc in wfn.occupation_b().to_array()))

    LH1flag = True
    LHeflag = True
    electron_scatter = np.zeros((nelec, 2))
    i = 0
    for irrep, E in MO_energies:
        label = False
        if 'sigma' in irrep:
            # sigma orbital levels are centred and labeled.
            cx = 0.5
            label = True
        elif LH1flag:
            # Lefthand pi orbital level isn't labeled.
            cx = -0.1
            LH1flag = False
```

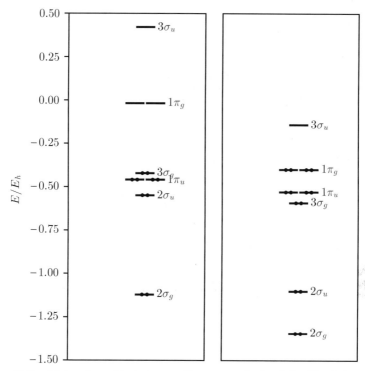

Figure 40.6 Molecular orbital energy diagrams calculated for N_2 (left) and F_2 (right).

```
    else:
        # Righthand pi orbital level is labeled.
        cx = 1.1
        LHlflag = True
        label = True
    # The level is plotted as a short black line.
    ax.plot([cx-0.5, cx+0.5], [E, E], c='k')

    if label:
        # Annotate with the numbered irrep label, to the right of the level.
        ax.annotate(f'${irrep}$', xy=(cx+0.7, E), va='center')

    # Add electrons to the scatter plot array as long as there are some
    # left to occupy the current orbital under consideration.
    nocc = 0
    while nelec and nocc<2:
        dx = 0.15 - nocc*0.3
        electron_scatter[i] = cx + dx, E
        LHeflag = not LHeflag
        nelec -= 1
        nocc +=1
        i += 1
ax.scatter(*electron_scatter.T, fc='r')
```

```
    ax.set_xlim(-4, 4)
    ax.set_ylim(-1.5, 0.5)
    ax.set_xticks([])

fig, axes = plt.subplots(nrows=1, ncols=2)
plot_MO_energies(axes[0], N2wfn, N2energies)
axes[0].set_ylabel('$E /E_h$')
plot_MO_energies(axes[1], F2wfn, F2energies)
axes[1].set_yticks([])
```

The molecular orbital diagram produced by the above code is given in Figure 40.6.

The molecular orbitals can be visualized using `fortecubeview`. First, create the directory `N2-cubes/`, then run the code cell below to generate cube files for the electron density and the first nine orbitals for N_2.

```
psi4.set_options({"writer_file_label":"N2"})
psi4.set_options({
    'CUBEPROP_TASKS': ['ORBITALS', 'DENSITY'],
    'CUBEPROP_ORBITALS': list(range(9)),
    'CUBEPROP_FILEPATH': 'N2-cubes'
})
psi4.cubeprop(N2wfn)
```

```
fortecubeview.plot(path='N2-cubes', width=500, height=500, sumlevel=0.7,
                   opacity=0.7)
```

An example of the `fortecubeview` widget visualization of the lowest-unoccupied molecular orbital is given in Figure 40.7.

MO 8a (1-B3g)

Figure 40.7   The $1\pi_g$ (LUMO) orbital of N_2.

Index

Printed in the United States
by Baker & Taylor Publisher Services